AF806223

Engineering Materials

This series provides topical information on innovative, structural and functional materials and composites with applications in optical, electrical, mechanical, civil, aeronautical, medical, bio- and nano-engineering. The individual volumes are complete, comprehensive monographs covering the structure, properties, manufacturing process and applications of these materials. This multidisciplinary series is devoted to professionals, students and all those interested in the latest developments in the Materials Science field, that look for a carefully selected collection of high quality review articles on their respective field of expertise.

More information about this series at http://www.springer.com/series/4288

Syed Shahabuddin · Adarsh Kumar Pandey ·
Mohammad Khalid · Priyanka Jagadish
Editors

Advances in Hybrid Conducting Polymer Technology

Springer

Editors
Syed Shahabuddin
Department of Science, School of Technology
Pandit Deendayal Petroleum University
Gandhinagar, Gujarat, India

Mohammad Khalid
Graphene and Advanced 2D Materials Research Group (GAMRG), School of Engineering and Technology
Sunway University
Petaling Jaya, Selangor, Malaysia

Adarsh Kumar Pandey
Research Centre for Nano-Materials and Energy Technology (RCNMET), School of Engineering and Technology
Sunway University
Petaling Jaya, Selangor, Malaysia

Priyanka Jagadish
Graphene and Advanced 2D Materials Research Group (GAMRG), School of Engineering and Technology
Sunway University
Petaling Jaya, Selangor, Malaysia

ISSN 1612-1317 ISSN 1868-1212 (electronic)
Engineering Materials
ISBN 978-3-030-62089-9 ISBN 978-3-030-62090-5 (eBook)
https://doi.org/10.1007/978-3-030-62090-5

© Springer Nature Switzerland AG 2021
This work is subject to copyright. All rights are reserved by the Publisher, whether the whole or part of the material is concerned, specifically the rights of translation, reprinting, reuse of illustrations, recitation, broadcasting, reproduction on microfilms or in any other physical way, and transmission or information storage and retrieval, electronic adaptation, computer software, or by similar or dissimilar methodology now known or hereafter developed.
The use of general descriptive names, registered names, trademarks, service marks, etc. in this publication does not imply, even in the absence of a specific statement, that such names are exempt from the relevant protective laws and regulations and therefore free for general use.
The publisher, the authors and the editors are safe to assume that the advice and information in this book are believed to be true and accurate at the date of publication. Neither the publisher nor the authors or the editors give a warranty, expressed or implied, with respect to the material contained herein or for any errors or omissions that may have been made. The publisher remains neutral with regard to jurisdictional claims in published maps and institutional affiliations.

This Springer imprint is published by the registered company Springer Nature Switzerland AG
The registered company address is: Gewerbestrasse 11, 6330 Cham, Switzerland

Preface

This is an advanced text, compiled and designed to apprise postgraduate and undergraduate researchers, students and research industries with the basic theory and advanced applications of conductive polymers and its nanocomposites. The authors of this book also believe that the content of this book would serve as introductory literature for those researchers and students who are not acquainted with the background of the subject.

The research of electrically conductive, conjugated and electroactive polymers has progressed in due course of time and continues to thrive. There has been a significant advance in the understanding of fundamentals and advanced applications in the field of conductive polymers and its nanocomposites. Almost 40 years has been passed since the discovery of conducting polymers. Still, this field of research continues to flourish with the development of wide variety of applications being explored and reported. The importance of conducting polymers can be visualised by the fact the Nobel Prize was awarded to Alan Heeger, Alan MacDiarmid and Hideki Shirakawa in the year 2000 for the discovery of conductivity in polyacetylene. With the advent of nanotechnology and development of nanomaterials, conducting polymers have been tailored into unique nanocomposites which has enhanced the properties and developed unique characteristics. This field of research is still growing steadily, and there is a need of complete understanding of the concept of conducting polymers and how they can address the modern-day issues faced by humanity.

The properties of conducting polymers such as conductivity, optical properties and various physical properties along with the advance applications such as photocatalysis, energy storage devices, nanoelectronics devices, energy harvesting devices, heat transfer applications, waste water treatment, EMI shielding, tissue engineering, antimicrobial properties and dye-sensitised solar cells were discussed throughout the book.

We hope this book will be useful and will serve as a guide for universities and industrial institutions, postgraduate and Ph.D. scholars, researchers working in the area of chemistry, biology, physics, polymer sciences, material sciences, nanoscience and so on.

Lastly, this book has been the final product of the outstanding work of many experts, distinguished researchers and renowned scientists. They have provided their expertise and efforts in compilation of the state-of-the-art chapters. We appreciate the valuable time, hard work and efforts of all the contributing authors.

As a matter of fact, perfection is an assumption. The authors and the editors can never be satisfied with their final efforts. Albeit the book chapters and final draft have been comprehensively reviewed by authors, reviewers and editors, there is always a window to enhance and modify various sections of the chapters, providing an opportunity for improvement in the quality and quantity of the contents. Therefore, the readers of this book are welcome to recommend the errors, omissions and lack of inclusion of some important information. The recommendations from students, researchers, scientists and academicians will be highly appreciated.

Gandhinagar, India — Syed Shahabuddin
Petaling Jaya, Malaysia — Adarsh Kumar Pandey
Petaling Jaya, Malaysia — Mohammad Khalid
Petaling Jaya, Malaysia — Priyanka Jagadish

Contents

Introduction to Conducting Polymers

Syed Shahabuddin, Nurul Aqilla Mazlan, Siti Nor Atika Baharin, and Kavirajaa Pandian Sambasevam

Abstract In recent years, the research about Conducting Polymers (CPs) have seen exponential growth due to their versatile applications. The widespread attention on CPs is due to its extraordinary properties such as simple preparation step, low cost of monomers, environmentally benign, and most importantly the high conducting properties like metals. In addition, lightweight of CPs and non-corrosive nature, have made it one of the versatile polymers in the materials group. These remarkable properties of CPs have made it to be easily integrated with the latest applications on photocatalyst, sensors, and actuators, solar cells, energy devices, and batteries. However, many have not realised the historical background of these versatile CPs. Hence, this chapter is an attempt to address the forgotten history of CPs with respect to certain selected well-known CPs.

Keywords Conducting polymers · Polyaniline · Polypyrrole · Polythiophene · History of conducting polymers

1 Introduction

Over the years, the concept of conducting polymers (CPs) continues to fascinate the scientific community as several research papers regarding this material have been annually increased. CP is an organic macromolecule that integrates both conventional polymers and metal properties into one system. This "synthetic metal" was

---

S. Shahabuddin (✉)
Department of Science, School of Technology, Pandit Deendayal Petroleum University, Knowledge Corridor, Raisan Village, Gandhinagar, Gujarat 382007, India
e-mail: syedshahab.hyd@gmail.com

N. A. Mazlan
Department of Orthopaedic Surgery, Faculty of Medicine, National Orthopaedic Centre of Excellence for Research & Learning (NOCERAL) University of Malaya, 50603 Kuala Lumpur, Malaysia

S. N. A. Baharin · K. P. Sambasevam
School of Chemistry and Environment, Faculty of Applied Sciences, Universiti Teknologi MARA (UiTM), 72000 Kuala Pilah, Negeri Sembilan, Malaysia

© Springer Nature Switzerland AG 2021

S. Shahabuddin et al. (eds.), *Advances in Hybrid Conducting Polymer Technology*, Engineering Materials, https://doi.org/10.1007/978-3-030-62090-5_1

formerly discovered by a group of Nobel laureates named Hideki Shirakawa, Alan MacDiarmid, and Alan Heeger with the recognition of new phenomena; "dawn of the new plastic age" [1]. Here, these scientists reported a dramatic increase in electrical conductivity of polyacetylene (PA) when doped with controlled amounts of arsenic pentafluoride (AsF_5) together with halogen group elements [2]. Subsequently, a series of CPs such as polyaniline (PANI), polypyrrole (Ppy) and polythiophene (PTh) have been widely developed (Fig. 1). Since the 1980, CPs have been used as the key principle in numerous potential applications such as electronics devices [1–4], energy storage devices [4–6], corrosion inhibitors [7, 8], degradation of pollutants [11–10] electromagnetic shielding materials [11] and sensors [12, 13]. Besides tough mechanical strength and remarkable rheological properties [14], this material possesses high magnetic properties, unique optical features, and good electrochemical reactions [18–14]. CPs also own an excellent light absorption ability where it acts as transporting material in visible-light excitation state, improving photovoltaic efficiency [11, 16].

Furthermore, CPs also show high conductive ability similar to those metals and semiconductors although originally, the polymer was classified as an insulator as for their dominant insulating properties. Figure 2 represents the comparison of conductivity between CPs with other materials. The conduction of electricity in these types of polymers is due to two factors; the conjugation of the double bonds in the polymeric chains and the doping process [17]. The conjugated π electron together with dopant ions allows the mobility of electrons in CPs to acts as charge carriers and to delocalise into the conduction band, thus increasing the metallic behavior of the polymer [17].

Generally, CPs can be doped via two methods widely known methods such as addition (n-doping) or reduction (p-doping) of an electron by using a different type of dopant as shown in Fig. 3. Though CPs own many promising characteristics, there are still few questionable issues that arise such as difficulties in the process, low

Fig. 1 Molecular structure of most studied CPs

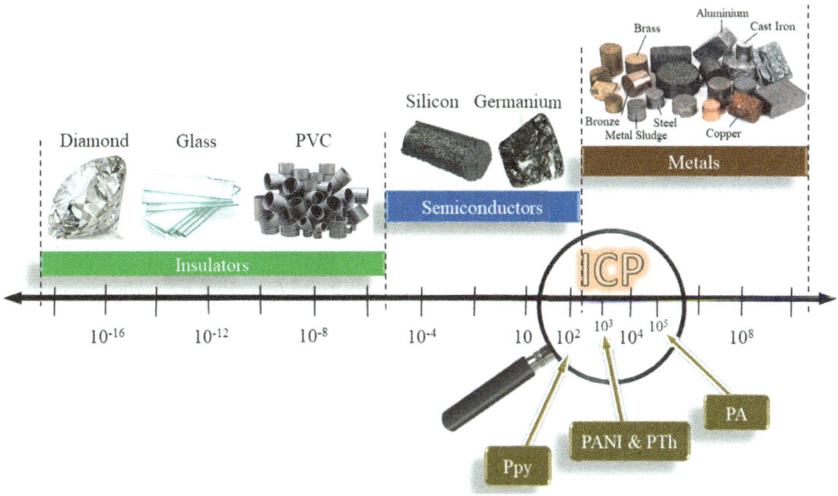

Fig. 2 The comparison of conductivity between CP and other material

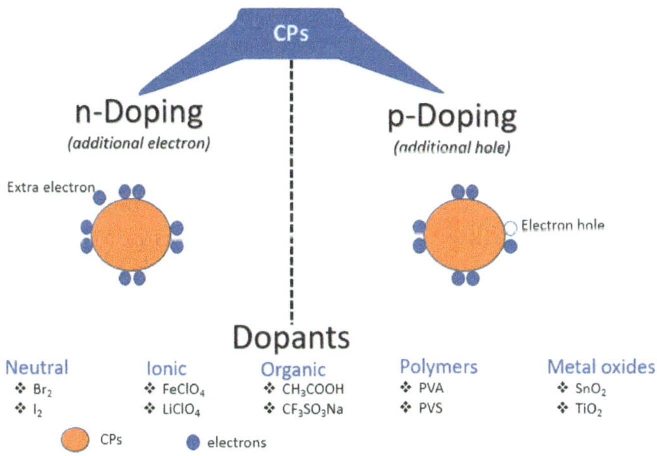

Fig. 3 Types of dopants used in n-doping and p-doping of CP materials

stability, and insolubility issues. Nevertheless, various modifications to the structure and functional group can be done to overcome those flaws in terms of blending, co-polymer, or hybrid composition.

1.1 *Historical Background of the Development of Conducting Polymers*

The initial work on conducting polymers have begun since the nineteenth century by Prof Dr Letheby [18]. In 1862, he attempted to investigate the properties of conducting polymers, and a study on electropolymerisation of aniline sulfate was published in the Journal of the Chemical Society [19]. However, the scientists of that time were still not clear about the chemistry behind the materials that they were investigating. Even though the research about CPs emerged 60 years ago, the possibility of producing those polymers was not recognised. A key discovery that changed the perspective was the findings of inorganic polymer polysulfur nitride $(SN)_x$ which was highly conducting in 1973 [20] by Macdiarmid and Heeger partnership. The researchers also reported that the conductivity of $(SN)_x$ was in the order of 10^3 S/cm which was close to the conductivity of copper (~10^5 S/cm). This discovery convinced the entire scientific community to produce more polymeric conductors.

Polyacetylene was first synthesised by Natta and co-workers which showed a range of conductivity between 10^{-11} and 10^{-3} S/cm in 1958. The conductivity of this polymer could be manipulated by the synthesis process. However, this polymer also did not get widespread attention, until a co-worker of Prof Dr Hideki Shirakawa accidentally added an excess of Ziegler–Natta catalyst which resulted in a silvery polyacetylene thin film in 1967. This film exhibited a higher conductivity than the graphite powder. Then after some time, MacDiarmid who was a visiting professor at Tokyo Institute of Technology, Japan gave a talk on $(SN)_x$ conducting properties where it brought together MacDiarmid and Shirakawa to sit down for a cup of green tea while sharing their fascination for conducting materials. Soon after, MacDiarmid invited Shirakawa to the University of Pennsylvania to work on the CPs with Alan Heeger. In 1976, the group of trios—Alan Heeger, Alan Macdiarmid, and Hideki Shirakawa announced the discovery of novel conducting polymers which triggered the interests of many scientific communities around the world to venture into the field of conducting polymers. The trios received Nobel Prize for conducting polymers in 2000 [21].

1.2 *Types of Conducting Polymers and Their Properties*

1.2.1 Polyacetylene

One of the well-known conducting polymers is polyacetylene. The structure contains C-H units that constructed C_2H_2. The conjugating of single and double bond C–C bond makes polyacetylene special with a stagnant carbon backbone with mobilised electron clouds around the structure as shown in Fig. 4. The structure of polyacetylene depended on the synthesis condition, with normal polymerisation condition cis polyacetylene predominates to trans polyacetylene [22].

Fig. 4 Polyacetylene

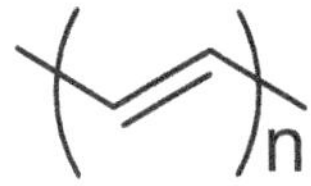

Fig. 5 Polyparaphenylene

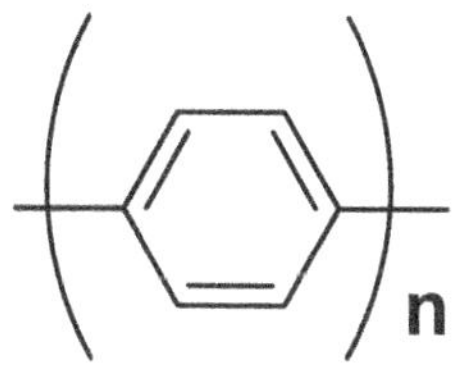

The conductivity of polyacetylene was known to be increased as the polymer chain increased. Doping action with a halogen such as bromine and iodine have greatly increased the conductivity than undoped polyacetylene (10^3–1.7×10^5 S/cm). Polyacetylene was found to be crystalline with x-ray diffraction of powder samples, which were found to be ~ 90–95% crystalline, with a small amount of amorphous content [23]. Insolubility, high thermal stability, sensitivity towards air become some of the drawbacks of polyacetylene which leads to low processability and conducting performance [24, 25].

1.2.2 Polyparaphenylene

Another significant class of conducting polymer is polyparaphenylene (PPP). The polymer chain linked through conjugated aromatic rings as illustrated in Fig. 5. PPP is known to have a tremendous thermal property which can go up to 600 °C. However, PPP shows some issues in terms of solubility, due to the polymeric chains made up by mainly benzene rings [26].

PPV can be processed into a crystalline film by vacuum deposition which in turn shown interesting electronic and optical properties when doped (10^2–10^3 S/cm). The main application of PPV is electroluminescence specifically in organic light-emitting diodes (OLED). Delayed fluorescent and fluorescent are normal behavior of PPP which is mainly contributed by electrons mobilisation in the conjugated chains [27].

1.2.3 Polyparaphenylene Vinylene

High crystallinity rigid-rod structure is a property for polypara-phenylene vinylene (PPV). The polymeric chain is connected via alternating of benzene and vinyl group; Fig. 6 Similar to other conducting polymers, PPV tends to have unusual electrical properties upon doping (3–5 × 10^4 S/cm) and can be tailored by the functionalisation inclusion. PPV can be processed with a greatly ordered crystalline thin film.

Fig. 6 Polyparaphenylene vinylene

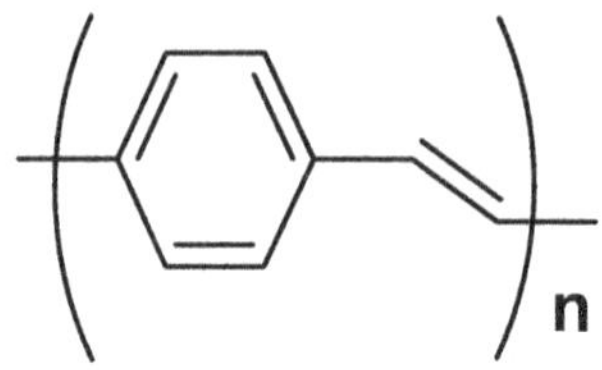

Dopants such as H_2SO_4 increase the conductivity of PPV with good metallic transport property. However, the H2SO_4-PPV dopant was revealed to have low stability in the presence of moisture [28].

Moreover, many approaches haven done in incorporating PPV with either organic or inorganic materials to enhance or tune the conducting properties of PPV was made in light emitting diode applications such as polysilicon-PPV [29], polyparaphenylene-PPV [30] and poly(2,5-dimethyl-para-phenylene vinylene)-PPV [31]. Another, interesting application of PPV, is solar cell device, for example blend of the poly 2,5-dimethoxy-1,4-phenylenevinylene-2-methoxy-5-(2-ethylhexyloxy-1,4-phenylenevinylene) (M3EH-PPV) and polyoxa-1,4-phenylene1,2-(1-cyano-ethylene-2,5-dioctyloxy-1,4-phenylene-1,2-(2-cyano-ethylene-1,4-phenylene) (CN-ether-PPV) [32] and TiO_2-PPV bilayer [33].

1.2.4 Polypyrrole

Pyrrole is a 5-membered ring that contains nitrogen (N) heteroatom. The conducting properties of polypyrrole (PPy) is a result of π electrons and a positive charge along the backbone [34]. PPy was first synthesised by using a common oxidative route with pyrrole monomer and hydrogen peroxide as the oxidising agent in 1916 [34]. The synthesis yielded a compound which was known as 'pyrrole black' and was not recognised as CPs. Then, in 1968 Dall'Ollio and co-workers prepared PPy via an electrochemical route on a platinum electrode which has given the recognition for PPy as CPs [35]. This discovery has led to explosive research that revolves around the synthesis and application of PPy materials [30–25].

PPy possessed enormous advantages such as large surface area due to its fibrous structure and it is a high capacity electrode material. PPy can be easily synthesised via chemical or electrochemical route, high stability in aqueous and air medium makes it a promising candidate interfacial material [37]. Therefore, it made PPy one of the extensively applied and investigated materials in the current state of the art.

1.2.5 Polythiophene

Owing to the interesting properties; stable towards environment perturbation, thermal, conductivity, ease of synthesis, and variety of application standpoint, polythiophene (PTh) becomes one of the most studied conducting polymers for the last

Fig. 7 Polythiophene

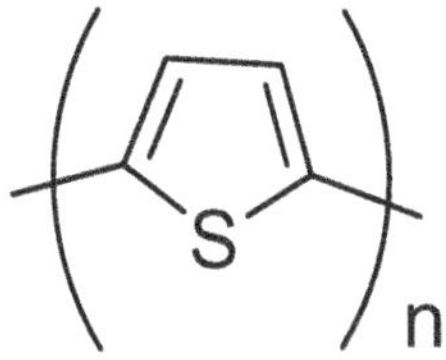

twenty years with relative conductivity range of 10–10^3 S/cm [38]. PTh belongs to a class of heterocyclic compounds containing a five-membered ring made up of one sulfur as heteroatom with the formula C_4H_4S. Thiophene and its derivatives isolated from petroleum or coal. Owing to their efficient light harvesting, structural versatility, and intrinsic charge transport behavior, thiophene-based π-conjugated systems have attracted much attention in developing high performance conducting cells. Figure 7 demonstrates the structure of thiophene repeating units PTh.

Property tailoring to match an explicit electronic application is likely by integrating electronic groups along with the conjugated framework in addition to copolymerising with other heterocyclic monomers. Copolymerisation and homopolymerisation of PTh are commonly prepared either electrochemically or via chemical polymerisation [39]. have been the focus of many applications due to its outstanding optoelectrical properties including field-effect transducers (OFET), light-emitting diodes (OLED), photovoltaics (OVD), and nonlinear optical devices (NLO) [40, 41]. Moreover, PTh also has been applied as materials for the electrochromic device in which by principle the material owns the capacity to transform the color by tuning the redox state; for example, polyalkyl thiophene can appear blue color in oxidised and red in the reduced state [42].

1.2.6 Polyaniline

Among the CPs, polyaniline (PANI) has attracted widespread attention around the world due to its versatile properties such as ease of synthesis by both chemical and electrochemical route, thermal stability up to 250 °C [43], unique acid–base chemistry and tunable electrical conductivity [44]. These extraordinary properties opened a wider window of application for PANI in solar cells [45], corrosion inhibitors [46], organic light-emitting diodes [47], electromagnetic interfering devices [11] and bio/chemical sensors [48]. These widespread applications of PANI have made it one of the versatile CP in its group.

Historically, PANI was prepared and used centuries ago, before it was realised that it could be functionalised as a semiconductor. The research on PANI was started by Runge (1834) and was extensively continued by Prof Dr Letheby [49]. Between 1907 and 1911 a Nobel Laureate Richard Willstatter characterised the PANI derivatives via his methodic way of research. Then, in the 1960 a group of researchers from

Czechoslovak developed an iodine doped PANI which produced a 1 S/cm conductivity for the PANI-iodine complexes. However, at that time, the whole global attention was diverted to the discovery of polyacetylene by the Nobel Prize trios. Then, the conducting properties of PANI were reaffirmed by another study conducted in parallel with PPy research. Interestingly, the study reported a higher conductivity for PANI ranging from 5–30 S/cm.

Then, the research of Willstatter was continued and more oxidation states of PANI were revealed. Figure 8 shows the chemical structure PANI in various oxidation states. PANI possessed different colors in the acid–base environment [50] and reversible protonation-deprotonation which made it an exclusive CP compare to other CPs in the group [51].

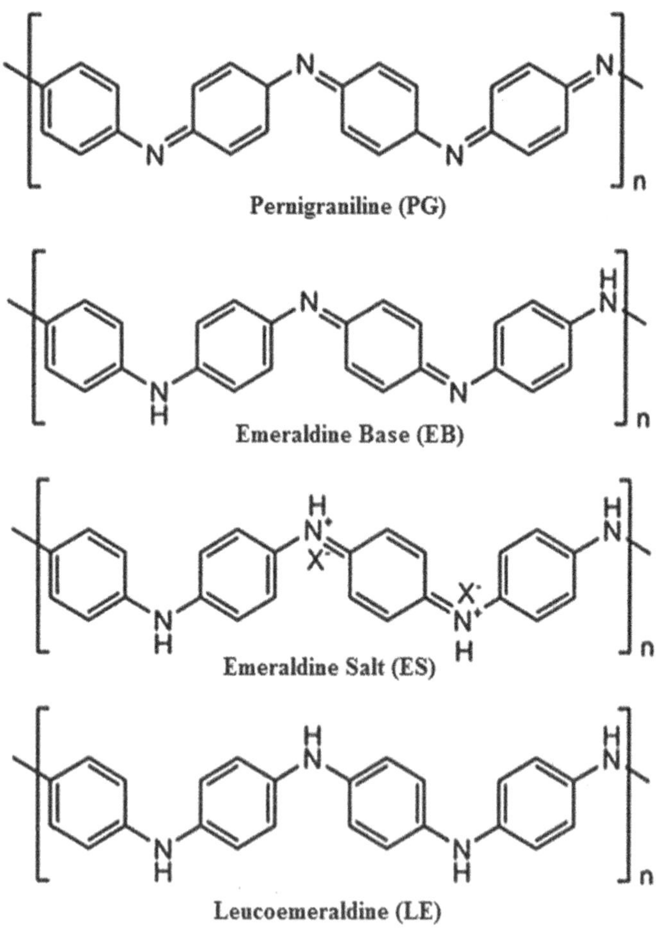

Fig. 8 Various oxidations states of PANI [52]

2 Synthesis Pathways and Polymerisation Mechanism

2.1 *Synthesis Method for Conducting Polymers (CPs)*

Polymerisation is a process where the monomer undergoes chemical reaction to form a polymeric chain. Theoretically, the monomer that has been oxidised will produce radical cation, which will be reacted with another monomer, resulting in the formation of an oligomer [53]. In general, three methods that commonly used to synthesise CPs are (i) chemical polymerisation, (ii) electrochemical polymerisation, and (iii) photochemical polymerisation. Each of the synthesis methods possesses its advantages and disadvantages.

2.1.1 Chemical Synthesis

In general, this method uses dopant ions such as ammonium persulfate or potassium nitrate to react with the targeted monomer through the oxidation process [23, 24]. Chemical synthesis can be divided into two main classes which are condensation polymerisation with step-growth mechanism and addition polymerisation with chain-growth mechanism (Fig. 9).

Chemical synthesis is the most widely used technique for CPs due to mass production, small expenditure, processability and it offers various routes in accordance with the type of CPs used [54, 24, 25]. However, chemical synthesis can only produce thick film and have a problem of large activation barriers in its reaction, making it inconvenient for certain application [53].

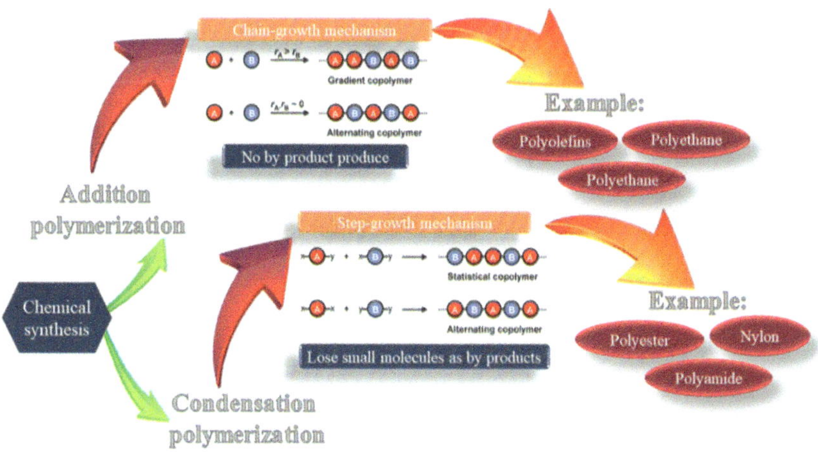

Fig. 9 Classification of chemical synthesis in CPs' polymerization

Another chemical technique commonly used in this chemical synthesis are sol–gel polymerisation and emulsion polymerisation. Sol–gel polymerisation (Fig. 10) is a colloidal precipitation process of dispersing polymeric molecules into a solid–liquid state [55]. This technique offers several advantages such as it occurs at relatively low temperature, has high homogeneity of organic moieties into the inorganic materials and it is an effortless fabricating process [56].

Despite that, sol–gel polymerisation often faces problems in porosity and weak intermolecular bonding [56]. On the other hand, emulsion polymerisation (Fig. 11) is a hetero-phase process involving initiator, monomer, and surfactant where it is instantaneously added to the solvent (water), resulting in the formation of micelles [16]. The monomers get polymerised within the micelles and lead to the formation of stable colloidal dispersions [55]. This polymerisation technique lowers the viscosity of the reaction making good reaction kinetics between the hybrid materials [55]. Moreover, this technique produces non-toxic and inflammable product [55], thus making it an environmentally friendly approach, however, the usage of surfactants limit its usage in other applications [57].

Besides, the selection of an appropriate oxidising agent is crucial to yield CPs with desired properties. Besides that, the use of oxidising agents in solution polymerisation is feasible, less time consuming, common, and economic method for the synthesis of CPs. Various oxidising agents have been utilised by the researchers in the synthesis art of CPs such as ammonium peroxydisulfate [48], hydrogen peroxide [58], ferric chloride [59] and ceric nitrate and sulfate [60] to obtain certain desired properties in CPs.

The use of chemical polymerisation has been widely reported by many authors. In a study by Peng et al. [61] polythiophene grafted poly(methacrylate) (PTh-g-PMA)

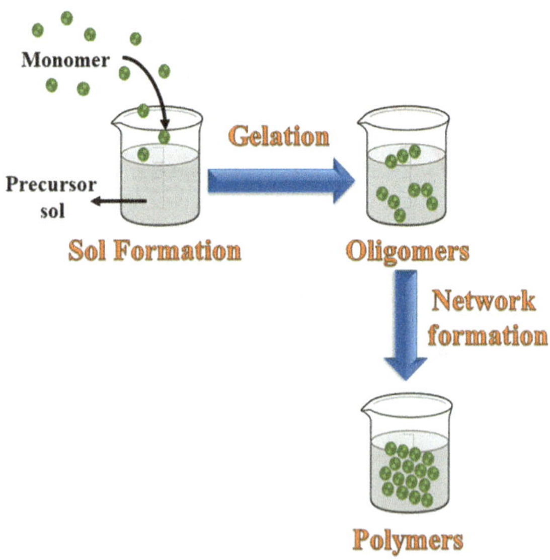

Fig. 10 Basic mechanism of sol gel polymerization

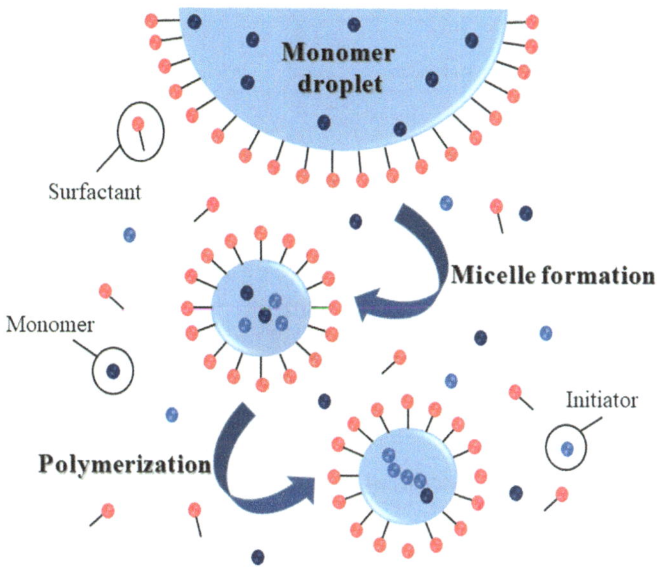

Fig. 11 Basic mechanism of emulsion polymerization

was synthesised using in-situ chemical polymerisation to investigate the optical properties of this material. This material has a great photon efficiency whereby it can reduce 85% of photoluminescence, making it suitable for photovoltaic applications. Similarly, Kwon et al. [62] have also investigated PPy composite by sol–gel technique as suitable photovoltaic materials.

The authors found that the higher the PPy concentration, the higher the photoelectric power conversion. This is due to the uniform chain interpretation, which intensifies the charge transfer and thus resulting in increased power conversion. Another study conducted by Dai et al. [63] have reported the synthesis of PEDOT using a high concentration of poly[2-(3thienyl)- ethoxy-4-butylsulfonate] (PTEB) has boosted the thermal stability, crystallinity and electrical conductivity of the hybrid material. Clement et al., have reported that PTh hybrid silica nanomaterial synthesised via chemical oxidative method [64] has an excellent performance in terms of optical and thermal features whereby it could be the primary material for LED, photoactive and conducive materials.

2.1.2 Electrochemical Synthesis

Electrochemical synthesis is a process using a three-electrode system that allows manipulation of molecular oxidation states and capable of generating high reactive species [65]. The principle of electrochemical synthesis includes a system where it makes use of working, counter, and reference electrodes in the presence of monomer, suitable electrolyte, and proper additives in the system [53]. The polymerisation can

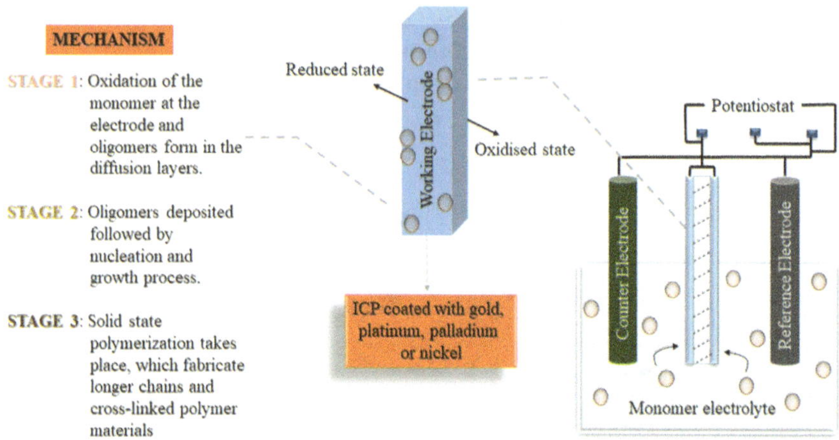

Fig. 12 Electrochemical polymerization cell setup and its basic mechanism

be done in three different techniques; with constant current (galvanostatic), constant applied potential (potentiostatic) or potential scanning or sweeping method [66]. Figure 12 shows the basic mechanism in electrochemical synthesis. This technique is a simple and reproducible method for CPs polymerisation. Additionally, electrochemical synthesis with sufficient electrical potential allows the film to have a good control in the film thickness.

Electrochemical synthesis can produce high-quality films with a desired thickness, which makes this route is much preferable to compare to chemical synthesis polymerisation in producing a thin film of CPs. Besides, the electrochemical route has been proven to offer a better polymerisation mechanism for CPs, which easily controls the initiation and termination steps during polymerisation. Moreover, electrochemical polymerisation provides a cleaner and cheaper route because it does not require the addition of oxidants, surfactants, and so on. Hence, it makes this method one of the facile and environmentally benign procedures [67].

However, this method has complications in terms of producing bulk sample as it is difficult to discharge the film that has been deposited on the electrode surface [53]. The use of gold and platinum as an electrode in electrochemical synthesis also becomes a conflict as the material is expensive. Moreover, this method is also restricted to the only monomer that can undergo oxidation through applied potential [65].

Despite its drawback, many authors have reported that electrochemical synthesis brings out an excellent electrocatalytic and conductive activity of the hybrid material whether in acid, alkali, or neutral medium, which makes them a suitable candidate for various applications such as sensors, catalysis, and energy conversion devices [55]. Wang et al. [68] reported PANI that has been synthesised via electrochemical polymerisation with constant potential (0.75 V) possesses excellent gravimetric and volumetric capacitance to compare to the existing carbon-based electrode, which enables it to become a suitable electrode in supercapacitors. Another study conducted by Liu et al. [69] also observed an increased conductivity of PPy when synthesised using electrochemical polymerisation. This is because of the surface morphology of this hybrid material is denser and more compact, which is responsible for enhancing conductivity. This hybrid material was synthesised using electrochemical polymerisation under UV irradiation constant applied potential of 0.85 V. Cai et al. showed a similar result., [70] where synthesised PANI via solvothermal and electrochemical polymerisation exhibit unique nanotubules structure, which leads to high electrical conductivity. PEDOT synthesised by electrochemical polymerisation also showed good electrochemical ability with maximum faradaic interaction [71].

2.1.3 Photochemical Synthesis

Photochemical synthesis is a process that utilised light sources such as ultraviolet (UV) light, visible light, or laser generating radicals as the initiator for monomer polymerisation [53]. This technique reduces the time consuming experimental procedures, economical method and it is not destructive to the surroundings [16]. Moreover, in comparison with chemical synthesis, photochemical polymerisation produces its radical via hydrogen abstraction by irradiation which gives higher efficiency rather than direct fragmentation via thermal reaction in the chemical polymerisation method.

Additionally, this method provides great control over the shape, size, and physical properties of the CPs. This is because in this method, many parameters, such as light intensity and temperature, can be easily controlled. For example, the initiation rate can be easily adjusted by turning the light source on or off [53]. The photochemical synthesis method also enhanced the electrical conductivity of nanomaterial as illumination of the semiconductor electrode embedded in this method employs band gap irradiation, which will help in increasing the conductivity via photoconductivity. The basic mechanism is illustrated in Fig. 13 below.

Photochemical synthesis can be divided into two which are direct polymerisation (where this method directly absorbs the photon energy to decompose the monomer to form radicals) and photosensitised polymerisation (where this method use dye-sensitised semiconductor (metal oxide or chalcogenides) as a photosensitiser to transfer the photon energy into the corresponding excited states. For CPs, it should be noted that it is impossible to use direct photochemical polymerisation. This is because direct polymerisation gives more positive oxidation peak potential compare to redox potential by photosensitiser in photosensitise polymerisation [72].

The working principle of photochemical polymerisation is as follows; when photosensitiser being hit with photon energy greater than the bandgap energy, it will make photosensitiser undergo electron excitation. The electron from the valence band jumps to the conduction band of the semiconductor, leaving a hole that is highly oxidising. The polymerisation takes place when the holes produced reacts with the monomer. A primary aspect of photochemical polymerisation is that the oxidation

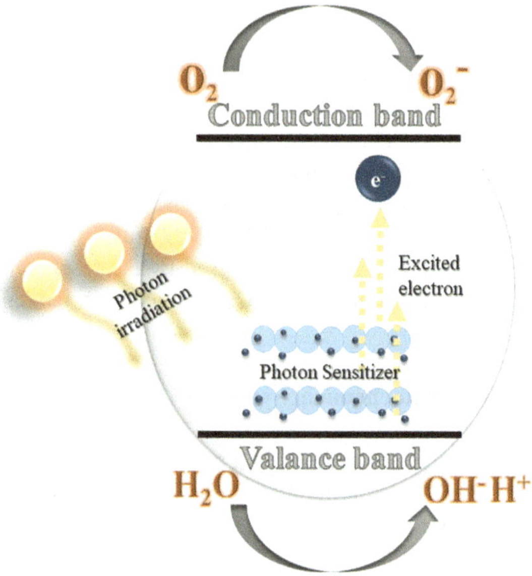

Fig. 13 Basic mechanism of photochemical synthesis

step is guided by the difference between the potential of the valence band edge of the inorganic species and the oxidation potential of the monomer [53].

Lee and co-workers have carried out an extensive study of using the photochemical route to synthesise, [73] polythiophene (PTh) with self-assembled $RuL_2(NCS)_2$/Di(3-Aminopropyl)Viologen on Indium Thin Oxide (ITO) as working electrode using 10% thiophene in acetonitrile as the electrolyte medium and platinum as the counter electrode. Here, they reported that the PTh showed a thickness of ~360 nm, a dense surface, and possessed an optical gap of 2.38 eV. Zhang et al. [74] have conducted a photochemical synthesis of poly(3,4-ethylenedioxythiophene) as hole-transporting material in dye-sensitised solar cell (DSSC) application. The material showed a better dye regeneration along with the high charge recombination energy levels. Besides, it also showed excellent light absorption capability where it exhibits a high power conversion efficiency of 5.2%.

Another study in the same DSSC application also showed a similar result where the use of PEDOT as the hole transporting material enhanced proficient power conversion efficiency of 3.4%. The procedure of in-situ photochemical polymerisation involved the use of stainless steel as counter electrode and two electrolyte medium which are organic electrolyte (10 mM EDOT and 0.1 M $LiN(CF_3SO_2)_2$ in acetonitrile) and aqueous micellar electrolyte (10 mM EDOT in an aqueous solution containing 0.1 M $LiN(CF_3SO_2)_2$ and 50 mM TritonX-100 surfactant) [75]. The summary of each method presented in this section is listed in Table 1.

Table 1 Different polymerisation methods of CPs

Routes	Details	Technique	Advantages	Limitation
Chemical	A method that uses a dopant ion to react with targeted monomer through the oxidation process	Sol-gel polymerisation Emulsion polymerisation	Low cost Easy process Produce bulk quantity of sample	Large activation barriers Poor control of film thickness
Electrochemical	A method that uses a three-electrode system with the applied potential to generate high reactive species	Galvanostatic Potentiostatic Potential scanning/sweeping	Great control on film thickness Reproducible method	Difficult to discharge film deposited on the electrode Consume a lot of time
Photochemical	A method that utilised light intensity in generating radicals	Direct polymerisation Photosensitiser	Great control of the morphology of the material Green process Fast reaction	Oxygen inhibition Need to use a high resolution of light intensity

3 Conclusion

CPs are a versatile group of macromolecules. Even though it was challenging to understand the polymerisation and the conducting properties of CPs back in the 1970, the current industrial revolution with sophisticated instrumentation techniques has opened the door for a better understanding of these macromolecules. Hence, it is always important to track the evolution of CPs in terms of its origin and synthetic route to tailor a better polymer engineering in the prospect. In conclusion, this chapter covered most of the notable CPs and its polymerisation pathway which would bridge the knowledge gap to enhance the design of new CPs.

References

1. Shirakawa, H., Louis, J., Macdiarmid, A.G.: Synthesis of electrically conducting organic polymers: halogene derivatives of polyacetylene, (CH)x. J. C. S. Chem. Comm. **578**, 578–580 (1977)
2. Chiang, C., et al.: Electrical conductivity in doped polyacetylene. Phys. Rev. Lett. **39**(17), 1098–1101 (1977)
4. Kumar, L., Rawal, I., Kaur, A., Annapoorni, S.: Flexible room temperature ammonia sensor based on polyaniline. Sens. Actuat. B Chem. **240**, 408–416 (2017)
5. AlSalhi, M.S., Alam, J., Dass, L.A., Raja, M.: Recent advances in conjugated polymers for light emitting devices. Int. J. Mol. Sci. **12**(3), 2036–2054 (2011)

6. Yun, S., Freitas, J.N., Nogueira, A.F., Wang, Y., Ahmad, S., Wang, Z.S.: Dye-sensitised solar cells employing polymers. Prog. Polym. Sci. **59**(February), 1–40 (2016)
7. Ng, H.M., Ramesh, S., Ramesh, K.: Quasi-solid polymer electrolyte composed of poly(1-vinylpyrrolidone-co-vinyl acetate) copolymer and the influence of its composition on electrochemical properties and the performances of dye-sensitised solar cells. Polym. Plast. Technol. Eng. **57**(2), 98–107 (2018)
8. Kausar, A.: Overview on conducting polymer in energy storage and energy conversion system. J. Macromol. Sci. Part A **54**(9), 640–653 (2017)
9. Deshpande, P.P., Jadhav, N.G., Gelling, V.J., Sazou, D.: Conducting polymers for corrosion protection: a review. J. Coatings Technol. Res. **11**(4), 473–494 (2014)
10. Shahabuddin, S., Sarih, N.M., Ismail, F.H., Shahid, M.M., Huang, N.M.: Synthesis of chitosan grafted-polyaniline/Co_3O_4 nanocube nanocomposites and their photocatalytic activity toward methylene blue dye degradation. RSC Adv. **5**(102), 83857–83867 (2015)
11. Koh, Y., Sambasevam, K.P., Yahya, R., Phang, S.: Improvement of microwave absorption for PAni/HA/TiO2/Fe3O4 nanocomposite after chemical treatment. Polym. Compos. **34**(7), 1186–1194 (2013)
12. Weder, C.: Functional polymer blends and nanocomposites. Chimia (Aarau) **63**(11), 758–763 (2009)
13. Opoku, F., Kiarii, E. M., Govender, P. P., Mamo, M. A.: Metal oxide polymer nanocomposites in water treatments. Descr. Inorg. Chem. Res. Met. Compd. August 2017
14. Syed, S.: Polyaniline based nanocomposites as adsorbents and photocatalysts in the removal of organic dyes. pp. 1–179 (2016)
16. Kumar, R., Singh, S., Yadav, B.C.: Conducting polymers: synthesis, properties and applications. Int. Adv. Res. J. Sci. Eng. Technol. **2**(11), 595–604 (2015)
17. Kaur, G., Adhikari, R., Cass, P., Bown, M., Gunatillake, P.: Electrically conductive polymers and composites for biomedical applications. RSC Adv. **5**(47), 37553–37567 (2015)
18. Scott, J.C.: History of conductive polymers, In: Eftekhari, A. (ed.) Nanostructured Conductive Polymers, John Wiley & Sons Ltd, pp. 1–17 (2010)
19. Essays, U.: History of conducting polymers engineering essay. UKEssays.com (2018) (Online). Available: https://www.ukessays.com/essays/engineering/history-of-conducting-polymers-engineering-essay.php?vref=1. Accessed 29 Apr 2020
20. Walatka, V., Labes, M., Perlstein, J.: Polysulfur Nitride—a one-dimensional chain with a metallic ground state. Phys. Rev. Lett. **31**(18), 1139–1142 (1973)
21. Bengt, N.: The Nobel Prize in Chemistry, 2000: Conductive polymers. Kungl, Vetenskapsakademien (2000)
22. MacDiarmid, A.G., Heeger, A.J.: Organic metals and semiconductors: the chemistry of polyacetylene, (CH)x, and its derivatives. Synth. Met. **1**(2), 101–118 (1980)
23. Etemad, S., Heeger, A.J.: Polyacetylene, (CH)x: the prototype conducting polymer. Annu. Rev. Phys. Chem. **33**(1), 443–469 (1982)
24. Saxman, A.M., Liepins, R., Aldissi, M.: Polyacetylene: Its synthesis, doping and structure. Prog. Polym. Sci. **11**(1–2), 57–89 (1985)
25. Chien, J. C. W.: Polyacetylene : Chemistry, Physics, and Material Science. Elsevier Science (1984)
26. Brédas, J.L., Street, G.B., Thémans, B., André, J.M.: Organic polymers based on aromatic rings (polyparaphenylene, polypyrrole, polythiophene): evolution of the electronic properties as a function of the torsion angle between adjacent rings. J. Chem. Phys. **83**(3), 1323–1329 (1985)
27. Hertel, D., Setayesh, S., Nothofer, H.G., Scherf, U., Müllen, K., Bässler, H.: Phosphorescence in Conjugated Poly(para-phenylene)-Derivatives. Adv. Mater. **13**(1), 65–70 (2001)
28. Ahlskog, M., Reghu, M., Noguchi, T., Ohnishi, T.: Doping and conductivity studies on poly(p-phenylene vinylene). Synth. Met. **89**(1), 11–15 (1997)
29. Kimura, M., et al.: Low-temperature polysilicon thin-film transistor driving with integrated driver for high-resolution light emitting polymer display. IEEE Trans. Electron Devices **46**(12), 2282–2288 (1999)

30. De Carvalho, L.C., Dos Santos, C.N., Alves, H.W.L., Alves, J.L.A.: Theoretical studies of poly(para-phenylene vinylene) (PPV) and poly(para-phenylene) (PPP). Microelectron. J. **34**(5–8), 623–625 (2003)
31. Kiebooms, R., Resel, R., Vanderzande, D., Leising, G.: Polymer leds based on N-Alkylsulfinyl Ppv precursor polymers. MRS Proc. **558**, 409 (1999)
32. Melzer, C., et al.: Hole transport in poly(phenylene vinylene)/methanofullerene bulk-heterojunction solar cells. Adv. Funct. Mater. **14**(9), 865–870 (2004)
33. Van der Zanden, B., Goossens, A.: Oxygen doping of TiO_2/poly(phenylene-vinylene) bilayer solar cells. J. Appl. Phys. **94**(10), 6959–6965 (2003)
34. Heeger, A.J.: Semiconducting and Metallic Polymers: the fourth generation of polymeric materials (nobel lecture). Angew. Chem. Int. Ed. Engl. **40**(14), 2591–2611 (2001)
35. Papeo, G., Pulici, M.: Italian chemists' contributions to named reactions in organic synthesis: an historical perspective. Molecules **18**(9), 10870–10900 (2013)
37. Song, X., et al.: In situ pPy-modification of chitosan porous membrane from mussel shell as a cardiac patch to repair myocardial infarction. Appl. Mater. Today **15**, 87–99 (2019)
38. Kutsche, C., Targove, J., Haaland, P.: Microlithographic patterning of polythiophene films. J. Appl. Phys. **73**(5), 2802–2804 (1993)
39. Xu, Q., An, L., Yu, M., Wang, S.: Design and synthesis of a new conjugated polyelectrolyte as a reversible pH sensor. Macromol. Rapid Commun. **29**(5), 390–395 (2008)
40. Huo, L., Guo, X., Zhang, S., Li, Y., Hou, J.: PBDTTTZ: a broad band gap conjugated polymer with high photovoltaic performance in polymer solar cells. Macromolecules **44**(11), 4035–4037 (2011)
41. Mannerbro, R., Ranlöf, M., Robinson, N., Forchheimer, R.: Inkjet printed electrochemical organic electronics. Synth. Met. **158**(13), 556–560 (2008)
42. Nawa, K., Imae, I., Noma, N., Shirota, Y.: Synthesis of a novel type of electrochemically doped vinyl polymer containing pendant terthiophene and its electrical and electrochromic properties. Macromolecules **28**(3), 723–729 (1995)
43. Ren, L., Zhang, X.F.: Preparation and characterisation of polyaniline micro/nanotubes with dopant acid mordant dark yellow GG. Synth. Met. **160**(7–8), 783–787 (2010)
44. Sambasevam, K. P., Mohamad, S., Phang, S.-W.: Effect of dopant concentration on polyaniline for hydrazine detection vol. 33, Elsevier (2015)
45. Kawata, K., Gan, S.N., Ang, D.T.C., Sambasevam, K.P., Phang, S.W., Kuramoto, N.: Preparation of polyaniline/TiO_2 nanocomposite film with good adhesion behavior for dye-sensitised solar cell application. Polym. Compos. **34**(11), 1884–1891 (2013)
46. Zheng, L., Song, J.: Curcumin multi-wall carbon nanotubes modified glassy carbon electrode and its electrocatalytic activity towards oxidation of hydrazine. Sens. Actuat. B Chem. **135**(2), 650–655 (2009)
47. Moulton, S., Innis, P., Kane-Maguire, L.A., Ngamna, O., Wallace, G.: Polymerisation and characterisation of conducting polyaniline nanoparticle dispersions. Curr. Appl. Phys. **4**(2–4), 402–406 (2004)
48. Sambasevam, K.P., Mohamad, S., Phang, S.W.: Sensor kimia untuk mengesan hidrazin dengan menggunakan polyanilina berbentuk filem nipis. Malaysian J. Anal. Sci. **21**(4), 762–769 (2017)
49. Letheby, H.: On the production of a blue substance by the electrolysis of sulphate of aniline. J. Chem. Soc. **15**, 161 (1862)
50. Bai, H., Chen, Q., Li, C., Lu, C., Shi, G.: Electrosynthesis of polypyrrole/sulfonated polyaniline composite films and their applications for ammonia gas sensing. Polymer (Guildf) **48**(14), 4015–4020 (2007)
51. Pang, Z., Fu, J., Lv, P., Huang, F., Wei, Q.: Effect of CSA concentration on the ammonia sensing properties of CSA-Doped PA6/PANI composite nanofibers. Sensors **14**(11), 21453–21465 (2014)
52. Kavirajaa Pandian, S.: Synthesis and characterisation of polyaniline for hydrazine detection/Kavirajaa Pandian Sambasevam (2015)
53. Nguyen, D. N., Yoon, H.: Recent advances in nanostructured conducting polymers: from synthesis to practical applications. Polymers (Basel). **8**(4) (2016)

54. Bakhshi, A.K., Bhalla, G.: Electrically conducting polymers: materials of the twentyfirst century. J. Sci. Ind. Res. **63**(September), 715–728 (2004)
55. Iqbal, S., Ahmad, S.: Recent development in hybrid conducting polymers: Synthesis, applications and future prospects. J. Ind. Eng. Chem. (2017)
56. Danks, A.E., Hall, S.R., Schnepp, Z.: The evolution of 'sol–gel' chemistry as a technique for materials synthesis. Mater. Horiz. **3**(2), 91–112 (2016)
57. Yamak, H. B.: Emulsion polymerization : effects of polymerisation variables on the properties of vinyl acetate based emulsion polymers. Emuls. Polym. 35–73 (2013)
58. Wang, Y., Jing, X., Kong, J.: Polyaniline nanofibers prepared with hydrogen peroxide as oxidant. Synth. Met. **157**(6–7), 269–275 (2007)
59. Yasuda, A., Shimidzu, T.: Chemical and electrochemical analyses of polyaniline prepared with FeCl3. Synth. Met. **61**(3), 239–245 (1993)
60. Hand, R.L.: The Anodic decomposition pathways of ortho- and meta-substituted anilines. J. Electrochem. Soc. **125**(7), 1059 (1978)
61. Peng, X., Zhang, L., Chen, Y., Li, F., Zhou, W.: In situ preparation and fluorescence quenching properties of polythiophene/ZnO nanocrystals hybrids through atom-transfer radical polymerisation and hydrolysis. Appl. Surf. Sci. **256**(9), 2948–2955 (2010)
62. Kwon, J.D., Kim, P.H., Keum, J.H., Kim, J.S.: Polypyrrole/titania hybrids: Synthetic variation and test for the photovoltaic materials. Sol. Energy Mater. Sol. Cells **83**(2–3), 311–321 (2004)
63. Matsumura, S., et al.: Ionomers for proton exchange membrane fuel cells with sulfonic acid groups on the end-groups: Novel branched poly(ether-ketone)s. Am. Chem. Soc. Polym. Prepr. Div. Polym. Chem. **49**(1), 511–512 (2008)
64. Clément, S., et al.: Synthesis and characterisation of π-conjugated polymer/silica hybrids containing regioregular ionic polythiophenes. J. Mater. Chem. **21**(8), 2733 (2011)
65. Yan, S., Wu, Q., Chang, A., Lu, F., Xu, H.C., Wu, W.: Electrochemical synthesis of polymer microgels. Polym. Chem. **6**, 3979–3987 (2015)
66. Sadki, S., Schottland, P., Sabouraud, G., Brodie, N.: The mechanisms of pyrrole electropolymerisation. RSC Adv. 283–293 (2000)
67. Bhadra, S., Khastgir, D., Singha, N.K., Lee, J.H.: Progress in preparation, processing and applications of polyaniline. Prog. Polym. Sci. **34**(8), 783–810 (2009)
68. Paper, C., et al.: Fabrication of Graphene/Polyaniline **3**(7), 1745–1752 (2009)
69. Liu, Y., Huang, J., Tsai, C., Chuang, T. C., Wang, C.: Effect of TiO_2 nanoparticles on the electropolymerisation of polypyrrole. 387, 155–159 (2004)
70. Cai, G.F., Tu, J. P., Zhou, D., Zhang, J. H., Wang, X. L., Gu, C. D.: Solar energy materials & solar cells dual electrochromic fi lm based on WO 3/polyaniline core/shell nanowire array **122**, 51–58 (2014)
71. Jacob, D., Mini, P.A., Balakrishnan, A., Nair, S.V., Subramanian, K.R.V.: Electrochemical behaviour of graphene – poly (3, 4-ethylene- dioxythiophene) (PEDOT) composite electrodes for supercapacitor applications. Bull. Mater. Sci **37**(1), 61–69 (2014)
72. Yamada, K., Yamada, Y., Sone, J.: Three-dimensional photochemical microfabrication of poly(3,4-ethylene- dioxythiophene) in transparent polymer sheet. Thin Solid Films **554**, 102–105 (2014)
73. Lee, W., Hyung, K.H., Hwang, Y., Cho, B.W., Lee, S.H., Han, S.H.: Photoelectrochemical polymerisation of thiophene on self-assembled RuL 2(NCS) 2/Di(3-aminopropyl)viologen on indium thin oxide. J. Nanosci. Nanotechnol. **11**(5), 4501–4505 (2011)
74. Zhang, J., et al.: Poly(3,4-ethylenedioxythiophene) hole-transporting material generated by photoelectrochemical polymerisation in aqueous and organic medium for all-solid-state dye-sensitised solar cells. J. Phys. Chem. C **118**(30), 16591–16601 (2014)
75. Zhang, J., Jarboui, A., Vlachopoulos, N., Jouini, M., Boschloo, G., Hagfeldt, A.: Photoelectrochemical polymerization of EDOT for solid state dye sensitized solar cells: role of dye and solvent. Electrochim. Acta **179**, 220–227 (2015)

Intrinsically Conducting Polymer Based Nanocomposite in Photocatalytic Study

Syed Shahabuddin, Nurul Aqilla Mazlan, Siti Nor Atika Baharin, Kavirajaa Pandian Sambasevam, and Adarsh Kumar Pandey

Abstract It is well noted that Intrinsically Conducting Polymer (ICP) have been the main focus for many researchers for many applications involving energy storage devices and system. It is due to its remarkable properties, simple preparation technique, environmentally stable and safe as well as low expenditure cost. Integrating ICP with nanocomposite have also widened its capabilities in venturing into other application such as photocatalysis process. The photocatalytic study is a photoreaction process involving the use of nanomaterials and sunlight. Using ICP nanocomposite (ICP) as the sole material in the photocatalysis process increases the performance for photocatalytic study in various applications, water remediation, and production of Hydrogen gas. This chapter is an attempt of combining the recent performance of ICP nanocomposites mainly in the photocatalytic study that involves the preparation routes, fundamentals, comparison on recent literature with prospects and recommendation.

Keywords Conducting polymer · Nanocomposite · Photocatalyst · Photocatalytic process · Polyaniline

---

S. Shahabuddin (✉)
Department of Science, School of Technology, Pandit Deendayal Petroleum University, Knowledge Corridor, Raisan Village, Gandhinagar, Gujarat 382007, India
e-mail: Syedshahab.hyd@gmail.com

N. A. Mazlan
Department of Orthopaedic Surgery, Faculty of Medicine, National Orthopaedic Centre of Excellence for Research and Learning (NOCERAL), University of Malaya, 50603 Kuala Lumpur, Malaysia

S. N. A. Baharin · K. P. Sambasevam
School of Chemistry and Environment, Faculty of Applied Sciences, Universiti Teknologi MARA (UiTM), 72000 Kuala Pilah, Negeri Sembilan, Malaysia

A. K. Pandey
Research Centre for Nano-Materials and Energy Technology (RCNMET), School of Engineering and Technology, Sunway University, No. 5, Jalan Universiti, Bandar Sunway, 47500 Petaling Jaya, Selangor, Malaysia

© Springer Nature Switzerland AG 2021

S. Shahabuddin et al. (eds.), *Advances in Hybrid Conducting Polymer Technology*, Engineering Materials, https://doi.org/10.1007/978-3-030-62090-5_2

1 Introduction

In general, polymers that can conduct electricity almost like metal are known as conducting polymers. These conducting polymers can be further classified into three categories such as intrinsically conducting polymers (ICPs), extrinsically conducting polymers (ECPs) and co-ordination, or inorganic conducting polymers (CICPs). ECPs are prepared by simply adding conducting fillers or blending insulating polymers with conducting polymers either physically or chemically [1]. CICPs are prepared via the addition of charge transfer complexes such as metal atoms with polydentate ligands [2]. Interestingly, ICPs are designer polymers where one can architecture the modeling of the polymer via an appropriate synthetic route [3]. ICPs usually possess a solid backbone with extensive conjugation which results in its fascinating conductivity. Besides that, ICPs can be achieved via the doping process namely, p-doping or n-doping (refer to Chap. 1; Fig. 3).

Recently, ICP based nanocomposite (ICPn) has received great attention and become a prominent area of current research as well as development. ICPn is a hybrid material between polymer matrix and different types of nanofillers reinforced within it [4]. The nanofillers can be in one-dimensional (nanotubes and fibers), two-dimensional (layered materials like clay), or three-dimensional (spherical particles). ICPn have been recognized by various application where it: (i) fasten the electrolyte diffusion in supercapacitors, (ii) give effective excitation dissociation in solar cells (iii) enhancing the sensitivity of sensors and (iv) reduce responses time in electrochromic devices [5]. Other applications associated with ICPn is illustrated in Fig. 1.

Hybridizing ICP with nanoscale material contributes to many advantageous features including controlled morphology, film-forming ability, dimensional variability, and activated functionalities [6]. Moreover, ICPn gives a greater performance

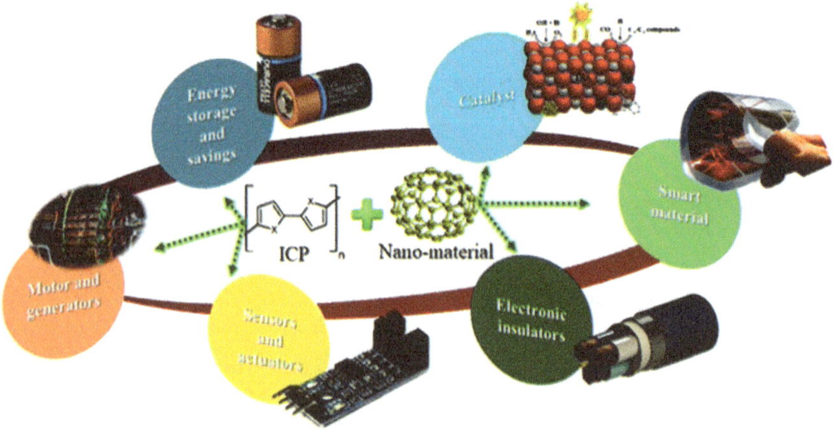

Fig. 1 Application of ICPn

in electrical conductivity, electrochemical, and mechanical properties in comparison with macro-scale materials [7, 8]. In comparison with micro filler composites like glass or carbon fibers, ICPn attributes substantial advantages due to the properties of the nanomaterial itself which are lightweight, provide large surface area, size confinement, dominance in interfacial phenomena and good quantum effects [9]. Although nanofiller composite shows remarkable features, the polymer matrix also plays a vital role in determining the whole performance of ICPn [10]. Typically, ICP has usually been combined with inorganic species such as quantum dots (semiconductors metal oxides and sulfide), carbonaceous material (nanotubes or nanowires), and chalcogenides [5]. Table 1 shows some examples of ICPn together with its highlighted properties.

This article is concentrating on the different types of ICP with nanocomposite for the photocatalytic process. Owing to low cost, facile methods of fabrication, and environmentally friendly nature, conducting polymer have believed to be the most promising material for many applications. Research regarding ICPn gives rise in interest towards many researchers. This can be related to the increasing number of publications of ICP doped nanomaterials that are illustrated in Fig. 2. This article is categorized into the following subsection which are the different synthesis pathways and polymerization mechanism of ICPn, fundamentals on photocatalysis with its principles, and the performance of the different types of ICPn for photocatalytic study. The review provides the reader with a systematic study about different kinds of blended ICP with their commercial impacts mainly as photocatalysts and problems related to its performance by recent literature examples.

2 Synthesis Method for ICPn

The properties of ICPn do not depend on the properties of individual components but also depends on the whole system. The most common feature of ICPn is the existence of phase border between matrix and filler material and the development of an interface layer between the two. The properties, composition, and microstructure at the interface vary across the interface region and are different from both matrix and filler. Most of the interphase properties depend on the bound surface and therefore the nanocomposite properties can be tailored by optimizing the interfacial bond between the nanofiller and polymer matrix. Thus, to modify its properties, ICPn can be fabricated through various methods available. In this review, ex-situ and in-situ polymerization, solution mixing, electrospinning, and some other method will be briefly introduced.

Table 1 List of ICPn with its properties and application

No.	ICPn	Morphology	Properties	Application	References
1	PANI/HA/TiO_2/Fe_3O_4	Nanorods/nano-tubes	• Good dielectric and magnetic properties • High conductivity and heterogeneity • Possess better microwave absorption ability	Stealth technology	[11]
2	PANI/CNT	Nanotubes	• High thermal ability • Great capacitance performance	Supercapacitor	[12]
3	Ppy/NiO	Nano-crystalline	• High sensitivity and stability • Increase in hole density	Gas sensor	[13]
4	Ppy/$SrCO_3$–$Sr(OH)_2$	Core shell	• Improve catalytic efficiency • Efficient photo catalyst for pollutant degradation • Easy preparation	Photocatalysis	[14]
5	Pth/ZnO	Nano-sphere	• Great adsorption capacity • Possess good efficiency in pollutant degradation	Photocatalysis	[15]

(continued)

Table 1 (continued)

No.	ICPn	Morphology	Properties	Application	References
6	Pth/Graphene	Core shell	• Enhanced conductivity • High specific capacitance • Provide large effective surface	Supercapacitors	[16]

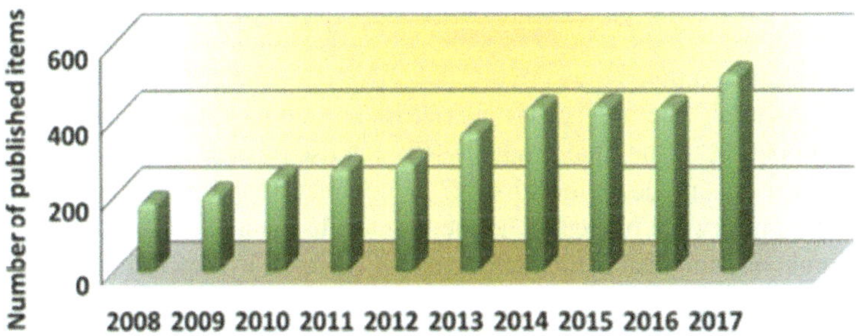

Fig. 2 Number of publication for ICP doped material from 2008 to 2017 [17]

2.1 *In-Situ and Ex-Situ Synthesis*

By general definition, in-situ means on-site or in the position of its original place/group in contrast with ex-situ. In chemistry, in-situ means a reaction that happens in the mixture while ex-situ is vice versa. In-situ synthesis is a component that will be generated from the corresponding precursors [5]. This synthesis uses radiation, heat, initiator diffusion, or organic initiator in its process [18]. The monomer is then polymerized between interlayers, thereby forming either exfoliated or intercalated nanocomposites. Figure 3 shows the example mechanism of in-situ polymerization for hybrid ICP with TiO_2 nanomaterial.

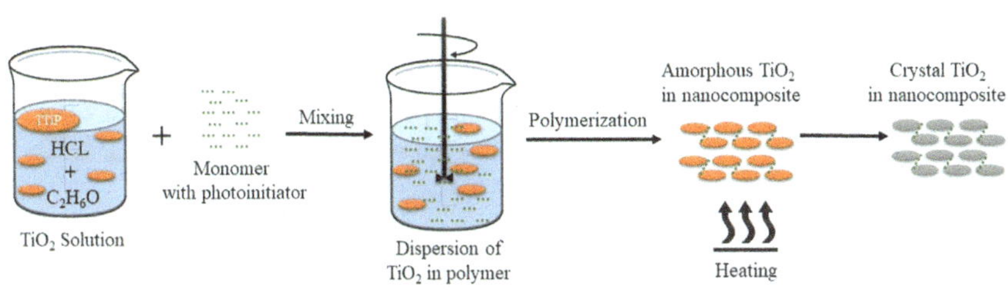

Fig. 3 In-situ polymerization of Titanium (IV) isopropoxide (TTIP)/TiO_2

In-situ synthesis gives advantages to hybrid ICPn as it has a better and easy process control of the material at a molecular level. For instance, thin-film ICPn can be easily obtained through in-situ electrochemical polymerization as the hybrid heterogeneous *material is deposited on the electrode surface.* The in-situ synthesis also enables excellent spatial distribution where it can avoid the particle from agglomeration [19]. This is due to the formation of a covalent bond between the nanomaterial and the matrix. The presence of the polymeric chain onto the nanomaterial's surface further facilitates their dispersion while providing a strong interface at the same time [20]. Furthermore, the structure and properties of ICPn material can also be easily altered by just tuning the desired parameters such as deposition time and current density [21].

The in-situ synthesis also said to be a versatile technique as it allows the preparation of composites to be diluted by other techniques such as chemical and electrochemical polymerization. However, this polymerization method could not ensure full converted material. The unreacted material in the polymer matrix can easily influence the properties of the final product [19]. The use of high temperature in in-situ synthesis also causing decomposition of the polymer. Oppositely, ex-situ synthesis is not a straightforward process. The method needs to undergo several subsequent steps (simple or complex) before achieving the material hybridization. Both of ICP and the nanomaterial species are usually being synthesized separately.

Interfacial tension of the hybrid material is the most important aspect of the ex-situ synthesis method. This is because the interfacial tension will determine the resulting major properties of the hybrid material. High interfacial tension will block the polymeric material to infiltrate into nanomaterial's pores [5], leading to incomplete filling, making it insufficient in certain applications. Low dispersibility and low stability are other limitations of ex-situ synthesis. Despite that limitations, ex-situ synthesis is usually easy, simple, and produces large mass production which makes it is most suitable for industrial processing [19]. Figure 4 shows the example mechanism of in-situ polymerization for hybrid ICP with TiO_2 nanomaterial. The differences between both of these syntheses are tabulated in Table 2.

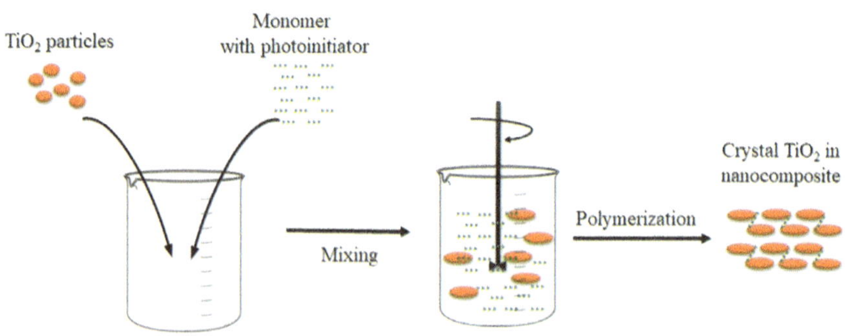

Fig. 4 Ex-situ polymerization of TTIP/TiO_2

Table 2 Differences between in-situ and ex-situ synthesis

In-situ synthesis	Ex-situ synthesis
• One step fabrication within the reaction mixture • Straightforward process	• Consists of subsequent step (simple or complex); synthesized separately • Not straightforward
Advantages	
• Good control of material properties • Excellent spatial distribution • Low agglomeration	• Large mass production • Suitable for solution-based processing • Easy and simple method
Disadvantages	
• Material not fully converted • The unreacted material will influence the final product's properties	• Low dispersability and agglomerated • Low long term stability

2.2 *Direct Mixing Method*

Direct mixing is a method for the fabrication of ICPn that implicates the breaking of aggregated nanofillers during the mixing process [18]. This method usually classified into two processes which are (i) mixing polymer matrix with a nanomaterial in the presence of organic solvents that is called solution mixing method and (ii) mixing polymer matrix with nanomaterial above the glass transition temperature of the polymer that is called melt compounding method. The ICPn that was fabricated by one of the above methods are then will be processed by conventional manufacturing methods like injection molding, calendaring, casting, compression molding, blow molding, rotational molding, extrusion molding, thermoforming, etc. [18]

2.2.1 Solution Mixing

This method also is known as 'solvent synthesis', where it is based on in-situ reaction to form organic or inorganic phases simultaneously in a single-phase [5]. The solution mixing process is the most used technique by many authors especially for the preparation of ICPn [20]. The fact that it can facilitate the separation and dispersion efficiently for most nanocomposite makes this process to be mostly chosen [22–25]. Similar to in-situ synthesis, the ICP solution and the nanomaterial are separately done but both of the material will use the same organic solvent. The nanomaterial will be dispersed and homogenous through the sonication process into an organic solvent to breakdown the nanofiller aggregates [18], prior before mixing with ICP solution, which was previously dissolved in the same solvent. The solvent coagulation method will take place to remove the solvent, lowering the shear stresses induced in the polymer matrix, remaining only the ICPn [22, 26]. The solution mixing method also allows both chemical and electrochemical polymerization as its preparation method same as in-situ synthesis. Though this method is less time consuming and an easy technique, it is irreversible. Solution mixing also has limited control over the structure and morphology of the nanomaterial.

2.2.2 Melt Compounding

Melt compounding is a method that involves internal mixer techniques that come along with simplicity, availability, and speed characteristics. Melt compounding is preferably chosen by the industry as this technique is environment friendly; free from solvent and contamination [27]. This technique is said to have quite an advantage for nanomaterials addition compare to other processing techniques as it can maintain the filler high aspect ratio, reduce the fiber breakage, and avoid aggregations [28]. In general, the dispersion of inorganic fillers in the matrix polymer depends largely on the internal shear stresses induced by viscous drag on the fillers during melt-compounding. In this kind of method, the shear stress-induced in the polymer melt by melt compounding is employed for the breakdown of aggregated fillers to the nano-scale. According to a study made by Xiao et al. [7] reported that melt compounding can minimize the filler aggregates by applying an appropriate shear during the process. Therefore, using this method much probably will give a better compounding effect which will result in better mechanical tensile properties. The conceptual illustration of the mechanism that happened in the melt compounding technique is simplified in Fig. 5.

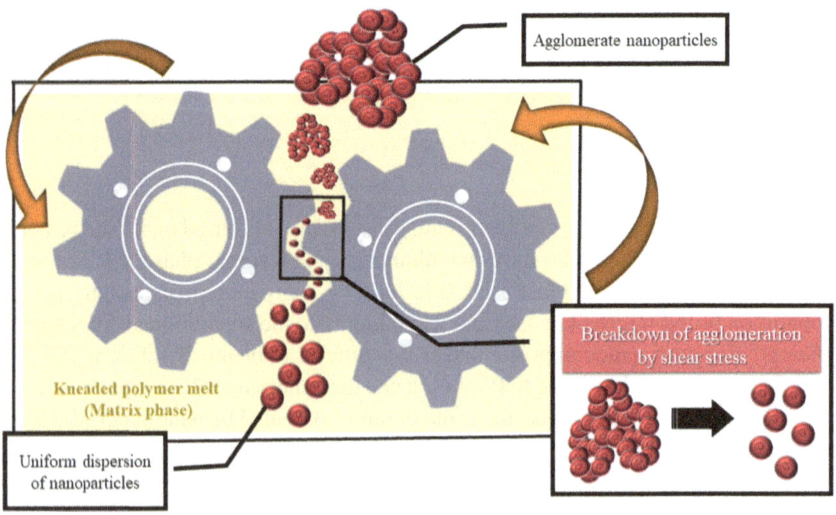

Fig. 5 Conceptual illustration of direct melt compounding technique for ICPn

2.3 *Intercalation Method*

The intercalation method is a commonly used method to create nano dimension filler [8]. In this method, the exfoliation of the layered silicates used as the inorganic filler occurs by intercalating ICP into the interlayer space, resulting in the uniform dispersion of plate-like nanofillers [8]. Intercalated dispersion is the key element in using the intercalation method for the diffusion of nanoparticles into polymer matrices. Uniform dispersion of nanoparticles can be achieved by the intercalation method that involves proper surface modification techniques which are either by (i) chemical or (ii) mechanical technique [29]. The chemical intercalation method is an indirect method that incorporates in-situ polymerization. The nanoparticle will be dispersed in monomer and intercalated between the sheets to form the polymer chains. As for mechanical technique, it is a direct intercalation method that occurs through solution mixing. Both monomer and nanoparticles will be dissolved in a solvent before been mix together. The intercalation of polymeric chains into the nanoparticle layers will displace the solvent used.

2.3.1 Melt Intercalation

Melt intercalation is a non-solvent method that is commonly used by the industry where it blends the nanofiller with polymer at molten state in intensive mixing [8]. The thermal motion of polymer chains that resulting from the mixing will form nanomaterial layers by either intercalated or exfoliated polymer nanocomposites (Fig. 6) [30]. In this method, both polymer and nanofiller are been annealed either statically or under shear. The polymer powder will be melted to form a viscous solution before nanofillers are added into it under a high shear rate and high-temperature diffusion.

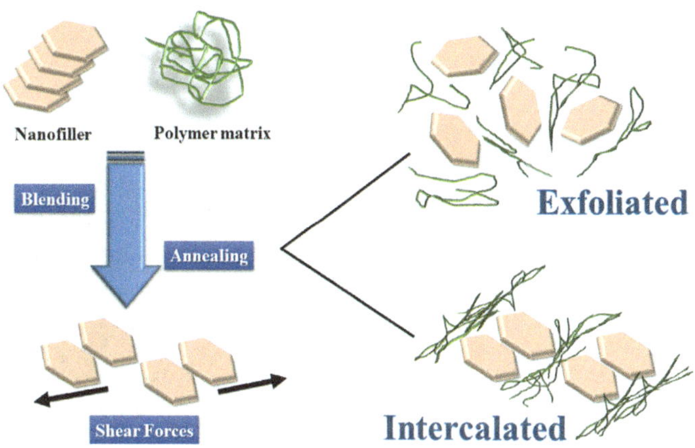

Fig. 6 Schematic illustration of melt intercalation method for ICPn

Although this method has the advantages of not using any solvent, the use of high temperatures in the process may damage the properties of the material.

2.4 Other Methods

Several other methods have been reported by other publications for the synthesis of ICPn. These include electrospinning, template synthesis, soft lithography, etc. Electrospinning is a method that involves a high electrical charge where it consists of three main parts; (i) the ICPn solution inside tube capillary, (ii) high voltage supplier and (iii) a charged collector. When undergoes the electrospinning process, the surface of the polymer solution will form an electrical field, coming from the voltage supplier along with the development of a jet. This electrical field will then create repulsion which induces a force between the polymer/composite solution and subsequently overcomes the surface tension that creates the formation of nanofiber after the solvent evaporated [30].

Electrospinning is the most chosen technique to produce nanofibers as it possesses a simple process, inexpensive, high speed, and provides good control on the particle's diameter [31]. A study conducted by Katepalli et al. had used this method to synthesize poly(acrylonitrile) (PAN)-activated carbon fibers (ACF) for the adsorption of organic pollutants existed in the atmosphere. Here, the authors stated that this composite material is a good adsorbent for air pollution control as this material can absorb SO_2 and NO [32]. The experimental setup is shown in Fig. 7a where the electrospinning technique was used to coat 8% (wt/wt) of PAN solution in DMF for 10 min. This technique produces a hierarchical fabric ACF-PANS of rod-like shape fibers with a size approximately 360 ± 38.1 nm when been analyzed under FESEM, SUPRA 40VP (Fig. 7b).

Similarly, a study conducted by Homaeigohar et al. also had used the electrospinning method. This technique was used for the deposition of polyethersulfone (PES)

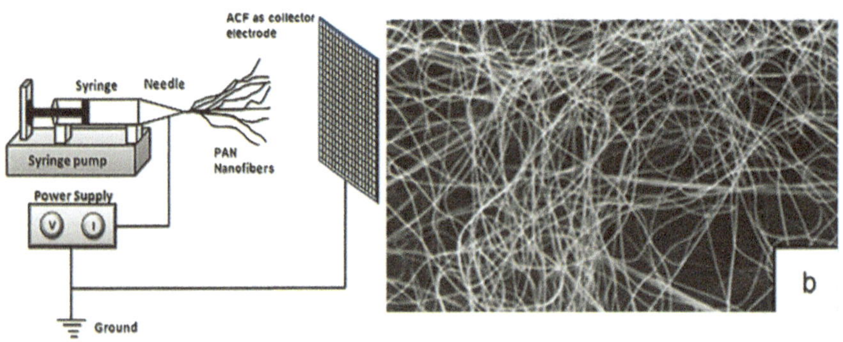

Fig. 7 **a** Experimental setup for PAN deposited on ACF Substrate. **b** FESEM image of PAN-ACF fibers after electrospinning technique. Adapted from Katepalli et al. [32]

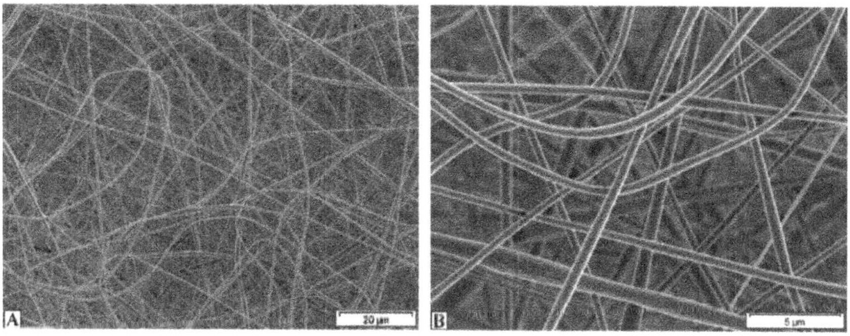

Fig. 8 SEM micrograph of PES morphology under **a** 20 um and **b** 5 um. Adapted from Homaeigohar et al. [33]

on poly(ethylene terephthalate)(PET) as a filtration material in wastewater removal [33]. The surface morphology in Fig. 8 also shows a similar rod-like pattern of PES fibers, similar to the previous study, when analyzed under SEM after electrospun. In this study, it was found that using the electrospinning technique, would avoid the difficulty of handling and accumulation of nanofibrous mat. Typically, 20 wt% of PES in the DMF solution was prepared and fed with a constant rate of 0.5 ml/h into a needle using a syringe pump. PES was electrospun on top of an aluminum foil with 20 kV energy by Heinzinger Electronic device.

The template synthesis method is a method that in-cooperate inorganic filler in the form of aqueous solution or gel. The polymer, which acts as the nucleating agent, enhances the growth of the crystals of the inorganic filler. The polymer mainly will be intercalated with the nanomaterial after the growth of the crystals and forming the polymer nanocomposite [34, 35]. This method is usually used for the preparation of double-layered nanocomposites [36]. A synthesized TiO_2/PDMS nanocomposite films by template synthesis was made by Iketani et al., where they studied photocatalytic behavior of the nanocomposite films for the decomposition of organic pollutants [37]. Soft lithography also is an example of an alternative lithography method for ICPn synthesis. Soft lithography is a technique that replicates the material structures using a micro-mold with the presence of vapors solvent and control temperature. This method is convenient as it is a low cost, gives high resolution, and possesses good control over nanoscale patterns [5].

3 Conductive Polymer-Based Nanocomposite for Photocatalysis

Photocatalysis is an advanced oxidation process that uses natural light or artificial illumination as its main source to decompose especially organic pollutants present in wastewater [38]. This process also is an eco-friendly approach as it terminates pollutants rather than merely transferring them to another phase without the use of potentially hazardous oxidants such as ozone and chlorine [39]. Novel photocatalysts for example ZnO, ZnS, ZrO_2, semiconductor-graphene, perovskites, MoS_2, WO_3, CdS, and Fe_2O_3 were actively studied this past few years [40, 41]. Despite the successful TiO_2 material to degrade various kinds of pollutants the degradation only can be done under UV light due to the large bandgap of TiO_2 material (3.2 eV). Recently, utilizing renewable energy sources (solar) for photodegradation has attracted more attention due to its pollution-free and low cost.

To utilize the visible light, 43% of the solar spectrum, the bandgap must be in the range of 1.7–2.2 eV. This value may be in the range of many other semiconductor materials, including oxides, oxynitrides, and oxysulfides [42]. However, their practical uses have been limited by their low photocatalytic activity under solar light, short-term stability against the photo, and chemical corrosion. Heterogeneous photocatalysis oxidations performed with light irradiated semiconductors dispersions have been extensively investigated owing to their high efficiency to completely mineralize the harmful organic and inorganic ions species to CO_2 and water.

3.1 Process and Fundamental of Photocatalysis

A photocatalyst is known as a material that has the capability of absorbing light and producing electron–hole pairs that enable chemical transformations. To control the performance of the photocatalysis process, some measures need to be highlighted. Photocatalysis employs hydroxyl radicals (OH·) as a very reactive oxidizing agent for the degradation of organic compounds [38]. The OH· can oxidize a wide spectrum of organic matters in water quickly and non-selectively [43]. In general, photocatalysis is a photo-induced reaction initiated in the presence of a photocatalyst [44]. Photocatalysts also can regenerate its chemical composition after each cycle of interaction [45, 46]. The illustration of the photocatalytic system is shown in Fig. 9 by using Polypropylene with TiO_2 nanoparticles as the photocatalyst to degrade Alizarin Red S from wastewater [47]. The important features of the photocatalytic system are the desired bandgap, suitable morphology, high surface area, stability, and reusability [48]. Based on a study by Van Gerven et al., to maintain the photocatalyzed reaction, the recombination of the electron with the hole must be prevented [44]. The presence of the hole is highly oxidizing, thus when organic molecules come in contact with the hole, it will be readily oxidized, ideally to carbon dioxide. While the excited electron then can be scavenged by the oxygen from the environment or react with photon to produce hydrogen. Both of the electron/hole pairs need to be recaptured by the targeted molecules quickly because if it does not, it will recombine and it will not generate the desired reaction. The physicochemical properties of the metal oxides are also crucial to maintain a good photocatalytic performance. These include its size, shape, morphology, and composition [49]. Moreover, the source and type of light used are also important to be controlled as it can also affect the performance of this process [50].

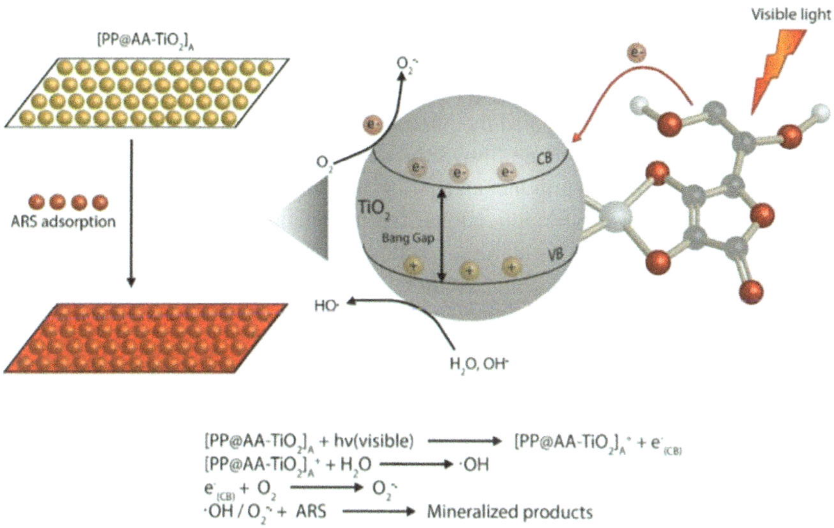

Fig. 9 The schematic diagram of photocataysis process of PP/TiO_2 for degradation of ARS compound. Adapted from D'Amato et al. [47]

3.2 *Properties of Photocatalyst Material*

The properties of the materials play a crucial role in making it as an effective photocatalyst. Metal oxides catalysts such as oxides (TiO_2, ZnO, CeO_2, ZrO_2, WO_3, V_2O_5, and Fe_2O_3) and sulfides (CdS, and ZnS) is commonly used for photocatalyst. It possesses the characteristics of light absorption, which induces a charge separation process with the formation of positive holes that can oxidize organic substrates [51]. In this photocatalysis process, a metal oxide is activated using either UV light, visible light, or a combination of both. The energy from the light source makes the electrons excited and promoted them from the valence band to the conduction band, forming an electron/hole pair (e−/h+) [52]. The photogenerated pair (e−/h+) can reduce and/or oxidize a compound adsorbed on the photocatalyst surface. According to Khan et al., the photocatalytic activity of metal oxide comes from two sources which are first, from the generation of OH^- radicals by oxidation of OH^- anions and the other one is from the generation of O_2^- radicals by reduction of O^{2-} ions [52]. Both of these radicals and anions can react with pollutants to degrade or otherwise transform them into lesser harmful by-products. Most of the photocatalytic activity also can completely mineralize the pollutant into CO_2 and H_2O with the use of a suitable catalyst [38].

3.3 Role of ICPn in Photocatalysis

In 1995, Tennakone and his team have established the very first hybrid polymer nanocomposite. They investigated the photocatalytic degradation of phenol by TiO_2 that was deposited on polythene films. Since then, this topic has been numerously explored by using different polymer and modifying different nanocomposite to obtain the best efficiency in various applications. To understand the mechanistic aspects of visible light-responsive photocatalysis, cyclic voltammetry (CV) measurements for ICPn were used to calculate the experimental HOMO and LUMO energy levels from the ionization potential and the electronic affinity, respectively and determine the bandgap of PEDOT nanospindles [53] (Fig. 10).

The main p-doping (oxidation) and n-doping (reduction) are irreversible processes, and the values of the peak potentials versus Ag/AgCl are +0.139 V (oxidation) and −1.552 V (reduction) for PEDOT nanospindles. This reveals the onset of oxidation and reduction processes occurring at lower potentials having a much lower energy gap around 1.69 eV than TiO_2 (3.2 eV). When illuminated with photons of energy exceeding (or equal to) the bandgap ($E \geq 1.69$ eV or $\lambda \leq 733$ nm), excess electrons and holes are formed in the conjugated polymer chains. Based on the bandgap structure of the as-prepared polymer nanostructures and the effects of scavengers, a possible photocatalysis mechanism for degradation of organic pollutants and dyes can be proposed. Due to the low bandgap of the PEDOT, less energy is required to promote an electron to the conduction band and as a result, more electronic transitions occur.

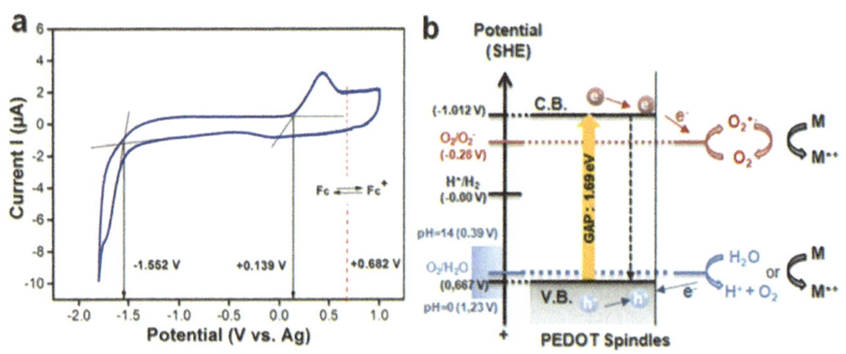

Fig. 10 **a** Cyclic voltammograms of PEDOT nanospindles obtained at 20 mV/s in acetonitrile, 0.1 M Tetrabutylammonium Perchlorate. Ferrocenium/ferrocene (Fc/Fc +) redox potential has been measured at the end of the experiment in order to calibrate the pseudo reference electrode (0.63 V vs. Ag in the present study). The energetic levels of PEDOT nanospindles is determined as follows: EHOMO (eV) ~ ionization potential = −4.8 − e (Eox_onset − 0.63) and ELUMO (eV) ~ electronic affinity = −4.8 −e (Ered_onset − 0.63). **b** Photocatalysis mechanism with charge separation in PEDOT nanospindles, with electron reducing oxygen and hole oxidizing water. Adapted from Ghosh et al. [53]

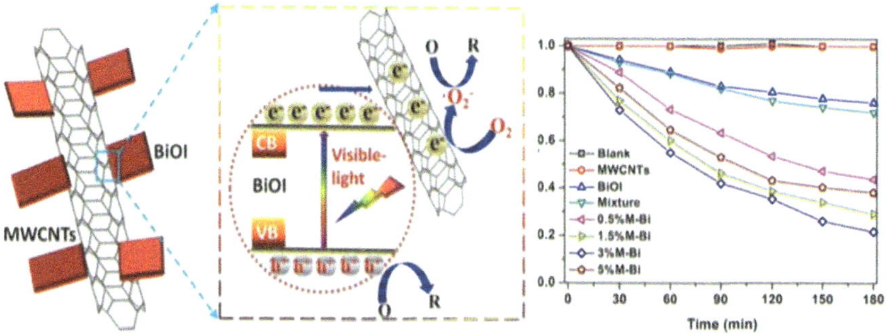

Fig. 11 The degradation efficiencies of BiOI Nanosheets Anchored onto MWCNTs towards RhB. Adapted from Li et al. [54]

The remarkable performance of ICPn in the photocatalytic application, for example, the degradation efficiencies shown by MWNTs in Fig. 11 [54], are now widely known and have been widely studied by many researchers [17, 55–57]. This is due to the electronic and optoelectronic characteristics that make it a very suitable material for a stable photosensitizer. ICPn photosensitizers enhanced the performance in the photocatalytic study as it owned a high electron–hole carrying efficiency [17].

Besides possessing high electroactive and excellent charge-transporting properties, ICPn also owing to a high thermal and photochemical stability [58]. According to several studies, they have reported that ICPn plays the role of p–n junctions. P-type ICP tailored to n-type nano-scale semiconductors compromise each other to overcome the drawbacks of semiconductor oxide that is typically used in photocatalytic applications. These disadvantages include the low visible light response, high tendency of electron–hole recombination, leaching, and decomposition at high temperatures. Particle size is strongly associated with the electrical and photophysical properties of ICPN.

This is proven by a study conducted by Jang et al. [59] where PANI nanoparticles (less than 4 nm) exhibited high electrical conductivity up to 85 S cm^{-1} after being fabricated by emulsion polymerization method. According to his research, smaller particle size contributed to highly compact and orderly structured in PANI chains thus making the electrical conduction to rise. Another study likewise highlighted that the conductivity of polydiphenylbutadyiene nanofibers (PDPBn) slightly increase from 10^{-11} to 10^{-4} S cm^{-1} with the decreasing diameter of nanofibers used when been doped with I_2 [26]. ICPn with controlled morphology contributed to photovoltaic efficiency where it enhanced the interchain charge, enable it to possess pseudocapacitance and light absorption ability [23]. Friedel et al. [24] found the functionalities of poly(3-methyl)thiophene as nanofibers donor in solar cells application has improved as this nanofibers provide high ion carrier mobility, excellent dispersion phase, and favorable optical absorption spectrum. Moreover, ICP also inherent the semiconducting bandgap, which makes it able to absorb photo-excited electrons to produce other elements such as H_2 gas in water splitting or decomposing organic pollutants through the oxidation process.

3.4 Application of Photocatalyst Based ICPn

The high percentage of pollutant existed in water reservoir have worsened day by day. This pollution not only endangered human health but also disturbing the ecosystem. It is reported that the major cause of the decreasing population of vulture species in Pakistan and India is due to water contamination [58]. Water pollution occurs when undesirable effluents discharged in a water system and reduce water quality. Water quality can be affected by many factors (Fig. 12) [25]. Some of the factors are precipitation, climate change, soil type, vegetation, geology, and also human activities [60]. Pollutants from industry, agriculture, or households contain emerging contaminants (Fig. 13). The emerging contaminants is a material which is left as a by-product during the manufacturing process and give adverse health effect to the living things [60].

Fig. 12 Different source of water pollution. Adapted from Chaudhry and Malik [60]

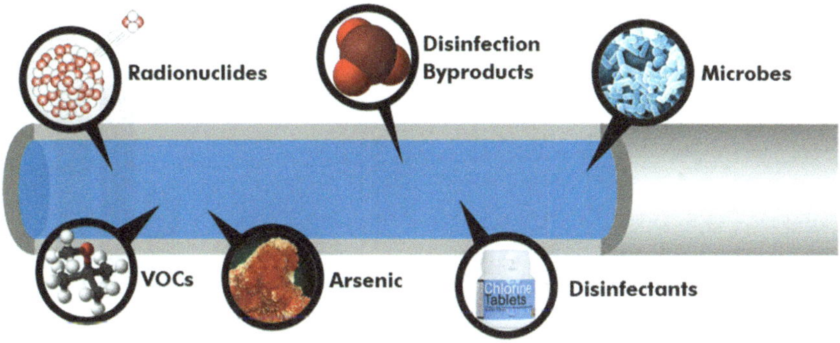

Fig. 13 The usual contaminants that existed in waste water

Thus, the excessive phenomena of water pollution worldwide have prompted the finding of possible solutions using clean energy routes. The best clean solutions to overcome these phenomena is by emerging photocatalysis for water and environmental detoxification. This is due to the unlimited source of solar energy where the sun effuses solar radiation up to 3.8 million exajoules per year, proving photocatalysis is the best choice for water pollution remediation. Using ICPn as the photocatalyst solved the limitations of common photocatalyst mainly semiconductor oxide. Semiconductor oxide could not give full utilization of solar energy as it can give efficiency only in the ultraviolet light region [61]. On the other hand, ICPn is a good light absorber and active in the visible light region [61]. This artificial photosynthesis process allowing the conversion of harvest sunlight into chemical reaction inside ICPn layer [62].

This makes ICPn an efficient visible-light-driven photocatalyst that can maximize the use of solar radiation and increase the performance of energy conversion [62]. Effective absorption of solar radiation can be obtained by the composition modifying process between ICP with the nanofiller composite. Moreover, ICPn provided many promising advantages in the photocatalytic study. These include excellent conductivity, good electrochemical activity, large specific surface area, short and direct pathways for charge transport.

3.4.1 ICPn Photocatalyst for Water Remediation

Experimental studies on hydrothermally synthesized polypropylene (Ppy) with zinc oxide nanorods and reduced graphene oxide (RGO) by Ariffin et al., has shown a tremendous catalytic performance as well as high stability for water purification [63], focussing on the degradation of Methylene Blue (MB). In his study, the photocatalytic degradation for MB when using PPy/ZnO with RGO has given the best performance as illustrated in Fig. 14. It showed significant improvement over the control samples as well as those containing the nanocrystals. The material also exhibited high stability and a good reusability cycle when being used for MB photodegradation (Fig. 15).

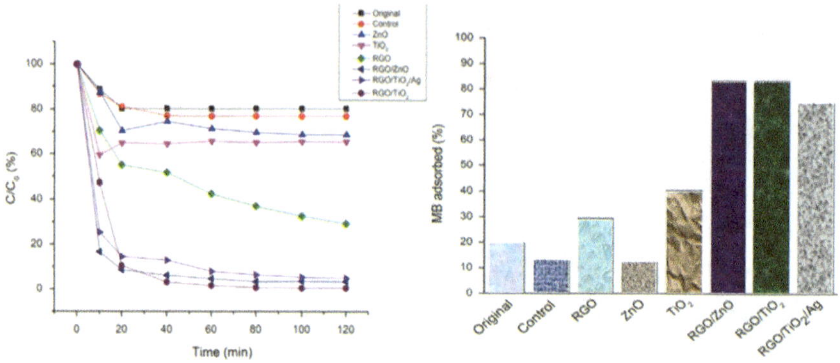

Fig.14 The photocatalytic study of filler for MB photodegradation. Adapted from Ariffin et al. [63]

Fig. 15 Number of reusability cycle for both filler. Adapted from Ariffin et al. [63]

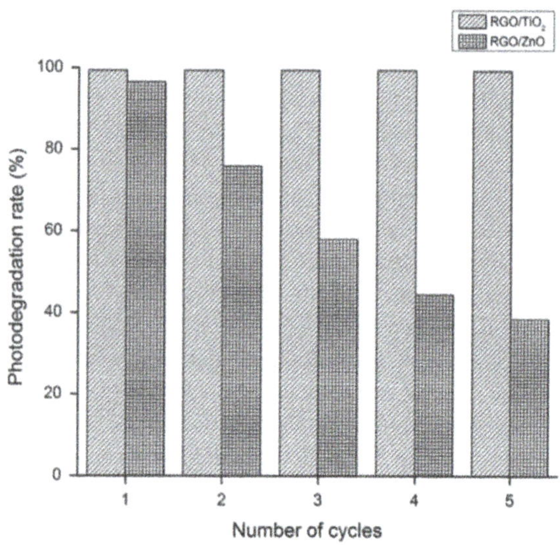

Additionally, the synergetic effect of the combination of ZnO and polymers allows for the protection of the PPy fiber against surface cracks and limits the well-known photo corrosion process of zinc oxide [63].

Another study of ZnO nanorods on polybutylene terephthalate (PBT) polymer fiber mats were synthesized by using hydrothermal growth was reported in the literature [64]. The thin films formed by the low temperature of vapor phase atomic layer deposition (ALD) a seed layer where it used ultraviolet (UV) radiation in the range of 320–390 nm providing 79 mW/cm^2 of energy flux. This ICPn was studied to absorb azo organic dyes namely Acid Red 40. Gong et al., stated that the degradation of the acid dye achieved up to 90% within 2 h of film immersion. The combination of the ALD and hydrothermal method here allows this film to obtain the best performance of the photocatalyst and may also be used for other crystal growth systems, such as TiO_2, Fe_2O_3, SnO_2 and V_2O_5, where high area and ready solution access are desired [64].

3.4.2 ICPn Photocatalyst for Degradation of Organic Pollutants

Recently, several studies were reported on the preparation of ICPs to curb organic compound pollution without producing secondary pollution. In a study conducted by Li et al., CeO_2–ZnO-polyvinylpyrrolidone (PVP) was evaluated for water treatment efficiency using an 8 W UV lamp with emission wavelengths of 254 nm for Rhodamine B (RhB) as its target compound [65]. The nanofibers of CeO_2–ZnO were synthesized using the electrospinning technique followed by thermal treatment and the nonwoven mat was prepared from the precursor solution of $PVP/Ce(NO_3)_3/Zn(CH_3COO)_2$. Here, Li et al., reported that 17.4% and 82.3% of Rhodamine B was decomposed catalyzed by pure CeO_2 and ZnO fibers, respectively, whereas almost 98% was decomposed applying the CeO_2–ZnO-composite fibers. It shows composite have the potential of treating organically polluted water as Rhodamine B was almost completely decomposed when it was catalyzed by CeO_2–ZnO nanofibers [65]. The result is shown in Fig. 16.

A similar study has been reported for the decomposition of Rhodamine B. by Zhang et al. [34]. ZnO/SnO_2-polyvinylpyrrolidone (PVP) was made by using the sol–gel process and electrospinning technique. The electrospun composite nanofibers

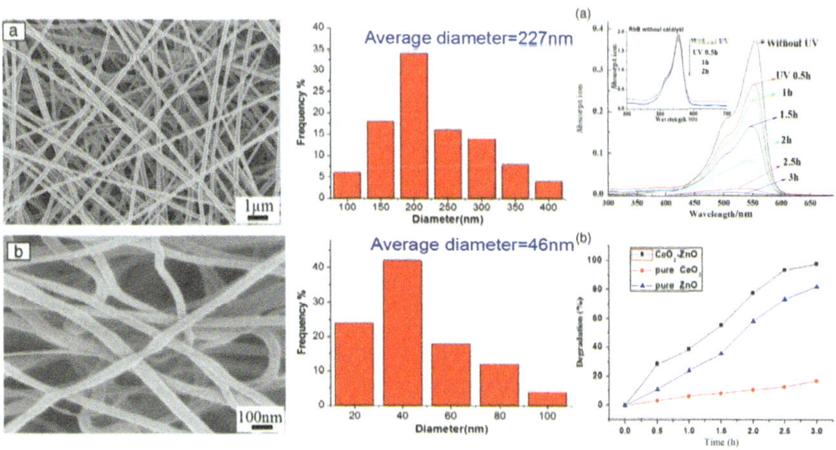

Fig. 16 SEM micrograph of PVP/CeO_2–ZnO with the absorption spectra of RhB solution under UV irradiation and without UV at different times and the curves of photocatalytic degradation on RhB with different catalysts. Adapted from [65]

were obtained by the precursor solution of PVP/$ZnCl_2$/$SnCl_2$. In this study, a high-pressure mercury lamp (50 W) with the main emission wavelength at 313 nm was used as its light source.

The degradation efficiency of RhB was equal to 75, 35, and 85% for ZnO, SnO_2, and TiO_2 fibers, respectively after 50 min of immersion. However, the time for complete decolorization of dye solution over the ZnO/SnO_2-nanofibers was 30 min. The photocatalytic activity of the ZnO−SnO_2 nanofibers for the degradation of rhodamine B (RB) was much higher than that of electrospun ZnO and SnO_2 nanofibers nevertheless, ZnO−SnO_2 nanofibers could be easily recycled without the decrease of the photocatalytic activity due to their one-dimensional nanostructural property [34].

Another photocatalytic activity was tested by using ZnO nanoparticles on wool and polyacrylonitrile (PAN) fiber [66]. The polymeric fibers were synthesized by using a sol–gel process with minor modification under ambient temperature. It is stated that 77% of Methylene Blue (MB) was decomposed after 6 h upon ZnO/PANI and 80% upon ZnO/wool fibers. Similar results of degradation were obtained for eosin yellowish (EY) dye, where the degradation ratios equal 64% and 50%, respectively [66]. Based on this result, both of the ZnO-coated polyacrylonitrile and ZnO-coated wool fibers prove to acquire photocatalytic activity toward dyes degradation. Another dye degradation study was reported by using ZnO nanowires on polyethylene (PP) made by [67]. The ZnO nanowires were grown from seed ZnO nanoparticles affixed onto the commercially available fibers by hydrothermal method and 6 W of light sources were used throughout the analysis. After 3 h of irradiation, ZnO/polyethylene fibers degraded 83% of the MB, whereas the fibers without ZnO degrade only 32%. 24% of MB was found to undergo self-degradation under the same UV light without using polyethylene fibers [67].

3.4.3 ICPn Photocatalyst for Water Splitting

Photocatalytic water splitting is a mimic photoelectrochemical method of the photosynthesis process. This procedure harvest sunlight or artificial light to produce photons, and dissociate water into its constituent part which is Hydrogen (H) and Oxygen (O_2) atoms with the help of a catalyst. This process holds many advantages as it utilizes the uses of water fully, inexpensive renewable resources, and simple technique where it only needs a light source, catalyst, and water to produce H out from the water contents. Photocatalytic water splitting is highly in endothermic conditions ($\Delta H > 0$). In this technology, the oxidative and reductive reactions take place concurrently, while utilizing solar light [68].

ICPn photocatalyst has stimulated the development of new conjugated photocatalysts where it exhibited an electronic band, much like conductive metals, with a high level of tunability with respect to chemical composition, electronic structure, and texture. The emerging photocatalytic active polymers enrich the family of photocatalysts for solar-energy conversion. In comparison with the inorganic photocatalysts, the structures and properties of which are difficult to systematically adjust at the molecular level, the physicochemical properties of ICP can be easily tuned [69]. The advantages of ICP such as accessibility, chemical versatility, and tunable properties have made it more convenient for photocatalytic study. Further development of more efficient ICP photocatalysts for enhanced H_2 production is thus envisaged.

3.5 *Different Type of ICPn for Photocatalytic Study*

3.5.1 Polyaniline Based Photocatalysts

PANI composites are versatile conducting polymer that has shown remarkable photocatalytic activities. A study conducted by Bu and Chen [40] have shown a good performance, significantly increase both the photocatalytic degradation performance and the photocatalytic degradation (Fig. 17) by using Polyaniline/silver/silver phosphate (PANI/Ag/Ag_3PO_4) in Rhodamine B solution.

The PANI/Ag/Ag_3PO_4 composite was prepared by using in situ polymerization, depositing silver phosphate (Ag_3PO_4) nanoparticles on the surface of PANI. Based on his reports, PANI/Ag/Ag_3PO_4 that shows the best degradation is 20%

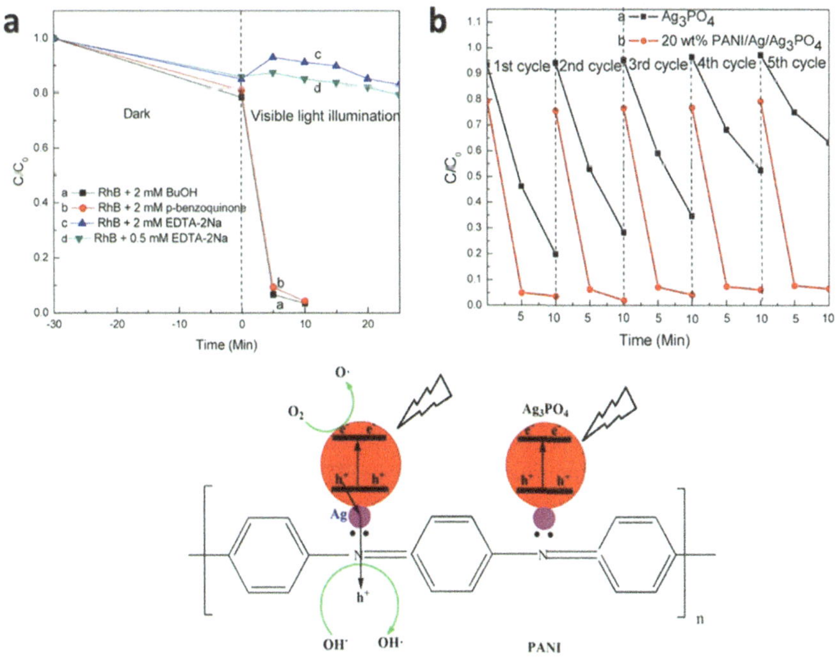

Fig. 17 **a** 20 wt% of PANI/Ag/Ag_3PO_4 degradation on Rhodamine B **b** stability graph for PANI/Ag/Ag_3PO_4 photocatalyst. Adapted from Bu and Chen [40]

by weight percentage. PANI/Ag/Ag_3PO_4 composite possesses the optimal photocatalytic degradation efficiency and only 5 min is needed to degrade more than 95% RhB dye [40]. With this, the composite contains approximately 4 times higher degradation efficiencies in comparison with pure Ag_3PO_4.

The excellent performance on the degradation of organic pollutants as well as the stability is attributed due to the formation of a heterojunction electric field between PANI and Ag_3PO_4, which is approximately 90 mV and points from Ag_3PO_4 to PANI [40]. The existence of this electric field can dramatically enhance the separation efficiency of the photogenerated electron–hole pairs, accelerate the transfer of photogenerated holes from Ag_3PO_4 to PANI and therefore inhibit the self-oxidation of Ag_3PO_4. Moreover, the photogenerated holes produced by Ag_3PO_4 are swiftly transferred to PANI under the formed electric field at the interface of Ag_3PO_4 and PANI and therefore, effectively slowing down the photo corrosion of Ag_3PO_4 [40].

Another photocatalytic degradation study of Rhodamine B into less dangerous compounds was reported by Gottam et al. [41]. In this study, the PANI salt and a porphyrin (PANI-H_2SO_4– TKHP) were synthesized by a chemical oxidative polymerization technique. Gottam et al. [41] stated that the oxidation of aniline in the presence of TKHP and ammonium persulfate results in the formation of the polyaniline salt (PANI-H_2SO_4-TKHP) and also a charge-transfer complex between PANI and TKHP via $\pi-\pi$ stacking [41]. PANI-H_2SO_4-TKHP was observed in a nanotubes morphology (Fig. 18) with a reasonably good yield of 88 wt%, the conductivity of 1.2 Ω^{-1} cm^{-1}, and thermal stability at 200 °C [41]. The photocatalytic performance of PANI-H_2SO_4-TKHP exhibited a 78% dye degradation, showing a good photocatalytic activity, compared to free-standing TKHP (2.5%) and PANI-H_2SO_4 (13%) aggregates under visible light irradiation [41].

Subsequently, a novel PANI/β-Bi_2O_3 composite was prepared by Dai et al. [70] and the composite was prepared via the chemical oxidative polymerization. The photocatalytic activity of the composite was evaluated by degrading Acid orange 7(AO 7) and PB-3 showed the highest photocatalytic activity which is by 5% higher compared to pure β-Bi_2O_3 [70]. According to the study, 30% of photocatalytic activity can be improved by the PB-3 when using PANI/β-Bi_2O_3 composite [40]. The PANI modification can effectively enhance the photocatalytic activity of the β-Bi_2O_3 [70]. The possible photocatalytic mechanism of the PANI/β-Bi_2O_3 composite was proposed (Fig. 19).

Another PANI based photocatalyst was prepared by Dhanavel et al. [71] which is the polyaniline/molybdenum trioxide composite (PANI/MoO_3). This composite was synthesized by using the chemical oxidative polymerization method. The composite was used as photocatalyst for the photodegradation of methylene blue under visible light irradiation. Based on this report, 46% of the methylene blue was adsorbed on to the composite [71]. This phenomenon occurs as the electrostatic interactions between cationic dye and negatively charged metal oxide surface between the composite and dye molecules. Also, the presence of a negatively charged sulphonic group in the system as well as by the presence of $\pi-\pi$ stacking interactions between the aromatic ring of PANI and MB is reasonable for the migration of dye molecules onto the surface

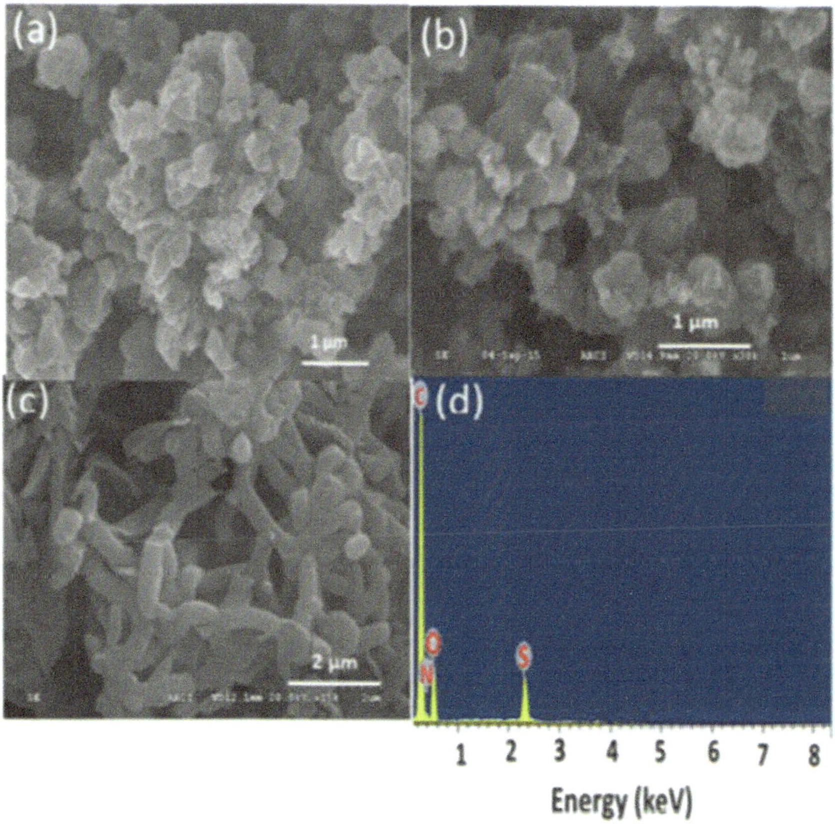

Fig. 18 FE-SEM images **a** PANI-H_2SO_4—TKHP-50, **b** PANI-H_2SO_4—TKHP-100 and with **c** PANI-H_2SO_4 and **d** EDAX spectrum of PANIH_2SO_4 -TKHP-100. Adapted from Gottam et al. [41]

of composites [71]. Figure 20 shows the MB removal percentage at a regular interval of time.

3.5.2 Polypyrrole Based Photocatalysts

PPy often exhibits a good conductivity, high polarizability, electrochemical stability and ease of preparation that makes it as a promisable photocatalytic material. Density functional theory study (DFT) of pyrrole-$Ti_{16}O_{32}$ bounded systems are made by Ullah et al. [72] for the photodegradation of environmental pollutants and solar water splitting. Inter-molecular interaction energy in nPy-$Ti_{16}O_{32}$ composites is simulated in the range of −41 to −72 kcal/mol which confirmed the existence of strong covalent and electrostatic type bonding [72]. This energy is simulated with the help of single-point energy and EgCP-D3 methods. So, after confirming the

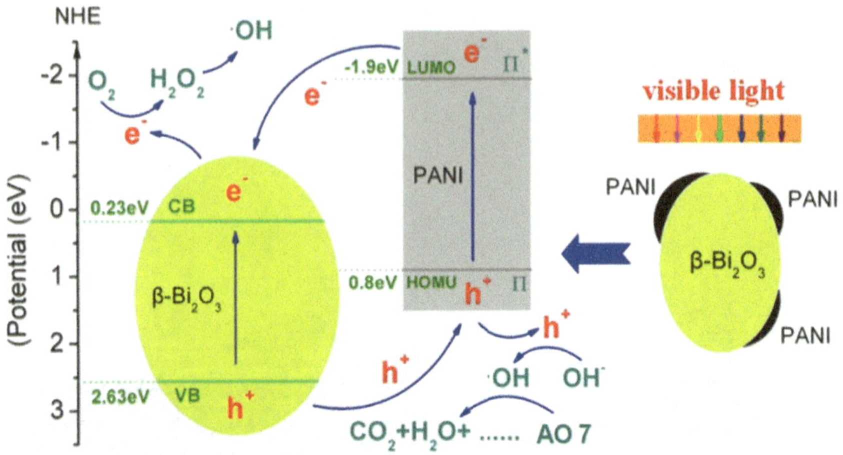

Fig. 19 Proposed mechanism of for the degradation of Acid Orange 7. Adapted from Dai et al. [70]

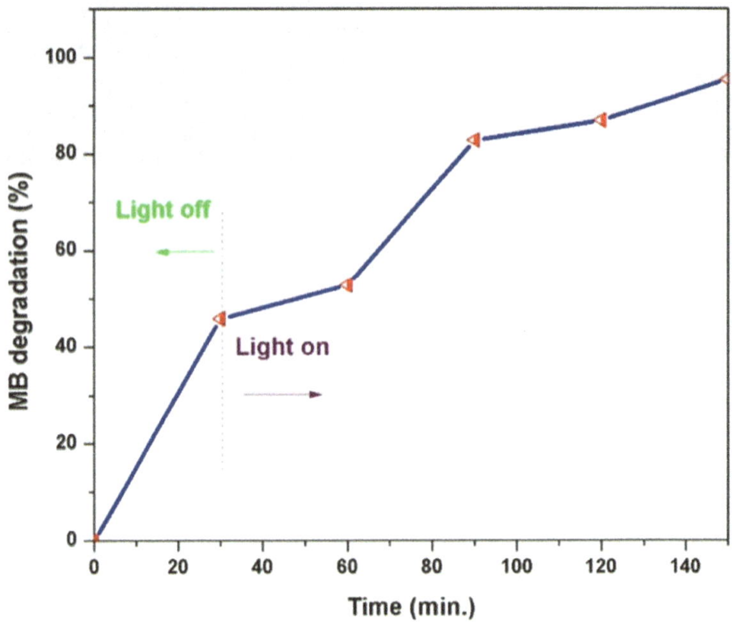

Fig. 20 Photodegradation for PANI/MoO_3 composite under visible light. Adapted from Dhanavel et al. [71]

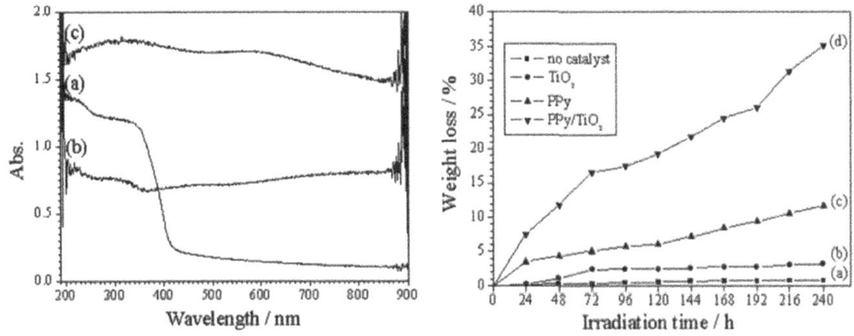

Fig. 21 UV–Vis absorption spectra of TiO_2 (**a**), pure PPy (**b**) and PPy = TiO_2 (**c**) photocatalysts and degradation curves. Adapted from [45].

composite formation, bandgap narrowing, and better visible light absorption capability is observed compared to the nPy and $Ti_{16}O_{32}$ constituents. Electronic properties such as HOMO and LUMO of nPy, $Ti_{16}O_{32}$, and nPy-$Ti_{16}O_{32}$ composites estimated using the B3LYP/LanL2DZ level, indicate excellent visible light absorption and charge transport efficiency of nPy-$Ti_{16}O_{32}$ [40]. Other electronic properties such as ESP, DOS, bandgap, and UV–vis spectra also support the formation of efficient photoactive nPy$Ti_{16}O_{32}$ composites.

A PPy/TiO_2 photocatalyst was prepared by another study where this time it uses sol–gel and emulsion polymerization methods [45]. The ability of PPY/TiO in visible light-capturing has enhanced and it can be shown in the UV absorption spectrum. The spectrum can be shown in Fig. 21. The photocatalytic degradation process of PE plastic use PPy = TiO_2 as photocatalyst was much faster than that use PPy or TiO_2 as photocatalyst. The escape of volatile organics came from the degradation of PE resulted in the formation of the cavities on the PE plastic surface. The photocatalytic reaction was initiated by reactive oxygen species generated on the PPy = TiO_2 surface. Further reactions involving these reactive oxygen species resulted in some by-products containing carbonyl and hydroxyl groups.

In another work conducted by Kumar et al. [46], an Ag-Ag_2O/TiO_2@polypyrrole (Ag/TiO_2@PPy) heterojunction has been synthesized by assembling a self-stabilized Ag-Ag_2O (p-type) semiconductor (denoted as Ag) and polypyrrole (π-conjugated polymer) on the surface of rutile TiO_2 (n-type). Ag/TiO_2@PPy was synthesized through simultaneous oxidation of pyrrole monomers and reduction of $AgNO_3$ in an aqueous solution containing well-dispersed TiO_2 particles. The photocatalytic activity of synthesized heterojunction was investigated for the decomposition of methylene blue (MB) dye under UV and visible light irradiation. The results are shown in Fig. 22 and it revealed that π-conjugated p–n heterojunction formed in the case of Ag/TiO_2@PPy significantly enhanced the photodecomposition of MB compared to the p–n type Ag/TiO_2 and TiO_2@PPy (n−π) heterojunctions [46]. A synergistic effect between Ag–Ag_2O and PPy leads to higher photostability and a better electron/hole separation leads to enhanced photocatalytic activity of

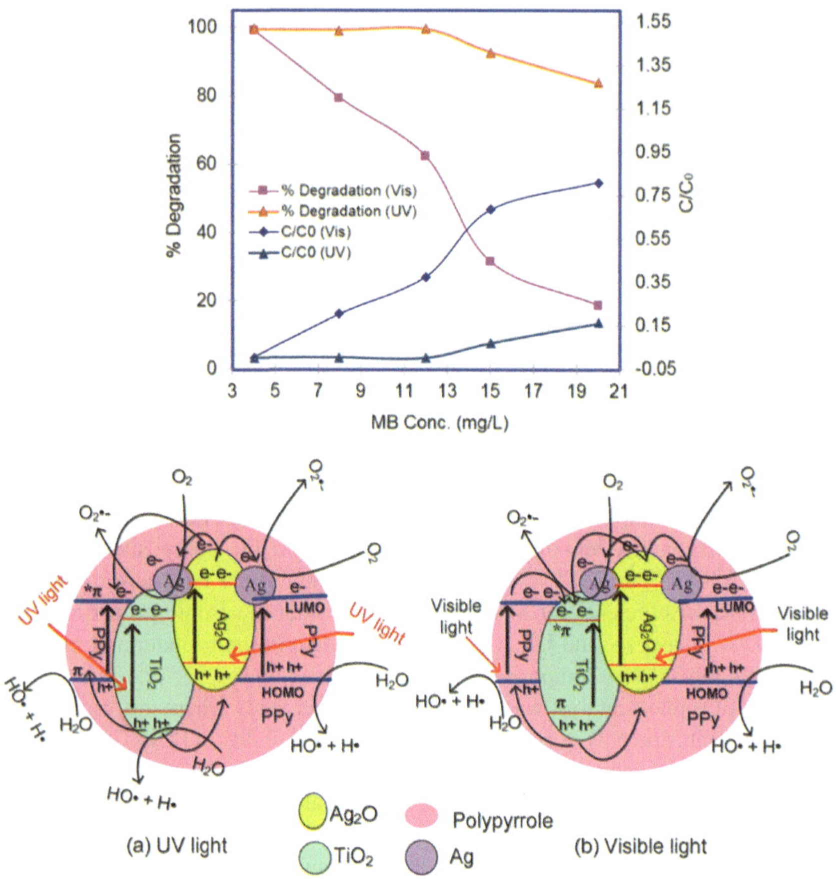

Fig. 22 The Effect of initial dye concentration on MB photocatalytic degradation and the schematic representation of possible electron/hole pair separation mechanism of the Ag–Ag_2O/TiO_2@PPy heterostructure under light irradiation. Adapted from [46]

Ag/TiO_2@PPy under both UV and visible light irradiations. The maximum photocatalytic activity of Ag/TiO_2@PPy for MB has observed at pH 9 and the pseudo-first-order rate constant values were 0.044 and 0.003 min^{-1} under UV and visible light irradiation, respectively [46]. The effect of initial dye concentration revealed that the degradation of MB decreased from 100 to 18.85% under UV light and from 100 to 83% in visible light when the MB concentration increased from 4 to 20 mg L^{-1} [46].

3.5.3 Polythiophene Based Photocatalysts

An effective photocatalytic process depends on the photocatalyst's surface as it depends on the adsorption capacity towards the target molecules. PTh could be a promising candidate for this instance due to the presence of π-conjugated electronic system that covers wide spectrum of applications. In a study by Ravi Chandra et al. [56] reported that hybrid Cu–TiO_2/polythiophene nanorods (HNRs) with the rutile phase were synthesized by a modified sol–gel method at low temperatures. It can be noticed that the significant change in the hypsochromic shift of Rhodamine B by using HNRs when compared to Cu–TiO_2 nanorods without polymer (CTNRs) photocatalyst. The Rhodamine B was removed 99.4% and 70.5% in 75 min and Orange G was removed 98% and 70.2% in 90 min, respectively under visible light irradiation. The results suggested that HNRs exhibited higher photocatalytic activity than CTNRs for the degradation of Rhodamine B and Orange G under visible light irradiation. There is no eloquent downturn of photocatalytic activity even after three cycles implying the good recyclability of photocatalytic activity of HNRs. Based on the remarkable photocatalytic performance, and reusability, it shows that the HNRs is a promising material in the field of photocatalysis and environmental remediation for industrial applications.

Figure 23a, b show the evolution of UV–vis spectra of the RhB solution in the presence of HNRs and CTNRs during visible light irradiation. The absorption peak

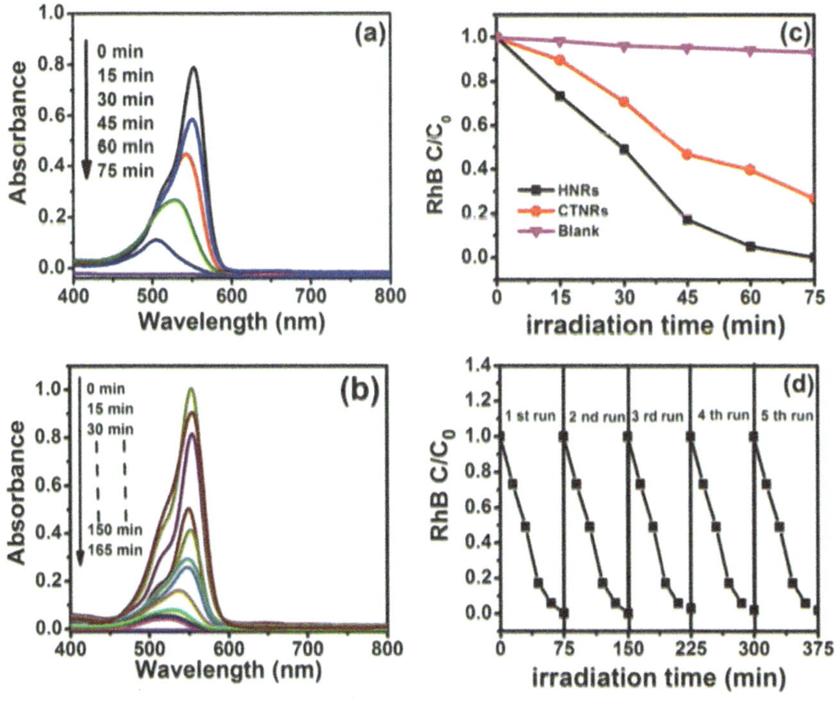

Fig. 23 UV–vis spectra of RhB collected at different intervals during the photodegradation.in the presence of **a** HNRs; **b** CTNRs; **c** comparison of photocatalytic activities for RhB photodegradation over different catalysts under visible light irradiation and **d** the profiles of cyclic tests of HNRs catalyst in reaction. Adapted from [56].

intensity of RhB decreased gradually for both photocatalysts. As shown in the figure, the 0 min spectrum shows that there is no degradation of dye along with the catalyst was observed. Further experiments were carried out along with catalyst and dye by exposing visible light degradation has been observed, which indicates that catalyst absorbed light and activated the catalyst to produce the reactive species leads to degradation.

In another study by Yang et al. [55], PThOD, a typical D–A type of conjugated polymer, was synthesized through a one-step polymerization. It was found that PThOD could be used as an effective photocatalyst to degrade various dyestuffs in the treatment of wastewater from dyeing. The species $O_2\cdot$ could be formed by the irradiation of PThOD with UV light and was the dominant ROS the degradation rate could be changed by adding H^+, which might be attributed to a change of adsorption capacity of PThOD [55]. RBB reusability experiments showed that PThOD was very stable and maintained its high efficiency during the treatment of dye solution [56]. Considering the simple synthesis procedure, effective photocatalytic ability in the treatment of dye solutions, and good reusability, PThOD may have a potential application in wastewater treatment.

Polythiophene nanoparticles prepared by water in oil (W/O) microemulsion method were found to be photocatalytically active for degradation of Orange II dye and methyl Orange by Cheng et al. [57]. During the photodegradation of Orange II and Methyl Orange as organic dyes in UV/H_2O_2 and UV irradiation systems using polythiophene nanoparticles as a photocatalyst, it exhibited significant photocatalytic activity towards both dyes in UV and UV/H_2O_2 system. Its photocatalytic efficiency for degradation of Orange II and Methyl Orange under UV irradiation is 2.54 and 1.95 times higher than normal polythiophene composites relatively. According to the testament by Accelerated Surface Area and Porosimetry System, the true cause lay in the fact that the specific surface area of polythiophene nanoparticles is 106.6 m^2/g compared to normal polythiophene is 24.2 m^2/g [57]. In the UV/H_2O_2 system, polythiophene nanoparticles exhibited the highest efficiency towards both dyes when the concentration of hydrogen peroxide is 3.0 wt%.

3.5.4 Others ICPn Photocatalysts

Copolymerization routes to carbon nitrides have been shown to dramatically increase the rate of H_2 evolution because of the extended delocalization of the p-conjugated system [10] which is advantageous for capturing more visible photons for the generation of charge carriers involved in the water splitting redox reactions. A remarkable advantage of organic conjugated semiconductors is that copolymers can be produced over a continuous range of monomer compositions, thus achieving systematic control over physical properties. This control allows the function-led design of conjugated semiconductors at the molecular level.

In principle, Cooper and co-workers prepared a series of pyrene-based CMPs with tunable optical gaps by statistical copolymerization [13]. All the synthesized copolymers (CP-CMPn, n = 1–15) possessed high specific surface areas that vary from 597 to 1710 m^2 g @1 according to the different monomer ratios and greatly favor the heterogeneous photocatalytic reactions because of the more exposed surface active sites. Interestingly, the optical properties of the polymers can also be facilely tuned by simple copolymerization with different amounts of featured monomers and are evidenced by the optical bandgap changing from 2.95 to 1.94 eV with optical absorption edges from l = 445 to 588 nm.

Lotsch and co-workers further synthesized a series of water- and photostable 2D azine-linked covalent organic frameworks (COFs) from hydrazine and triphenylene aldehydes, having a varying number of nitrogen atoms, by using a solvothermal method at 1208 °C [36]. These 2D azine-linked COFs were crystalline and presented high surface areas with a porous structure. The BET surface area of the as-prepared optimal COFs could be enlarged to as much as 1537 m^2 g @1, which contributed to improved heterogeneous catalytic activities. Apart from the surface area, the optical and electronic properties of the COFs could also be well controlled by the systematic change of the nitrogen content in the units. The diffuse reflectance spectrum reveals that all the COFs absorb light in the ultraviolet and blue sections of the visible region and show similar absorption profiles with an absorption edge at l = 465–475 nm thereby suggesting an optical gap of 2.6–2.7 eV as determined by the Kubelka–Munk function. These values are red-shifted by l = 40–60 nm in comparison to the solid-state absorption spectra of the precursor aldehydes and can be attributed to the introduction of the azine group and a higher degree of conjugation resulting from delocalization along, as well as across the plane in the extended framework.

In principle, the optimum N_3-COF exhibits a steady photocatalytic HER of 1703 mmol @1 g @1, thus suggesting that the activities are closely related to the surface area, and optical and electronic properties of the polymers. Further investigation revealed that nitrogen containing polymers tend to show higher activities for H_2 evolution. For instance, it was determined that the azine-based COF showed an improvement in the activities when the nitrogen content was increased.

A hydrazone-based COF with an optimized structure presented a higher activity for H_2 evolution. CTFs with triazine activities of the conjugated photocatalysts are closely related to the composition, structure, morphology, electronic, and optical properties. Thus, there is much room for future development by applying strategies, such as hybridization, heterojunction, nanostructure, and sensitization, for enhancing the activities of organic conjugated photocatalysts.

4 Conclusion and Recommendations

Researchers have explored extensively, promising, and well-defined work in the field of photocatalysis to create green remediation. Exploiting ICPn in environmental processes witnesses a continuing need to bridge the divide between the engineering advancements in green technology components and configurations, and the emerging needs. Hence, in conclusion, knowledge of the different synthetic procedures employed for the preparation of photocatalytic materials should be considered necessary. The search for new photocatalysts having desired properties to induce the oxidation reaction of organic substrates or the pollutants under visible light irradiation should be encouraged.

Acknowledgements The authors would like to thank the School of Technology, Pandit Deendayal Petroleum University and Sunway University (through internal grant INT-2018-SST-RCNMET-04) for the research facilities and financial support.

References

1. Chou, L.H.C., et al.: Semiconducting small molecule/polymer blends for organic transistors. Polym. (Guildf) **191**, 122208 (2020)
2. Du, Y., Shen, S.Z., Cai, K., Casey, P.S.: Research progress on polymer-inorganic thermoelectric nanocomposite materials. Progr. Polym. Sci. (Pergamon) **37**(6), 820–841 (2012)
3. Stenger-Smith, J.D.: Intrinsically electrically conducting polymers. Synthesis, characterization, and their applications. Prog. Polym. Sci. **23**(1), 5779 (1998)
4. Jusoff Albar, M.M., Jamion, N.A., Baharin, S.N.A., Raoov, M., Sambasevam, K.P.: Preparation of novel commercial polyaniline composites for ammonia detection. Solid State Phenom. **301**, 124–131 (2020)
5. Nguyen, D.N., Yoon, H.: Recent advances in nanostructured conducting polymers: from synthesis to practical applications. Polym. (Basel) **8**(4), 118 (2016)
6. Pandey, N., Shukla, S.K., Singh, N.B.: Water purification by polymer nanocomposites: an overview. Nanocomposites **0324**, 1–20 (2017)
7. Xiao, K.Q., Zhang, L.C., Zarudi, I.: Mechanical and rheological properties of carbon nanotube-reinforced polyethylene composites. Compos. Sci. Technol. **67**(2), 177–182 (2007)
8. Tanahashi, M.: Development of fabrication methods of filler/polymer nanocomposites: with focus on simple melt-compounding-based approach without surface modification of nanofillers. Mater. (Basel) **3**, 1593–1619 (2010)
9. Singh, P., Abdullah, M.M., Ikram, S.: Role of nanomaterials and their applications as photocatalyst and senors: a review abstract. Nano Res. Appl. **2**(1), 1–10 (2016)
10. Accepted Manuscript: J. Mater. Chem. A (2015)
11. Koh, Y., Sambasevam, K.P., Yahya, R., Phang, S.: Improvement of microwave absorption for PAni/HA/TiO_2/Fe_3O_4 nanocomposite after chemical treatment. Polym. Compos. **34**(7), 1186–1194 (2013)
12. Nguyen, V.H., Shim, J.J.: Green synthesis and characterization of carbon nanotubes/polyaniline nanocomposites. J. Spectrosc. **2015**, 1–9 (2015)
13. Nalage, S.R., Mane, A.T., Pawar, R.C., Lee, C.S., Patil, V.B.: Polypyrrole–NiO hybrid nanocomposite films : highly selective, sensitive, and reproducible NO_2 sensors. Ionics **20**(2), 1607–1616 (2014)
14. Márquez-herrera, A., Ovando-medina, V.M., Castillo-reyes, B.E., Zapata-torres, M., Meléndez-lira, M., González-castañeda, J.: Nanocomposite with enhanced photocatalytic activity under visible light. Materials **9**(1), 1–30 (2016)
15. Khatamian, M., Fazayeli, M., Divband, B.: Materials science in semiconductor processing preparation, characterization and photocatalytic properties of polythiophene-sensitized zinc oxide hybrid nanocomposites. Mater. Sci. Semicond. Process. **26**, 540–547 (2014)
16. Melo, J.P., Schulz, E.N., Horswell, S.L., Camarada, M.B.: Synthesis and characterization of graphene/polythiophene (GR/PT) nanocomposites: evaluation as high-performance supercapacitor. Electrodes **12**, 2933–2948 (2017)
17. Wang, G., Morrin, A., Li, M., Liu, N., Luo, X.: Nanomaterial-doped conducting polymers for electrochemical sensors and biosensors. J. Mater. Chem. B **6**(25), 4173–4190 (2018)

18. Reddy, R.J.: © Copyright 2010 by Ramya Jannapu Reddy all rights reserved, Dec 2010
19. Guo, Q., et al.: Comparison of in situ and ex situ methods for synthesis of two-photon polymerization polymer nanocomposites. Polym. (Basel) **6**(7), 2037–2050 (2014)
20. Kaur, G., Adhikari, R., Cass, P., Bown, M., Gunatillake, P.: Electrically conductive polymers and composites for biomedical applications. RSC Adv. **5**(47), 37553–37567 (2015)
21. Xia, X., et al.: Controllable growth of conducting polymers shell for constructing high-quality organic/inorganic core/shell nanostructures and their optical-electrochemical properties. Nano Lett. **13**(9), 4562–4568 (2013)
22. Guo, C., Zhou, L., Lv, J.: Effects of expandable graphite and modified ammonium polyphosphate on the flame-retardant and mechanical properties of wood flour-polypropylene composites. Polym. Polym. Compos. **21**(7), 449–456 (2013)
23. Wang, Y., et al.: Dopant-enabled supramolecular approach for controlled synthesis of nanostructured conductive polymer hydrogels. Nano Lett. **15**(11), 7736–7741 (2015)
24. Friedel, B., Mcneill, C.R., Greenham, N.C.: Influence of alkyl side-chain length on the performance of poly(3-alkylthiophene)/polyfluorene all-polymer solar cells. Chem. Mater. **2**, 3389–3398 (2010)
25. Reddy, D.H.K., Lee, S.M.: Water pollution and treatment technologies. Environ. Anal. Toxicol. **2**(5), 5–6 (2012)
26. Nakanishi, H., Matsuda, H., Kato, M.: Preparation of a variety of polymeric conductors from diacetylenes. Mol. Cryst. Liq. Cryst. **105**(1), 77–88 (1984)
27. Salleh, M.A.M., Razak, J.A., Ibrahim, N.A., Fakhru, A., Razi, L., Suraya, A.R.: The influences of melt-compounding parameters on the tensile properties of low filler loading of untreated-MWCNTs-polypropylene (PP) nanocomposites. J. Eng. Sci. Technol. J. Eng. Sci. Technol. **3**(31), 97–108 (2008)
28. Breuer, O., Sundararaj, U.: Big returns from small fibers: a review of polymer/carbon nanotube composites. Polym. Compos. **25**(6), 630–645 (2004)
29. Khan, W.S., Hamadneh, N.N., Khan, W.A.: Polymer nanocomposites—synthesis techniques, classification and properties Waseem. In: Science and Applications of Tailored Nanostructures, pp. 50–67. One Central Press (OCP) (2017)
30. G. Ucankus, M. Ercan, D. Uzunoglu, and M. Culha, Methods for preparation of nanocomposites in environmental remediation. Elsevier Inc., 2018.
31. Huang, Z., Zhang, Y., Kotaki, M., Ramakrishna, S.: A review on polymer nanofibers by electrospinning and their applications in nanocomposites. Compos. Sci. Technol. **63**, 2223–2253 (2003)
32. Katepalli, H., Bikshapathi, M., Sharma, C.S., Verma, N., Sharma, A.: Synthesis of hierarchical fabrics by electrospinning of PAN nanofibers on activated carbon microfibers for environmental remediation applications. Chem. Eng. J. **171**(3), 1194–1200 (2011)
33. Homaeigohar, S.S., Buhr, K., Ebert, K.: Polyethersulfone electrospun nanofibrous composite membrane for liquid filtration. J. Memb. Sci. **365**(1–2), 68–77 (2010)
34. Zhang, Z., Shao, C., Li, X., Zhang, L., Xue, H., Wang, C.: Electrospun nanofibers of $ZnO–SnO_2$ heterojunction with high photocatalytic activity. J. Phys. Chem. C **114**, 7920–7925 (2010)
35. Liu, L., Ouyang, S., Ye, J.: Gold-nanorod-photosensitized titanium dioxide with wide-range visible-light harvesting based on localized surface plasmon resonance. Angew. Chem. **200**, 1–6 (2013)
36. Fawaz, J., Mittal, V.: Synthesis of polymer nanocomposites. In: Mittal, V. (ed.) Synthesis Techniques for Polymer Nanocomposites, 1st edn., pp. 1–30. Wiley-VCH Verlag GmbH & Co. KGaA, Weinheim (2015)
37. Iketani, K., Sun, R., Toki, M., Hirota, K., Yamaguchi, O.: Sol–gel-derived TiO_2/poly(dimethylsiloxane) hybrid films and their photocatalytic activities. J. Phys. Chem. Solids **64**, 507–513 (2003)
38. Zangeneh, H., Zinatizadeh, A.A.L., Habibi, M., Akia, M., Hasnain Isa, M.: Photocatalytic oxidation of organic dyes and pollutants in wastewater using different modified titanium dioxides: a comparative review. J. Ind. Eng. Chem. **26**, 1–36 (2015)

39. Ji, Q., Yu, X., Zhang, J., Liu, Y., Shang, X., Qi, X.: Photocatalytic degradation of diesel pollutants in seawater by using $ZrO_2(Er^{3+})/TiO_2$ under visible light. J. Environ. Chem. Eng. **5**(2), 1423–1428 (2017)
40. Bu, Y., Chen, Z.: Role of polyaniline on the photocatalytic degradation and stability performance of the polyaniline/silver/silver phosphate composite under visible light. Appl. Mater. Interfaces **6**, 17589–17598 (2014)
41. La, D.D., Gottama, R., Srinivasan, P., Bhosale, S.V.: Improv-ing the photocatalytic activity of polyaniline and a porphyrin via oxidation to obtain a salt and a charge-transfer complex. New J. Chem. **41**(23), 14595–14601 (2017)
42. Pasternak, S., Paz, Y.: On the similarity and dissimilarity between photocatalytic water splitting and photocatalytic degradation of pollutants. ChemPhysChem **14**, 2059–2070 (2013)
43. Ljubas, D.: Solar photocatalysis—a possible step in drinking water treatment. Energy **30**(10), 1699–1710 (2005)
44. Van Gerven, T., Mul, G., Moulijn, J., Stankiewicz, A.: A review of intensification of photocatalytic processes. Chem. Eng. Process. Process Intensif. **46**(9), 781–789 (2007)
45. Li, S., Xu, S., He, L., Xu, F., Wang, Y., Zhang, L.: Photocatalytic degradation of polyethylene plastic with polypyrrole/TiO_2 nanocomposite as photocatalyst photocatalytic degradation of polyethylene plastic with polypyrrole/TiO_2 nanocomposite as photocatalyst. Polym. Plast. Technol. Eng. **49**, 400–406 (2010)
46. Kumar, R., El-Shishtawy, R.M., Barakat, M.A.: Synthesis and characterization of Ag-Ag_2O/TiO_2@polypyrrole heterojunction for enhanced photocatalytic degradation of methylene blue. Catalysts **6**, 76–87 (2016)
47. D'Amato, C.A., Giovannetti, R., Zannotti, M., Id, M.M., Gunnella, R., Di Cicco, A.: Band gap implications on nano-TiO_2 surface modification with ascorbic acid for visible light-active polypropylene coated photocatalyst. Nanomater. MDPI **8**, 599 (2018)
48. Hisatomi, T., Kubota, J., Domen, K.: Recent advances in semiconductors for photocatalytic and photoelectrochemical water splitting. Chem Soc Rev **22**, 7520–7535 (2014a)
49. Sun, C., Li, H., Chen, L.: Nanostructured ceria-based materials: synthesis, properties, and applications. Energy Environ. Sci. **5**(9), 8475 (2012)
50. Pelaez, M., et al.: A review on the visible light active titanium dioxide photocatalysts for environmental applications. Appl. Catal. B Environ. **125**, 331–349 (2012)
51. Hisatomi, T., Kubota, J., Domen, K.: Recent advances in semiconductors for photocatalytic and photoelectrochemical water splitting. Chem. Soc. Rev. **43**(22), 7520–7535 (2014b)
52. Khan, M.M., Adil, S.F., Al-Mayouf, A.: Metal oxides as photocatalysts. J. Saudi Chem. Soc. **19**(5), 462–464 (2015)
53. Ghosh, S., et al.: Visible-Light Active Conducting Polymer Nanostructures with Superior Photocatalytic Activity. Nature Publishing Group (2015)
54. Li, S., Hu, S., Xu, K., Jiang, W., Liu, J., Wang, Z.: A novel heterostructure of BiOI nanosheets anchored onto MWCNTs with excellent visible-light photocatalytic activity. Nanomater. (Basel) **Vld**, 1–13 (2017)
55. Yang, X., Ran, X.: Photocatalytic degradation of organic dyes by a donor–acceptor type conjugated polymer: poly(thiophene-1,3,4-oxadiazole) and its photocatalytic mechanism. Polym. Compos. **67**(9), 1282–1290 (2018)
56. Ravi Chandra, M., Siva Rao, T., Kim, H.S., Pammi, S.V.N., Prabhakarrao, N., Manga Raju, I.: Hybrid copper doped titania/polythiophene nanorods as efficient visible light-driven photocatalyst for degradation of organic pollutants. J. Asian Ceram. Soc. **5**, 436–443 (2017)
57. Cheng, Y., Wang, J., Bao, Y., Ma, Y., Wang, G.: Synthesis of polythiophene nanoparticles by microemulsion polymerization for photocatalysis. Adv. Mater. Res. **401**, 1312–1319 (2012)
58. Riaz, U., Ashraf, S.M., Kashyap, J.: Enhancement of photocatalytic properties of transitional metal oxides using conducting polymers: a mini review. Mater. Res. Bull. **71**, 75–90 (2015)
59. Jang, J., Ha, J., Kim, S.: Fabrication of polyaniline nanoparticles using microemulsion polymerization. Macromol. Res. **15**(2), 154–159 (2007)
60. Chaudhry, F., Malik, M.: Factors affecting water pollution: a review. J. Ecosyst. Ecography **07**(01), 1–3 (2017)

61. Ghosh, S., Maiyalagan, T., Basu, R.N.: Nanostructured conducting polymers for energy applications: towards a sustainable platform. Nanoscale **8**(13), 6921–6947 (2016)
62. Wang, C., Xie, Z., Kathryn, E., Lin, W.: Light-harvesting cross-linked polymers for efficient heterogeneous photocatalysis. Appl. Mater. Interfaces **4**, 2288–2294 (2012)
63. Ariffin, S.N., et al.: Modification of polypropylene filter with metal oxide and reduced graphene oxide for water treatment. Ceram. Int. **40**, 6927–6936 (2014)
64. Gong, B., Peng, Q., Na, J., Parsons, G.N.: Applied catalysis a: general Highly active photocatalytic ZnO nanocrystalline rods supported on polymer fiber mats: synthesis using atomic layer deposition and hydrothermal crystal growth. Appl. Catal. A Gen. **407**(1–2), 211–216 (2011)
65. Li, C., et al.: Electrospinning of CeO_2–ZnO composite nano fi bers and their photocatalytic property. Mater. Lett. **65**, 1327–1330 (2011)
66. Moafi, H.F., Shojaie, A.F., Zanjanchi, M.A.: Semiconductor-assisted self-cleaning polymeric fibers based on zinc oxide nanoparticles. J. Appl. Polym. Sci. **121**(6), 3641–3650 (2011)
67. Baruah, S., Thanachayanont, C., Dutta, J.: Growth of ZnO nanowires on nonwoven polyethylene fibers. Sci. Technol. Adv. Mater. **9**(2), 025009 (2016)
68. Grabowska, E., Zaleska, A., Sorgues, S., Etcheberry, A., Colbeau-Justin, C., Remita, H.: Modification of titanium (IV) dioxide with small silver nanoparticles: application in photocatalysis. J. Phys. Chem. **117**(4), 1955–1962 (2013)
69. Zhang, G., Lan, Z.A., Wang, X.: Conjugated polymers: catalysts for photocatalytic hydrogen evolution. Angew. Chem. Int. Ed. **55**(51), 15712–15727 (2016)
70. Dai J., Chen, X., Yang, H.: Visible light photocatalytic degradation of dyes by bismuth oxide-reduced graphene oxide composites prepared via microwave-assisted method. J. Colloid Interface Sci. **1556** (2017)
71. Dhanavel, S., et al.: Visible light induced photocatalytic degradation of methylene blue using polyaniline modified molybdenum trioxide. Mech. Mater. Sci. Eng. **9** (2017). To cite this version: HAL Id: hal-01500533
72. Ullah, H., Tahir, A.A., Mallick, T.K.: Polypyrrole/TiO_2 composites for the application of photocatalysis. Sens. Actuators B Chem. **241**, 1161–1169 (2016)

Energy Storage Devices (Supercapacitors and Batteries)

Meenakshi Gusain, Poonam Singh, and Yiqiang Zhan

Abstract The realization of future energy based on safe, clean, sustainable, and economically viable technologies is one of the grand challenges faced by modern society. Electrochemical energy technologies underpin the potential success of this effort to divert energy sources away from fossil fuels, whether one considers alternative energy conversion strategies through photoelectrochemical (PEC) production of chemical fuels or fuel cells run with sustainable hydrogen, or energy storage strategies, such as in batteries and supercapacitors. This dissertation builds on recent advances in nanomaterials design, synthesis, and characterization to develop novel electrodes that can electrochemically convert and store energy. With the improvement of global economy, the fatigue of energy becomes inevitable in the twenty-first century. It is expected that the increase in world energy requirements will be triple at the end of this century. Thus, there is an imperative need for the development of renewable energy sources and storage systems.

Keywords Sustainable · Electrochemical · Energy storage · Nanomaterial · Supercapacitor

1 Introduction

The accelerated consumption of non-renewable sources of fuels (i.e. coal, petroleum, gas) along with the consequent global warming issues have intrigued immense research interest for the advancement and expansion of an alternate efficient energy conversion and storage technique in the form of clean renewable resource. The global demand for energy production is predicted to be at least double by 2050, while the rate at which the non-renewable fossil fuels are being consumed today; it will take not

M. Gusain (✉) · Y. Zhan
State Key Laboratory of ASIC and System, Centre of Micro-Nano System, SIST, Fudan University, Shanghai 200433, China
e-mail: meenakshigusain@fudan.edu.cn

P. Singh
Department of Applied Chemistry, Delhi Technological University, Delhi 110042, India

© Springer Nature Switzerland AG 2021

S. Shahabuddin et al. (eds.), *Advances in Hybrid Conducting Polymer Technology*, Engineering Materials, https://doi.org/10.1007/978-3-030-62090-5_3

more than 40 years to run all the known oil repositories dry leaving the entire world into an era of complete darkness. In addition to this, the combustion of fossil fuels at a faster rate also leads to the excessive emission of harmful gases such as CO_2, CH_4, N_2O etc. in the atmosphere because of which the temperature of the earth is getting raised. Therefore in order to overcome these issues, we need to focus at sustainable green energy origins such as wind, solar, hydroelectric, geothermal, biological, nuclear, etc. However, the infrequent availability of abovementioned sustainable green energy origins encourages the focus of research on efficient ways of storage systems for stockpile and provides energy in a steady mode. The research work in the direction of storing electrochemical energy has expanded significantly during the last few decades and a huge range of active materials have been reported, both for supercapacitor and battery type energy storage [1, 2]. But till today among all the systems for storing energy electrochemical energy storage/conversion system found to be prominent candidate to get rid of the prevailing energy crisis. Based on the energy conversion mechanisms electrochemical energy storage systems can be divided into three broader sections namely batteries, fuel cells and supercapacitors. In batteries and fuel cells, chemical energy is the actual source of energy which is converted into electrical energy through faradic redox reactions while in case of the supercapacitor, electric energy is stored at the interface of electrode and electrolyte material forming electrochemical double layer resulting in non-faradic reactions.

The selection of an energy storage device for various energy storage applications depends upon several key factors such as cost, environmental conditions and mainly on the power along with energy density present in the device. Basically an ideal energy storage device must show a high level of energy with significant power density but in general compromise needs to be made in between the two and the device which provides the maximum energy at the most power discharge rates are acknowledged as better in terms of its electrical performance. The variety of energy storage systems can be compared by the "Ragone plot". Ragone plot comprises of performance of energy storage devices, such as capacitors, supercapacitors, batteries, and fuel cells are shown in Fig. 1.

The relationship of specific energy (E) with specific power (P) is provided by the expression [3, 4]

$$E = 1/2(\mathrm{c}\Delta V)^2$$
$$P = E/t_{dis}$$

where c represents the specific capacitance (F g^{-1}), ΔV represents the operating potential window (V), and t_{dis} represents the discharge time (s).

Ragone plot is a plot in which the values of the specific power density are being plotted against specific energy density, in order to analyze the amount of energy which can be accumulate in the device along with the efficiency of the energy's release. According to the Ragone plot batteries and fuel cells both acquire large value of specific energy density with small value of specific power density in contrast capacitors have high value of specific power density with a small value of specific

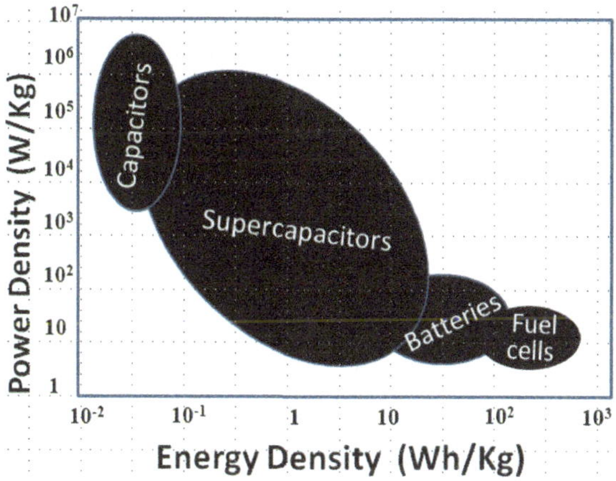

Fig. 1 Ragone plot representing varied energy storage devices (specific power vs. specific energy)

energy density. Whereas supercapacitor possess intermediate specific energy density together with power density and also possess a longer lifetime due to the absence of chemical reactions [5]. The low energy density of the supercapacitor is the only shortcoming in comparison to the batteries and fuel cell which is act as an obstacle for their commercialization. The type of material is being used with its structure for the preparation of electrode material of supercapacitor decides the performance of the supercapacitor.

Conducting polymers has immense electrical conductivity and undergoes reversible redox reactions due to its intrinsic tendency making it promising cathode materials for energy storage devices but suffers limitation owing to lesser specific capacity and tendency of large extent of self-discharge. The fusion of conducting polymer with other material having high specific capacity could create hybrid conducting polymer having beneficial properties of both component widening the applicability in storage devices with broadening application. In hybrid conducting polymers, conducting polymer can contribute towards the cyclic efficiency, high discharge capacity while other component contributes towards high capacity and ability to intercalate. The presence of inorganic materials in hybrid conducting polymer contributes towards the enhancement in the electron transfer tendency at the interface of the surface of the electrode and electrolyte material [6]. Hybrid conducting polymer can be classified into two major groups depending upon the conduction polymer and nanoparticle arrangement. The conducting polymer can serve as a matrix for inorganic nanomaterials or can be inserted inside the layer of the inorganic matrix.

The synthesis of hybrid conducting polymer can be classified into three categories [7]. In the first approach, the synthesis of nanoparticles occurs in the presence of a polymer matrix. It may cause aggregation which makes it difficult strategy to synthesize hybrid conducting polymer. In the second approach nanoparticles are generated in situ while the polymerization of monomer taking place resulting in homogeneous distribution of nanoparticles in the polymer matrix. In this, the polymerization reaction and the synthesis of nanoparticles proceed simultaneously leading to the synthesis of homogeneous hybrid conducting polymer. The third synthetic method includes polymerization of the monomer of the requisite polymer around the nanoparticles by means of chemically compatible ligands [8] or polymeric structures [9].

Controlling the size and shape of content phases, degree of mixing and synthetic protocol participates in the improvement of as-prepared nanocomposite properties. Different properties can be expected by varying attributes of the content phases and synthetic method used for hybrid conducting polymers.

2 Proposed Mechanism for Designing Hybrid Conducting Polymer

Conducting polymers behave as insulators or semiconductors in pure state. The doping of donor or acceptor substituents by using oxidizing or reducing agents respectively increases conductivity by several orders. The oxidation leads to p-doping in conducting polymers by using electron acceptor species such as I_2, H_2SO_4 etc. The reduction of conducting polymers leads to n-doping introduced by using electron donor species i.e. alkali metal ions (Na, K) and Li. Doping creates charged defects such as polaron, bipolaron and soliton acting as charge carrier for conduction. Hence increase in concentration of doping species enhances conductivity.

Although conductivity in conducting polymer has been explained and discussed by many research group but exact reason of conductivity in hybrid conducting polymer has been expected to be similar to conducting polymers and not explored extensively. The synergic effect arises due to the proper alignment of polymer chains may help in enhancement in conductivity. The interfacial interaction, photoinduced electron transfer, electrostatic interaction found to increase the conductivity of hybrid conducting polymers [9–11].

3 Batteries

Batteries have become the typical power source utilized for numerous purposes in industrial and consumer electronics because of its compactness, efficiency, reliability, and economical point of view. Battery maintains virtual instantaneous input and output response from the battery to network and vice-versa. Basically batteries are electrochemical devices exploiting redox reactions for converting the accumulated chemical energy of batteries into required electrical energy. Typically a battery consists of electrochemical cells containing electrically connected electrodes using a conductive electrolyte containing negatively charged ions and positively charged ions. The transportation of ions in the cell determines the polarity of a cell. The anion transporting electrode is defined as anode and cation transporting electrode defined as cathode taking place during charging process. Mostly, in batteries redox reaction takes place with oxidation and reduction reactions at the anode and cathode respectively.

On the basis of charging capabilities of batteries are classified in four classes: Primary batteries (non-rechargeable), secondary (rechargeable batteries), Grid-scale battery systems and Fuel cells.

Primary Batteries These batteries are not rechargeable, cannot be recycled and simple electric devices. These are simple and convenient to use involving irreversible processes. Primary batteries utilize aqueous and non-aqueous electrolytes. In these devices, electrolyte reacts with electrodes for creating flow of electric current along with formation of by-products that cannot be reused. The commonly employed primary batteries include zinc-carbon battery, alkaline battery and lithium primary batteries. It suffers from less energy density, reduced leakage resistance, and drop in voltage through discharge. These batteries commonly used in flashlight and many portable devices.

Secondary Batteries These batteries are rechargeable broadening the range of application for portable electronic devices. The longer charge–discharge cycles commercializes secondary batteries for residential power storage and for electric vehicles. Secondary batteries use reversible process having two distinct charge cycle and discharge cycles, marked by distinctive chemical reactions and peculiar electrical properties. In course of charging cycle, electrical energy transforms electrolyte storing electrical energy in form of chemical bonds. In discharge cycle, energy is released from chemical bonds and generates electrical energy by the transformation of electrolyte. Secondary rechargeable batteries comprise of lead-acid batteries, lithium-ion batteries, lithium-sulfur batteries, nickel-metal hydride batteries, and nickel-metal batteries depending upon their electrode component. The secondary batteries offer superior battery performance, high-quality performance in altering temperature range, elevated voltage, and fine charge retention. Amidst other secondary batteries, lithium–ion batteries found to show the highest storage efficiency valued nearly 83%, and have been installed in renewable energy systems widely along with micro-grid systems. The assets of using lithium-ion batteries includes the least maintenance, extended life-cycle, stability over a wide range of temperature, efficient charging-discharging ability, and elevated energy density. Secondary batteries are included in laptops and mobile phones.

Grid-Scale Battery Systems Grid scale storage provides peak power and stability for a sustained period. It includes red-ox flow batteries, Na–S batteries using advance level lead-A and Lithium-ion batteries.

Fuel Cells Fuel cells use either hydrogen, or indirect systems using fossil fuels such as methanol by the means of catalysed and thermal reaction. Fuel cells are resourceful in the output power supply, high reliability factor, and negligible amount of degradation process.

Thus batteries are storage option for the electrical energy providing smooth and steady electrical power for micro systems and are assembly of pseudocapacitive electrodes storing charge using faradic reactions. For various purposes batteries are preferred over supercapacitors due to their characteristics of slower discharge time providing lower energy densities available for much longer lifetime. The battery type electrodes exhibit non-reversible oxidation and reduction phenomenon. The galvanostatic charge-discharge curves presents typical non-linear behavior of the curve having flat discharge plateau reasoning for their ability to store large amount of energy. The flat discharge plateau represents phase transformation phenomenon occurring at the surface of electrode materials. The electrode materials due to phase transformation cause depletion in rate capability of the battery electrode limiting their use for longer run. The batteries used in industries for securing power in telecommunications, data networks etc. maintaining the continuous electricity supply. A range of battery chemistries is used for various types of energy storage applications. Extensive research has been performed to increase the capacitance and cyclic performance. Among various types of batteries, the commercialized batteries are lithium-ion batteries, sodium-sulfur batteries, lead-acid batteries, flow batteries and supercapacitors.

As we will be dealing with hybrid conducting polymer applicable for the energy storage devices in this chapter, here describing some important categories of hybrid conducting polymers consisting of conducting polymers with other constituents.

3.1 Metal Oxide/conducting Polymer

Metal oxides majorly oxides of iron, copper, cobalt and nickel have been investigated for the battery materials due to their high theoretical storage capacity shown for sodium. The reactions taking place reduces their life span due to large volume changes associated with it and therefore limited their exploitation in battery systems. However incorporation of conducting polymers not only increases electrical conductivity also improves its mechanical stress. The TiO_2 charge capacity is increased by sandwiching in between polyaniline (PANI) and graphene nanosheets serving in lithium ion batteries as anode material. It is accompanied by 99.19% efficiency after 100 cycles [12]. Metal oxides do not have higher capacity, however blending with conducting polymers can lead to the enhancement in the reversible capacity approximately ten times compared to the bare MnO_2 devices. The swelled layered structure of PANI intercalated MnO_2 possess high capacity along with excellent cycle stability and is therefore considered to be a promising cathode material for high-capacity lithium ion batteries [13]. The PANI/β-$AgVO_3$ triaxial nanowire indicates faster kinetics and better capacity compared to β-$AgVO_3$ nanowires which is expected owing to the decrease in charge transfer resistance.[14]. Polypyrrole (PPy) nanowires blended with silicon particles has been synthesized by mechanical milling showing improved reversibility and enhanced life cycle than that of silicon, because PPy nanowire act as a polymer matrix layer supporting active silicon granule since they have capability to continually get alloy with lithium ions during the charging-discharging process of lithium ion batteries [15]. PPy have been found to decrease the resistance of charge transfer reactions connected to the Li ions intercalation-deintercalation processes in PPy-metal oxide hybrid polymers such as MoO_3, V_2O_5, $LiCoO_2$ and LiV_3O_8 [16–20]. Hybrid composite of PPy/Polyethylene glycol (PEG) with $LiFePO_4$ showed higher discharge capacity than pristine $LiFePO_4$, PEG presence reduces structural defects due to stabilization effect provided by PEG [21].

3.2 Metal Chalcogenides/conducting Polymer

Metal chalcogenides owing to their tunable properties, unique compositional and structural features motivated for fundamental understanding and industrial advancement for varied applications. The two-dimensional metal chalcogenides have been explored extensively for lithium-ion batteries because of their aptitude to provide a platform by forming structures which support reversible intercalation of lithium-ion at very large applied potential and high surface-to-volume ratio. Due to structural instability and less electrical conductivity suffers these materials from low lifetime and small rate capability. Integration of metal chalcogenides with conducting polymer creating hybrid conducting polymer combine valuable properties of both. The remarkable electrochemical performance of the devices was observed due to the synergic effect involved in between metal sulfides and the conducting polymer. Molybdenum sulfide (MoS_2) in combination with conductive PANI proved the reduced resistance of charge-transfer at the interface of electrode and electrolyte resulting in high charge capacitance and cycling behaviour as seen from the Nyquist plots of bare MoS_2 and MoS_2/PANI. The MoS_2/PANI showed improved properties through large charging capacity valued 1063.9 $mAhg^{-1}$ with 90.2% retention in specific capacitance after long cycle-life of 95 cycles [22]. Hybridization of PANI with MoS_2 found to decrease the resistance of charge-transfer at the interface of electrode-electrolyte [22]. The PPy hybrid composite by SnS_2 exhibits specific capacitance of 1000 $mAhg^{-1}$ by retaining specific capacitance upto 703 $mAhg^{-1}$ after 500 cycles [23]. The lamellar structure of hybrid conducting polymer provides conductive network for the transportation of ions and enhancing the tendency of the expansion and contraction processes at the electrode material occurring in charge–discharge process [21]. The specific capacitance in batteries has also been found to increase by constructing MoS_2 nanocomposites with PPy and poly(3,4-ethylenedioxythiophene)

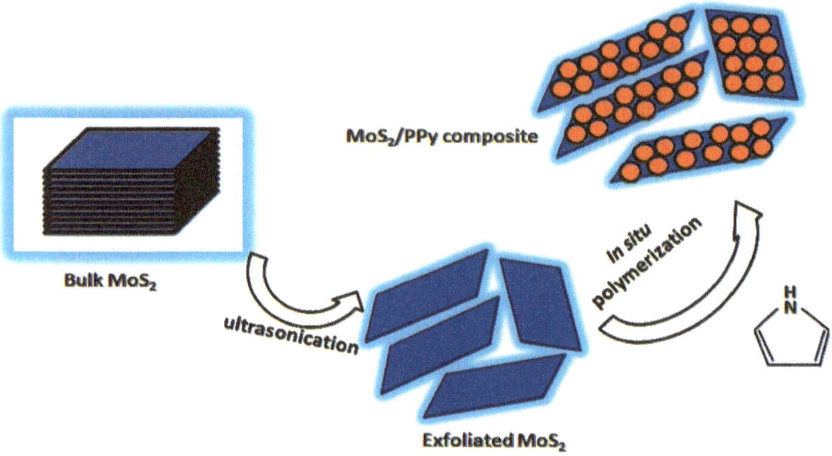

Fig. 2 Schematic diagram representing blending of MoS_2 with PPy forming MoS_2/PPy composite

polystyrene sulfonate (PEDOT: PSS) [24, 25]. Binary composite of PPy/MoS_2 illustrated the high capacitance with outstanding cycling stability that proves its excellence as an electrode material for supercapacitors and a schematic diagram for the blending has been presented in Fig. 2.

Similarly, The ternary hybrid conducting polymer are found to observe further improvement in the capacitance properties of conducting polymer reasoned out by blending the properties of three components present in hybrid conducting polymer [25–27].

3.3 Carbon Supported Hybrid Materials

During recent years, enormous efforts have been made to synthesize graphene hybrid materials as electrodes for novel energy storage devices. Graphene is two-dimensional layered material having total specific area of 2630 m^2/g along with 2000–5000 cm^2/V s of charge carrier mobility which is suitable for energy storage devices [28]. The principle of using graphene is to enhance the surface area which helps in allowing superior charge adsorption processes. Designing graphene using inorganic materials exploits the flexibility in graphene and reduces some stress off the inorganic material during operation cycles thereby improving specific capacitance as well as cyclic performance. Graphene oxide in reality is functionalized form of graphene obtained by chemically modification of graphene resulting in functional groups containing oxygen such as alcohols, epoxides, and acid groups dispersible in organic solvents, water, and different matrixes. The application of graphene in batteries is exploiting properties such as large surface area, remarkable electrical conductivity, excellent stability towards chemicals, thermal conductivity, electronic tunability, and mechanical strength. Graphene due to high mechanical strength and flexibility found to improve the storage of lithium ion in its hybrid material by reducing the stress cracking after repeated charge/discharge process which discontinues the connection between charge storing material and current collectors. According to the report, graphene nanosheets hybrid tin oxide exhibits a capacitance of 810 mAh/g retaining 70% efficiency after completing 30 charge-discharge cycles, whilst bare SnO_2 exhibited capacitance of only 550 mAh/g and also decreased to 60 mAh/g after completing 15th cycles [29]. The presence of graphene nanosheets in tin oxide creates enough void spaces buffering volume change occurs during the process of lithium insertion along with electronic conductive channels improves electrochemical performance. Sulfur based cathodic material has a theoretical specific capacity of 1672 mAh/g but has limited application due to their tendency to dissolve into electrolyte and experience swelling. However, sulfur hybrid material with graphene and PEG exhibit specific capacitance of 600 mAh/g reported for completing 100 cycles [30]. The metal oxide nanoparticles such as MnO and Fe_3O_4 integrated with graphene to achieve high specific capacity and stability [31, 32]. More than 90% of capacity retention exhibited by MoS_2/graphene oxide hybrid and rGO/SnS_2 shows 84% retention in charge capacity upto 500 cycles [33, 34]. $ZnMn_2O_4$–graphene hybrid nanosheets exploited in Li–ion battery as electrode exhibits higher rate capability and cycling stability compared to conventional graphite anode [35].

PANI based graphene hybrid material are examined for lithium ion batteries as electrode materials [36]. Graphene/PANI is the extensively studied hybrid materials as supercapacitor electrode attributed to their wide potential window, ease of its synthesis, and improved stability among other conducting polymers hybrid. A schematic diagram for the interaction showing between PANI and Graphene is presented in Fig. 3. MoS_2-graphene a combination of two-dimensional materials which has a high surface area, versatile electronic structure and high electrical conductance outperforms as an excellent candidate for energy storage application [37]. Although lithium-ion battery has been used mainly for practical purposes but sodium ion batteries have also been explored with MoS_2 composites with carbon and graphene materials [38].

Other than conducting polymer hybrids, nanoengineered two-dimensional MXenes and their derivatives attracted attention for energy storage devices improving limitations for practical applications [39]. MXene/polymer composites are widely used as electrode materials for hybrid supercapacitors. The frequently used polymers are electroactive in nature i.e. PVA, PPy, PANI, PEDOT and its derivatives, polyfluorene derivatives (PFDs), and polydiallyldimethylammonium chloride (PDAC). MXenes materials are including two-dimensional (2D) structure of metal carbides and nitrides behaving as hopeful electrode material having high electrical conductivity however suffer from low mechanical strength. MXenes suffers from the disadvantage of forming aggregation which is irreversible in nature and easily form stacking. The integration of graphene/conducting polymers linking the MXenes interlayer results in the formation of new types of material showing advanced functional and mechanical properties. The presence of conducting spacers in MXenes improves the kinetics and rate performance. For example, Ling et al. [40] have

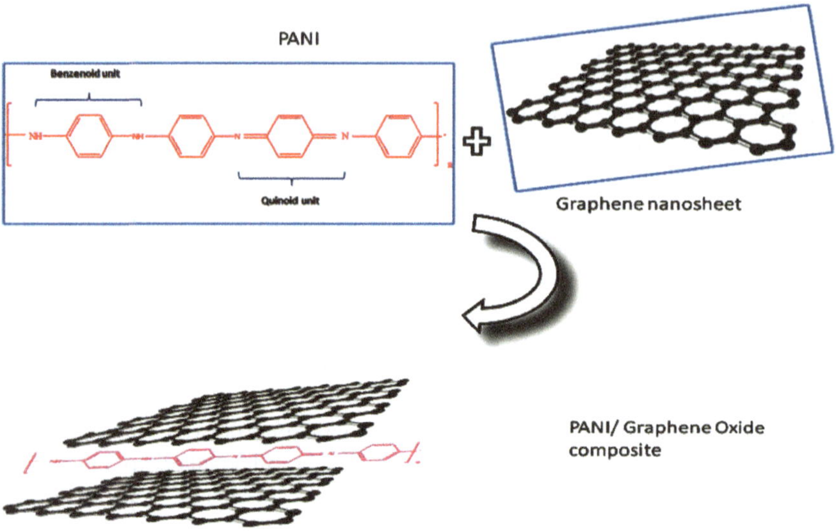

Fig. 3 Schematic diagram representing PANI/Graphene oxide composite

reported the synthesis of MXene/PVA nanocomposite composite having outstanding mechanical properties in comparison to their pure MXene/PVA counterparts [40]. Boota et al. [41] reported the addition of PPy (8 wt %) to MXene yields volumetric and gravimetric capacitances of 1000 F/cm^3 and 416 F/g, relatively higher than that of the pure MXene electrode [41]. Qin et al. [42] reported nanocomposite film based on MXene and PPy self-assembly as electrode-based supercapacitors, having excellent capacitance (69.5 mF cm^{-2}), ultrahigh energy density (250.1 mWh cm^{-3}) and outstanding cycling stability [42]. The enhanced electrochemical performance MXene/PPy nanocomposite may be attributed to the synergistic effect existing between MXene's (electric double layer capacitance mechanism) and PPy's (pseudocapacitance mechanism). Zhu et al. [43] also reported the synthesis of conducting hybrid films based on PPy/$Ti_3C_2T_x$ having excellent electrochemical performance as a supercapacitor electrode. However, still efforts need to be done on identifying novel organic materials that can be easily inserted in between the interlayer region of MXene to develop hybrid structures for high-performance energy storage devices [43].

Batteries have disadvantages in concern with the environment through hazardous waste and toxic fumes during manufacturing in addition with disposal and recycling. The use of heavy metal as electrode material when exposed causes serious effects on the health of animals and humans. There are some limitations in using batteries if not handled properly. A battery explosion is very common problem being faced and caused by misuse, short-circuit and excessive charging of batteries. The excessive charging or rate of charging leads to the formation of mixture of hydrogen and oxygen building up excessive pressure inside the battery. In extreme situation, battery chemicals may spray causing irreversible damage. The short-circuit generates large amount of current responsible for explosion. Another problem associated with batteries is leakage, releasing of dangerous chemical damaging the equipment or the environment.

4 Supercapacitors

In comparison to the batteries, supercapacitors are evolving as one of the most exciting innovative developments in the field of devices storing energy for future perspective. Supercapacitors fill the space having amid batteries quality and capacitors quality since its specific power density is higher compared to batteries and specific energy density is higher than that of the capacitor. Other significant features of supercapacitors include faster charge-discharge rate, longer cycling life time, simple fabrication with low maintenance, and without short circuit issues which are major concern in using available batteries. In addition these are exceptionally safe for storage since they can be easily discharged and do not release any toxic waste in the environment. Therefore supercapacitors are attractive and appropriate efficient energy storage devices mainly utilized in mobile electronic devices, hybrid electric vehicles, manufacturing equipment's, backup systems, defence devices etc. where the requirement of power density is high and cycling-life time required is longer are highly desirable [44–48]. Although the working of the supercapacitors is comparable to the conventionally used capacitors but supercapacitors are capable of storing greater charge through the presence of pores present within the electrodes having large surface area and also high charge separation between the electrolyte and an electrode occurs at a very small distance.

The components of a supercapacitor are two electrode system immersed in electrolyte having a separator. The electrodes possess high specific surface area and are separated by a separator i.e. membrane that permits the mobility of charged ions. The electrolyte is the mixture of positively and negatively charged ions dissolved in water. They are capable of storing a large amount of energy that can be released very fast. An ionic layer forms in between the electrodes sharing common electrolyte accumulate electric charge in the supercapacitor. Based on the mechanism involved in the charge storage and the active material of electrode, supercapacitors classified in three broader types, i.e. electrochemical double layer capacitors (EDLCs), pseudocapacitor and hybrid capacitors (Fig. 4). Each type has its own charge storage mechanism i.e. Faradic mechanism, Non-Faradic mechanism and the combination of Faradic and Non-Faradic mechanism respectively [44, 49, 50].

EDLC are storing their energy by non-Faradaic mechanism in which EDL charges at the interface of electrically conducting porous electrode such as carbon-based materials and ionically conducting electrolytes (Fig. 5). In EDLC, there do not occur any charge transformation at the interface of the electrode and electrolyte material. In EDLCs charges are distributed on the surface by physical mechanism without formation or cleavage of any chemical bond. Thus the electrode material remains

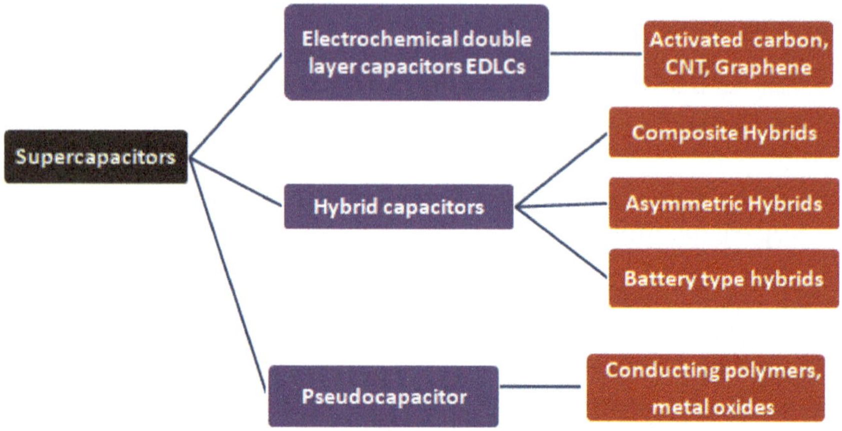

Fig. 4 Classification of supercapacitors

inert at all working voltage. The model of EDLCs was first proposed by Helmholtz in 1999 that was supplemented by Gouy and Chapman [51–53]. According to the theory, they proposed that the two opposite charge layers built up at the interface of electrode/electrolyte named as Inner Helmholtz Plane and Outer Helmholtz Plane. Stern modified the model presented by Gouy and Chapman, he united the model given by Helmholtz with Gouy-Chapman model to clearly distinguish between the two types of ion distribution layers i.e. the compacted layer and the diffused layer [54]. For ideal EDLCs, the specific capacitance, C (F/g), of electrode is generally given by the expression

$$C = \varepsilon_0 \varepsilon_r A/\mathrm{d}$$

where, ε_0 represents dielectric constant of the free space, ε_r represents dielectric constant of the insulating material present amid the electrodes, A is the total surface area of the electrode available for the electrolyte ions, and *d* represents the effective thickness of the EDLCs. Therefore, by making use of active materials with large surface area, significant increase in specific capacitances easily attained.

The pseudocapacitor or redox capacitor stores energy by Faradaic mechanism by means of the pseudocapacitive behaviour of the used redox-active material. The charge transfer reaction occur between the interface of electrode and electrolyte exploits redox-reactions, electro-sorption and intercalation processes. The rapid as well as reversible faradic reactions occur at the surface as well as in the vicinity of the surface of active electrode material. In general, pseudocapacitors are based on the materials possessing variable oxidation states. Pseudocapacitors possess superior capacitances along with energy densities compared to EDLCs (Fig. 6).

Additionally, the hybrid capacitors operate by means of both mechanisms i.e. using faradic and non-faradic mechanism. Commonly, the electrode materials of hybrid capacitor are composite of carbon based porous material blended with either

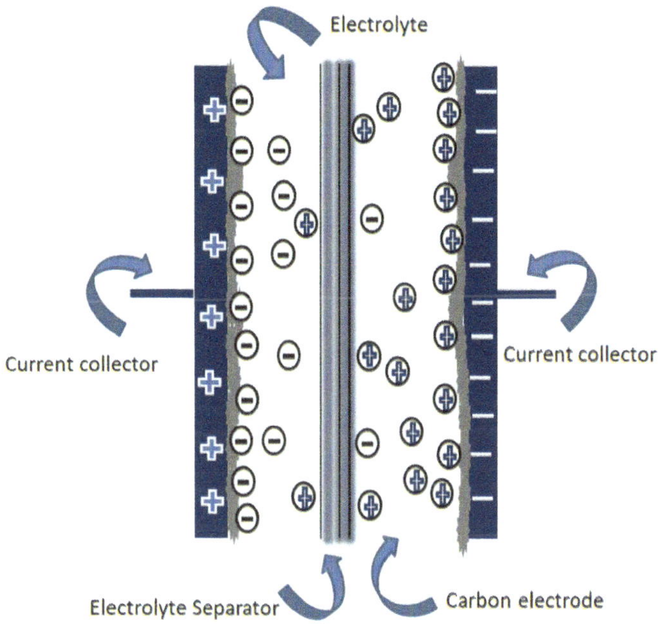

Fig. 5 Schematic representation of EDLCs

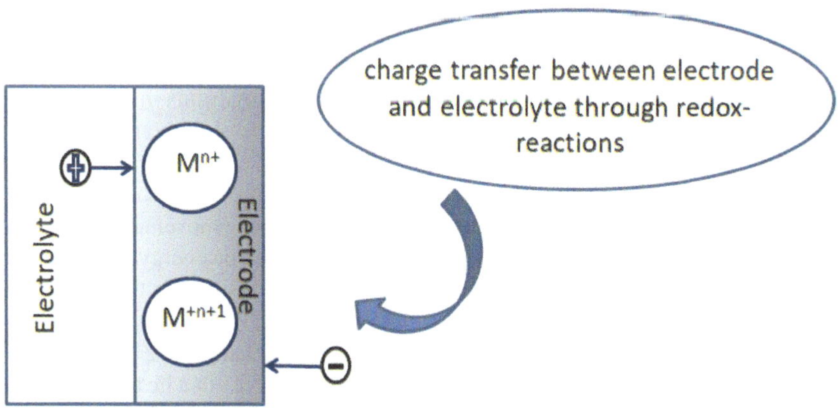

Fig. 6 Schematic diagram presenting mechanism of charge storage involved in pseudocapacitor

conducting polymers or metal oxide or both. Electrode materials employed in pseudocapacitors are usually made up of metal oxides and conducting polymers while EDLCs make use of large surface area based carbon electrode [55, 56].

The commonly employed electrode material in supercapacitor is carbon based materials attributed to their exceptional properties mainly superior conductivity, controlled pore size, variable allotropic forms, superior corrosion resistance, large surface area, high temperature stability, relatively low cost, environmental friendliness, easy and processability. Carbon based materials exists in variety of forms such as fine particles, fibre, composites, textile, etc. and showing various dimensional variety from zero-dimensional to three dimensional. The different forms of carbon that are mainly used as electrode for enhancing the performance of supercapacitor are discussed below:

Activated Carbon Activated carbon is usually employed in EDLCs as electrode material by virtue of their low cost, easy processing and large surface area (500–3000 m^2g^{-1}). Specific capacitance (SC) offered lies from 25 to 150 F/gm in aqueous electrolyte as well as organic electrolyte in the activated carbon electrodes. Along with large surface area the capacitive performances in activated carbons also depends upon several other factors such as pore structures and pore size distributions. Large pore size results in high power densities where as the small pore size results in high energy density. Zhang and Zhao [47] have reported that activated carbon makes use of complex porous structure consisting of variable size macropores (more than 50 nm), mesopores (from 2 to 50 nm) and micropores (less than 2 nm) wide to achieve a high value of surface area (BET) [47]. Also the surface functionality has profound effect on the capacitance and is exercised as additive for enhancing the conductivity performance in supercapacitor. Tang et al. [57] have reported that the oxygen and nitrogen functionalities present at the surface of carbon can enhance the specific capacitance by enhancing the wettability and by pseudo-Faradaic reaction [57]. Apart from activated carbon, carbon black is also exploited as active electrode material for supercapacitors as a result of its large surface area to volume ratio and high conductivity.

Carbon Nanotubes (CNTs) CNTs were first discovered by Iijima in the year 1991 in carbon soot and are amongst the most important forms of carbon gained enormous interest of researchers worldwide owing to its unique pore size, high thermal stability, superior electrical and mechanical properties. In comparison to activated carbon, CNTs possess mesoporous structure that allows facile movement of ions and lesser series resistance resulting in the improvement of specific energy density along with specific power density. CNTs can principally be classified in two type's i.e. multiwalled CNT (MWCNTs) and single walled CNT (SWCNTs) which bend on themselves forming tubes with hollow internal core area (Zhang and Zhao [47]). The diameter of a tube is the most important characteristic of CNTs that lie from 1 to 3 nm order and in length it has tens of microns orders. Niu etal.1997 first suggested the use of MWCNT based electrode material in supercapacitors exhibiting a power density and specific capacitance of 8 kW kg^{-1} and 102 F g^{-1} respectively in presence of H_2SO_4 solution as an electrolyte [58]. But the utilization of CNTs for various roles is still limited owing to its huge manufacturing cost. Recently helical carbon nanotubes have also attracted enormous consideration because of comparatively high specific surface area and excellent elasticity. Zeng et al. [59] have reported the synthesis of high-purity helical carbon nanotubes (HCNTs) by using one-step chemical synthetic method resulting in superior specific capacitance of 95 F g^{-1} at a current density of 0.1 A g^{-1} [59].

Carbon Fiber Carbon fibers possess one dimensional structure resulting in superior charge transportation property and also the presence of numerous pores present at the surface of the carbon fiber offers immense adsorption sites for ions to get adsorb and thus proves to be the most promising candidate electrode material.

Graphene Recently, a considerable shift in the attention of researchers from other frequently used carbon based materials such as CNTs, activated carbon etc. towards graphene has been observed due to their large surface area, appropriate pore size distribution, superior chemical stability, high conductivity, thermal stability, high elasticity, superior mechanical and thermal properties and fast heterogeneous electron transfer. Gonzalez et al. [60] reported intrinsic properties of graphene are responsible for make them an ideal electrode material for use in supercapacitor [60].

Metal Oxides Ruthenium oxide is the most commonly used metal oxide in pseudocapacitors because of its wide potential window, excellent stability towards heat, longer life time, high conductivity, high energy density as well as high power density. However, main drawback that limits the application of it is its limited availability leading to the higher cost and toxic nature. Therefore various transition metal oxides have been explored for the utilization in pseudocapacitor to act as an electrode material, as they possess appreciable conductivity and variable oxidation state. Extensively researched transition metal oxides such as MnO_2, CoO, NiO, Fe_2O_3, V_2O_3, etc. are considered as a potential candidate as an electrode material for supercapacitor owing to owing to its low price, environmental benign nature, excellent theoretical specific capacitance and superior capacitive behavior. Manganese dioxide is considered to be a better substitute for RuO_2 owing to its small cost, environmental benign nature, percentage abundance and excellent theoretical specific capacitance. Nickel oxide, vanadium oxide, iron oxide also behave as a superior candidate in supercapacitor due to its large theoretical specific capacitance, environmental stability, thermal stability, accessible layered structure (V_2O_5), mixed oxidation states, low cost and simple synthesis. Additionally, iron oxide (Fe_3O_4) based materials in aqueous solutions shows broad potential window up to 1.2 V which is higher when compared to the potential window of RuO_2, MnO_2 and NiO (less than 1 V) there by will enhance energy density. However, in order to enhance to supercapacitive performance of the metal oxides, research have been done to study the metal oxide composites, with CNTs, activated carbon and conducting polymers, since best features of variety of materials can be integrated to form a novel material having advanced applications. Ideally, the composite materials have low weight with attractive mechanical, thermal, chemical, electrical, magnetic and optical properties. Zhang et al. [61] have reported ZnO-CNT as electrodes in presence of poly vinyl alcohol and phosphomolybdic acid as a gel polymer electrolyte to improve the capacitance of electrode material [61]. Hu et al. [62] shown the enhanced specific capacitance and energy density of 305.3 F/g and 42.4 Wh/kg, respectively for the PANI and tin oxide based nano-composite material [62].

Conducting Polymers Polymeric materials are much more cost effective than RuO_2 and can produce comparatively higher specific capacitance. The generally used

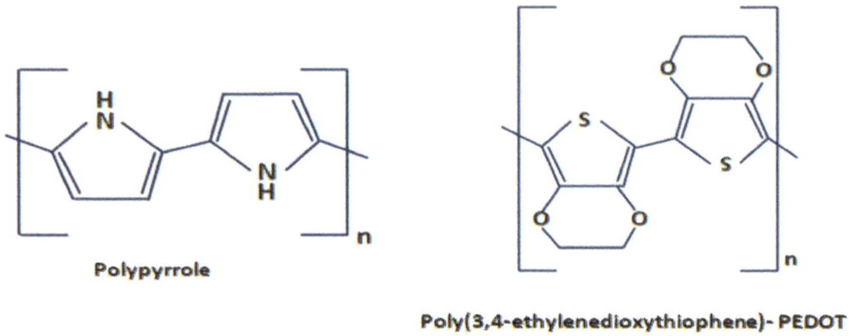

Fig. 7 Schematic representation of poly(pyrrole) and PEDOT

conducting polymers utilized in supercapacitor applications include PANI, PPy, PEDOT and their corresponding derivatives [63] (Fig. 7).

Conducting polymers exhibits red-ox reactions occurring both at the surface of polymer along with entire bulk polymer demonstrating excellent capacitive behaviour. The redox phenomenon in conducting polymers is highly reversible in nature since there is not any phase transformation or structural change. Among the various conducting polymers, PANI is the highly studied polymer due to its inherent exceptional properties for instance economical, chemical stability, environment stability, facile synthesis (chemical or electrochemical synthesis), doping process, dedoping process, mechanical flexibility and wide variety of applications in various fields. Conducting conjugated polymers and their derivatives, act as potential material for energy storage applications due to its exceptionally high electrical conductivity (up to 4.6×10^5 S m^{-1}) and excellent capacitance values (2000 F g^{-1} for PANI, 620 F g^{-1} for PPy, depending on the doping level).

PANI mainly occurs in three distinct oxidation states that make it a promising candidate as an electrode material for supercapacitor application. The reduced structure of PANI consists of only benzenoid units and is known as leucoemeralidine i.e. either faded yellow or colorless, where as the fully oxidized form of PANI is composed of quinoid units and is known as pernigraniline having blue/violet color. On the other hand semi oxidized of PANI is composed of both benzenoid and quinoid units and is known as emeralidine i.e. green or slightly blue in colour (Fig. 8).

Although conducting polymers exhibit unique properties such as low cost, light weight, corrosion resistance, large scale production, easy processing, fast redox reactions, and high conductivity but their reduced cycling stability along with the poor mechanical stability has declined the progress of conducting polymer based pseudocapacitors. Conducting polymers experience a range of physical properties variation with time such as swelling, shrinkage, doping etc. that deteriorates polymer's performance. However remarkable performance improvement of the conducting polymer based supercapacitor is obtainable using hybrid capacitors that store charge by exploiting Faradic and non-Faradic processes combining the best features of EDLCs and pseudocapacitors together into a unified supercapacitor resulting in advancement

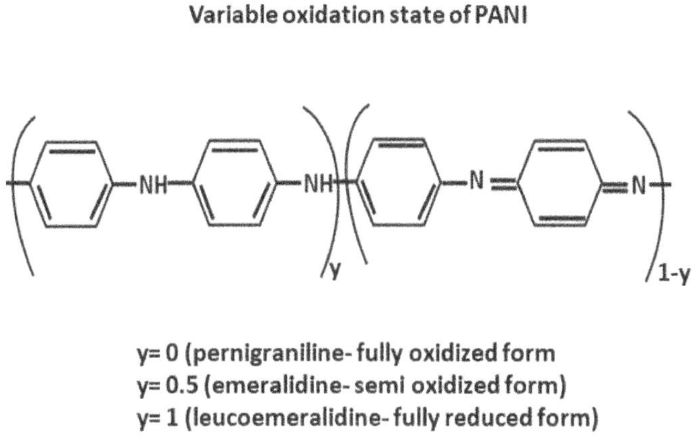

Fig. 8 Variable oxidation states of PANI

of the novel materials with outstanding properties. The key benefit in using these hybrid capacitors over the conducting polymer is that they present superior cyclic stability andexcellent power performance along with higher specific capacitance.

Du et al. [64] reported capacitance of 154.5 F g^{-1} for CNT/PPY composite (in organic electrolyte) which was considerably higher as compare to that of pure PPy and CNT [64]. In addition reinforcement of PPy with CNT reduces the swelling and shrinkage of polymer during the charge discharge process. Zhang et al. [65] reported synthesis of PPY/MnO_2 nanocomposite having higher conductivity (about 4–5 orders of magnitude) and specific capacitance (290 F g^{-1}) than that of MnO_2 (221 F g^{-1}) [65]. Huang et al. [66] reported ruthenium oxide/PEDOT:PSS composite having capacitance value of 1409 F g^{-1} [66]. However use of this composite material is still limited due to its exceptionally high cost. Therefore it's desirable to make use of cost effective metal oxide that can produce supercapacitor electrode of high specific energy, while being reasonably affordable. Xu et al. [67] reported synthesis of PPy/iron oxide nanohybrid having high capacitance value of 560 F g^{-1} and outstanding cycling stability [67]. Lee et al. [68] reported composites of PEDOT:PSS wrapped CNT/MnO_2 for flexible supercapacitors having capacitance of 428.2 F g^{-1} and high energy density of 63.8 Wh kg^{-1} [68].

Thus by using a proper combination of electrode, improvement in the cell voltage, directly resulting in the enhancement of energy and power densities is possible. In general, carbon based materials acts as best electrodes for EDLCs, while for pseudocapacitor the best candidate found to be transition metal oxides and conducting polymers. However, the rapid vanishing of the power density, reduced capacitance retention along with reduced cyclability at high power rates are the some of the major issues hindering the expansion of Hybrid supercapacitors. At present, researchers are focusing mainly on hybrid supercapacitors that are distinguishable by means of the configuration of their electrode and the hybrid supercapacitors are composite-type, asymmetric-type and battery-type.

Composite Electrodes It involves the combination of carbon-based materials together with either conducting polymer or with metal oxide forming a single electrode, showing charge storage mechanisms involving physical and chemical processes. The process of storage of charge involved in composite electrode is capacitive double layer mechanism obtained from carbon-based materials and the large surface area provided by carbon-based materials improves the contact between electrolyte and pseudocapacitive materials. Composite of MnO_2/CNT synthesized using simple hydrothermal approach revealed a large improvement in rate capability and capacitance in comparison to the bare MnO_2 or CNT electrode endorsed by the large surface area and porous structure of MnO_2 [69].

Asymmetric Hybrids It involves the coupling of EDLC with a Pseudocapacitor electrode by the integration of faradic and non-faradic processes. Asymmetric hybrids are fabricated with one electrode composed of a double-layer carbon material while another electrode made of a pseudo-capacitance material which can be metal oxide or conducting polymer. By opting an appropriate material for electrode, achieving a high working voltage along with high energy density is possible which contributes in raising the total energy density in supercapacitor. The main advantage of conductive polymers in asymmetric hybrids is their processability but the lack of efficiency has limited their application. This problem can be overcome using negatively charged, activated carbon electrode in place of negatively charged, conducting polymer. Although the conducting polymer electrodes exhibits high capacitance and low resistance compared with activated carbon electrodes, but shows poorer maximum voltage along with lesser cycling stability. Cheng et al. reported synthesis of MnO_2/CNT composite used as a asymmetric supercapacitor having MnO_2-coated/graphene as cathode electrode while anode electrode is formed by using pure graphene presented higher capacitances and energy densities exhibiting higher capacitances along with the energy and power densities [70].

Battery Type battery type hybrid is consists of dissimilar electrodes such as a supercapacitor electrode with battery electrode. The obtained combination utilises the properties of supercapacitors as well as batteries within single assembled cell. This specific configuration highlights the requirement of higher energy supercapacitors and higher power batteries, by merging the power, cycle life, energy qualities of batteries by the recharging time of supercapacitors. The first hybrid system was first proposed by Amatucci et al. [71] having $Li_4Ti_5O_{12}$ battery-type and AC supercapacitor-type, exhibiting energy density obtained 20 Wh kg^{-1} inside acetonitrile solution [71]. Positive electrodes of nickel hydroxide, lead dioxide, $LiCoO_2$, $Li_4Mn_5O_{12}$, $LiMn_2O_4$, $LiMn_2O_4$/activated carbon (AC) composite etc. and are used to create the battery type hybrid materials. They helped to accomplish the enhancement in energy density along with power density, high charge-discharge, enhance cycle life time, specific capacitance, etc. Zhang et al. reported hybrid capacitors having $LiMn_2O_4$ (LMO) as positive electrode with activated carbon (AC) as negative electrode having excellent performance [72]. Hu et al. [73] presented the battery-supercapacitor hybrid of $Li_4Ti_5O_{12}$ acting as an anode, while cathode is made by $LiMn_2O_4$/activated carbon composite with superior properties [73]. Although the available experimental figures utilizing hybrid battery type is comparatively lesser than that of other supercapacitors, but still there is ample opportunity to work so that gap present in between the supercapacitors and batteries can be filled in coming future.

5 Conclusion

Global warming and the inevitable depletion of fossil fuels, coupled with the growth of the human population and technology development has resulted in a rapidly increasing global energy demand. Therefore, it is of utmost importance to concentrate on an active search for efficient, rechargeable, renewable, electrical energy storage devices. The current chapter embodies an overview of the advanced hybrid conducting polymer for energy storage applications. The performance of these hybrid conducting polymers depends upon several factors i.e. environmental stability, surface area, conductivity, etc. In this chapter, we have discussed various conducting polymers based composites for energy storage applications. The improvement in the performance values of energy storage devices using these conducting polymer composites gives an indication that these hybrid conducting polymers are capable of bridging the gap existing between supercapacitor and batteries. In addition, they can also play a lead role in the development of smart, efficient, flexible and cost-effective energy storage systems in the coming future.

References

1. Zhang, C., Wei, Y.L., Cao, P.F., Lin, M.C.: Energy storage system: current studies on batteries and power condition system Renew. Sust. Energ. Rev. **82**, 3091 (2018)
2. Nitta, N., Wu, F., Lee, J.T., Yushin, G.: Li-ion battery materials: present and future Mater. Today **18**, 252 (2015)
3. Gunawardane, K.: Capacitors as energy storage devices—Simple basics to current commercial families. In: Energy Storage Devices for Electronic Systems, p. 137. Academic Press, Elsevier
4. Kularatna, N.: Capacitors as energy storage devices—simple basics to current commercial families. In: Energy Storage Devices—A General Overview, p. 1. Academic Press, Elsevier (2015)
5. Zuo, W., Li, R., Zhou, C., Li, Y., Xia, J., Liu, J.: Battery-Supercapacitor Hybrid Devices: Recent Progress and Future Prospects Adv. Sci. **4**, 1600539 (2017)
6. Ballarin B., Masiero S., Seeber R., Tonelli D.: Modification of electrodes with porphyrin-functionalised conductive polymers J. Electroanal. Chem. **449**, 173 (1998)
7. Iqbal, S., Ahmad, S.: Recent development in hybrid conducting polymers: Synthesis, applications and future prospects J. Ind. Eng. Chem. **60**, 53 (2018)
8. Mandal, T.K., Fleming, M.S., Walt, D.R.: Preparation of polymer coated gold nanoparticles by surface-confined living radical polymerization at ambient temperature Nano Lett. **2**, 3 (2002)
9. Corbierre, M.K., Cameron, N.S., Sutton, M., Mochrie, S.G.J., Lurio, L.B., Rühm, A., Lennox, R.B.: Polymer-stabilized gold nanoparticles and their incorporation into polymer matrices J. Am. Chem. Soc. **123**, 10411 (2001)

10. Girtan, M.: On the stability of the electrical and photoelectrical properties of P3HT and P3HT:PCBM blends thin films Org. Electron. Phys. Mater. Appl. **14**, 200 (2013)
11. Chen, B., Liu, C., Ge, L., Hayashi, K.: Electrical conduction and gas sensing characteristics of P3HT/Au nano-islands composite Sens. Actuators B Chem. **241**, 1099 (2017)
12. Zhang, F., Cao, H., Yue, D., Zhang, J., Qu, M.: Enhanced anode performances of polyaniline-TiO_2-reduced graphene oxide nanocomposites for lithium ion batteries Inorg. Chem. **51**, 9544 (2012)
13. Wang, Y.G., Wu, W., Cheng, L., He, P., Wang, C.X., Xia, Y.Y.: A Polyaniline-Intercalated Layered Manganese Oxide Nanocomposite Prepared by an Inorganic/Organic Interface Reaction and Its High Electrochemical Performance for Li Storage Adv. Mater. **20**, 2166 (2008)
14. Mai, L., Xu, X., Han, C., Luo, Y., Xu, L., Wu, Y.A., Zhao, Y.: Rational synthesis of silver vanadium oxides/polyaniline triaxial nanowires with enhanced electrochemical property NanoLett. **11**, 4992 (2011)
15. Zhou, X.Y., Tang, J., Yang, J., Zou, Y.L., Wang, S.C., Xie, J., Ma, L.L.: Effect of polypyrrole on improving electrochemical performance of silicon based anode materials Electrochim. Acta **70**, 296 (2012)
16. Wang, G.J., Yang, L.C., Qu, Q.T., Wang, B., Wu, Y.P., Holze, R.: An aqueous rechargeable lithium battery based on doping and intercalation mechanisms J. Solid State Electrochem. **14**, 865 (2010)
17. Tang, W., Liu, L., Zhu, Y., Sun, H., Wu, Y., Zhu, K.: An aqueous rechargeable lithium battery of excellent rate capability based on a nanocomposite of MoO_3 coated with PPy and $LiMn_2O_4$ Energy Environ. Sci. **5**, 6909 (2012)
18. Tang, W., Gao, X.W., Zhu, Y.S., Yue, Y.B., Shi, Y., Wu, Y.P., Zhu, K.: An aqueous rechargeable lithium battery of excellent rate capability based on a nanocomposite of MoO_3 coated with PPy and $LiMn_2$ O_4 J. Mater. Chem. **22**, 20143 (2012)
19. Liu, L.L., Wang, X.J., Zhu, Y.S., Hu, C.L., Wu, Y.P., Holze, R.: A hybrid of V_2 O_2 nanowires and MWCNTs coated with polypyrrole as an anode material for aqueous rechargeable lithium batteries with excellent cycling performance J. Power Sources **224**, 290 (2013)
20. Tian, F., Liu, L., Yang, Z., Wang, X., Chen, Q., Wang, X.: Polypyrrole-coated LiV_3 O_8-nanocomposites with good electrochemical performance as anode material for aqueous rechargeable lithium batteries Mater. Chem. Phys. **127**, 151 (2011)
21. Liu, H., Zhang, F., Li, W., Zhang, X., Lee, C., Wang, W., Tang, Y.: Porous tremella-like MoS /polyaniline hybrid composite with enhanced performance for lithium-ion battery anodes Electrochim. Acta **167**, 132 (2015)
22. Yang, L., Wang, S., Mao, J., Deng, J., Gao, Q., Tang, Y., Schmidt, O.G.: Hierarchical MoS_2 /Polyaniline Nanowires with Excellent Electrochemical Performance for Lithium-Ion Batteries Adv. Mater. **25**, 1180 (2013)
23. Liu, J., Gu, M., Ouyang, L., Wang, H., Yang, L., Zhu, M.: Sandwich-like SnS/Polypyrrole Ultrathin Nanosheets as High-Performance Anode Materials for Li-Ion Batteries ACS Appl. Mater. Interfaces **8**, 8502 (2016)
24. Zhao, X., Mai, Y., Luo, H., Tang, D., Lee, B., Huang, C., Zhang, L.: Nano-MoS_2 /poly (3,4-ethylenedioxythiophene): Poly(styrenesulfonate) composite prepared by a facial dip-coating process for Li-ion battery anode Appl. Surf. Sci. **288**, 736 (2014)
25. Xie, D., Wang, D.H., Tang, W.J., Xia, X.H., Zhang, Y.J., Wang, X.L., Gu, C.D., Tu, J.P.: Binder-free network-enabled MoS_2-PPY-rGO ternary electrode for high capacity and excellent stability of lithium storage J. Power Sources **307**, 510 (2016)
26. Azman, N.H.N., Mamat, M.S., Ngee, L.H., Sulaiman, Y.: Graphene-based ternary composites for supercapacitors Int. J. Energy Res. **42**, 2104 (2018)
27. Han, S., Ai, Y., Tang, Y., Jiang, J., Wu, D.: Carbonized polyaniline coupled molybdenum disulfide/graphene nanosheets for high performance lithium ion battery anodes RSC Adv. **5**, 96660 (2015)
28. Novoselov, K.S., Jiang, D., Schedin, F., Booth, T.J., Khotkevich, V.V., Morozov, S.V., Geim, A.K.: Two-dimensional atomic crystals Proc. Natl. Acad. Sci. U.S.A. **102**, 10451 (2005)

29. Paek, S.M., Yoo, E., Honma, I.: Enhanced Cyclic Performance and Lithium Storage Capacity of SnO_2 /Graphene Nanoporous Electrodes with Three-Dimensionally Delaminated Flexible Structure Nano Lett. **9**, 72 (2009)
30. Wang, H., Yang, Y., Liang, Y., Robinson, J.T., Li, Y., Jackson, A., Cui, Y., Dai, H.: Graphene-Wrapped Sulfur Particles as a Rechargeable Lithium–Sulfur Battery Cathode Material with High Capacity and Cycling Stability Nano Lett. **11**, 2644 (2011)
31. Luo, J., Liu, J., Zeng, Z., Ng, C.F., Ma, L., Zhang, H., Lin, J., Shen, Z., Fan, H.J.: Three-Dimensional Graphene Foam Supported Fe_3 O_3 Lithium Battery Anodes with Long Cycle Life and High Rate Capability Nano Lett. **13**, 6136 (2013)
32. Gao, F., Qu, J.Y., Zhao, Z.B., Dong, Y.F., Yang, J., Dong, Q., Qiu, J.S.: Easy synthesis of MnO-graphene hybrids for high performance lithium storage New Carbon Mater. **29**, 316 (2014)
33. David, L., Bhandavat, R., Singh, G.: MoS /Graphene Composite Paper for Sodium-Ion Battery Electrodes ACS Nano **8**, 1759 (2014)
34. Qu, B., Ma, C., Ji, G., Xu, C., Xu, J., Meng, Y.S., Wang, T., Lee, J.Y.: Layered SnS_2-reduced graphene oxide composite--a high-capacity, high-rate, and long-cycle life sodium-ion battery anode material Adv. Mater. **26**, 3854 (2014)
35. Yuan, F.W., Tuan, H.Y.: Scalable Solution-Grown High-Germanium-Nanoparticle-Loading Graphene Nanocomposites as High-Performance Lithium-Ion Battery Electrodes: An Example of a Graphene-Based Platform toward Practical Full-Cell Applications Chem. Mater. **26**, 2172 (2014)
36. Murugan, A.V., Muraliganth, T., Manthiram, A.: Rapid, Facile Microwave-Solvothermal Synthesis of Graphene Nanosheets and Their Polyaniline Nanocomposites for Energy Strorage Chem. Mater. **21**, 5004 (2009)
37. Wang, H., Tran, D., Qian, J., Ding, F., Losic, D.: MoS_2 /Graphene Composites as Promising Materials for Energy Storage and Conversion Applications Adv. Mater. Interfaces 1900915 (2019)
38. Li, Q., Guo, X., Zheng, M., Pang, H.: Some MoS_2-based materials for sodium-ion battery Funct. Mater. Lett. **11**, 1840004 (2018)
39. Nan J., Guo X., Xiao J., Li X., Chen W., Wu W., Liu H., Wang Y., Wu M., Wang G.: Nano-engineering of 2D MXene-Based Materials for Energy Storage Applications Small 1902085 (2019)
40. Ling, Z., Ren, C.E., Zhao, M.Q., Yang, J., Giammarco, J.M., Qiu, J., Barsoum, M.W., Gogotsi, Y.: Flexible and conductive MXene films and nanocomposites with high capacitance Proc. Natl. Acad. Sci. USA **111**, 16676 (2014)
41. Boota, M., Anasori, B., Voigt, C., Zhao, M.Q., Barsoum, M.W., Gogotsi, Y.: Pseudocapacitive Electrodes Produced by Oxidant-Free Polymerization of Pyrrole between the Layers of 2D Titanium Carbide (MXene) Adv. Mater. **28**, 1517 (2016)
42. Qin, L., Tao, Q., Liu, X., Fahlman, M., Halim, J., Persson, P.O., Rosen, J., Zhang, F.: Polymer-MXene composite films formed by MXene-facilitated electrochemical polymerization for flexible solid-state microsupercapacitors Nano Energy **60**, 734 (2019)
43. Zhu, M., Huang, Y., Deng, Q., Zhou, J., Pei, Z., Xue, Q., Huang, Y., Wang, Z., Li, H., Huang, Q., Zhi, C.: Highly Flexible, Freestanding Supercapacitor Electrode with Enhanced Performance Obtained by Hybridizing Polypyrrole Chains with MXene Adv. Energy Mater. **6**, 1600969 (2016)
44. Conway, B.E.: Electrochemical Supercapacitors: Scientific Fundamentals and Technological Applications. Kluwer-Plenum, New York (1999)
45. Winter, M., Brodd, R.J.: What Are Batteries, Fuel Cells, and Supercapacitors? Chem. Rev. **104**, 4245 (2004)
46. Wang, Y., Shi, Z., Huang, Y., Ma, Y., Wang, C., Chen, M., Chen, Y.: Supercapacitor Devices Based on Graphene Materials J. Phys. Chem. C **113**, 13103 (2009)
47. Zhang, L.L., Zhao, X.S.: Carbon-based materials as supercapacitor electrodes Chem. Soc. Rev. **38**, 2520 (2009)

48. Wu, Z., Li, L., Yan, J., Zhang, X.: Materials Design and System Construction for Conventional and New-Concept Supercapacitors Adv. Sci. **4**(6), 1600382 (2017)
49. Zhang J., Zhang L., Sun X., Liu H., Sun A., Liu R.S., Zhang, J.: Electrochemical Technologies for Energy Storage and Conversion, 1st edn. Wiley (2012)
50. Simon, P., Gogotsi, Y.: Materials for electrochemical capacitors Nat. Mater. **7**, 845 (2008)
51. Helmholtz, H.V.: Some laws concerning the distribution of electrical currents in conductors Ann. Phys. (Leipzig) **89**, 21 (1853)
52. Gouy, G.: Constitution of the Electric Charge at the Surface of an Electrolyte J. Phys. **4**, 457 (1910)
53. Chapman, D.L.: A contribution to the theory of electrocapillarity Philos. Mag. **6**, 475 (1913)
54. Stern, O.: The theory of the electrolytic double shiftThe theory of the electrolytic double shift Z. Electrochem. **30**, 508 (1924)
55. Lota, K., Khomenko, V., Frackowiak, E.: Capacitance properties of poly(3,4-ethylenedioxythiophene)/carbon nanotubes composites J. Phys. Chem. Solids **65**, 295 (2004)
56. Chen, L.M., Lai, Q.Y., Hao, Y.J., Huang, J.H., Ji, X.Y.: Pseudo-capacitive properties of $LiCoO_2$/AC electrochemical capacitor in various aqueous electrolytes Ionics **14**, 441 (2010)
57. Tang, C., Liu, Y., Yang, D., Yang, M., Li, H.: Oxygen and nitrogen co-doped porous carbons with finely-layered schistose structure for high-rate-performance supercapacitors Carbon **122**, 538 (2017)
58. Niu, C., Sichel, E.K., Hoch, R., Moy, D., Tennent, H.: High power electrochemical capacitors based on carbon nanotube electrodes Appl. Phys. Lett. **70**, 1480 (1997)
59. Zeng, Q., Tian, H., Jiang, J., Ji, X., Gao, D., Wang, C.: High-purity helical carbon nanotubes with enhanced electrochemical properties for supercapacitors RSC Adv. **7**, 7375 (2017)
60. González, A., Goikole, E., Barrena, J.A., Mysyk, R.: Review on supercapacitors: Technologies and materials Renew. Sust. Energ. Rev. **58**, 1189 (2016)
61. Zhang Y., Sun X., Pan L., Li H., Sun Z., Sun C., Tay B.K.: Carbon nanotube–zinc oxide electrode and gel polymer electrolyte for electrochemical supercapacitors J. Alloys Compd. **480**, L17 (2009)
62. Hu, Z.A., Xie, Y.L., Wang, Y.X., Mo, L.P., Yang, Y.Y., Zhang, Z.Y.: Polyaniline/SnO nanocomposite for supercapacitor applications Mater. Chem. Phys. **114**, 990 (2009)
63. Yan, Y., Wang, T., Li, X., Pang, H., Xue, H.: Noble metal-based materials in high-performance supercapacitors Inorg. Chem. Front. **4**, 33 (2017)
64. Du, B., Jiang, Q., Zhao, X.F., Huang, B., Zhao, Y.: Preparation of PPy/CNT Composite Applications for Supercapacitor Electrode Material Mater. Sci. Forum **610–613**, 502 (2009)
65. Zhang, X., Yang, W., Ma, Y.: Synthesis of Polypyrrole-Intercalated Layered Manganese Oxide Nanocomposite by a Delamination/Reassembling Method and Its Electrochemical Capacitance Performance Electrochem. Solid St. **12**, A95 (2009)
66. Huang, L.M., Wen, T.C., Gopalan, A.: Electrochemical and spectroelectrochemical monitoring of supercapacitance and electrochromic properties of hydrous ruthenium oxide embedded poly(3,4-ethylenedioxythiophene)–poly(styrene sulfonic acid) composite Electrochim. Acta **51**, 3469 (2006)
67. Xu, C., Puente-Santiago, A.R., Padron, D.R., Caballero, A., Balu, A.M., Romero, A.A., Muñoz-Batista, M.J., Luque, R.: Controllable Design of Polypyrrole-Iron Oxide Nanocoral Architectures for Supercapacitors with Ultrahigh Cycling Stability ACS Appl. Energy Mater. **2**(3), 2161 (2019)
68. Lee, H.U., Yin, J.L., Park, S.W., Park, J.Y.: Preparation and characterization of PEDOT:PSS wrapped carbon nanotubes/MnO_2 composite electrodes for flexible supercapacitors Synth. Met. **228**, 84 (2017)
69. Xia, H., Wang, Y., Lin, J., Lu, L.: Hydrothermal synthesis of MnO_2/CNT nanocomposite with a CNT core/porous MnO_2 sheath hierarchy architecture for supercapacitors Nanoscale Res. Lett. **7**, 33 (2012)
70. Cheng, Q., Tang, J., Ma, J., Zhang, H., Shinya, N., Qin, L.C.: Graphene and nanostructured MnO_2 composite electrodes for supercapacitors Carbon **49**, 2917 (2011)

71. Amatucci, G.G., Badway, F., Pasquier, A.D., Zheng, T.: An Asymmetric Hybrid Nonaqueous Energy Storage Cell J. Electrochem. Soc. **148**, A930 (2001)
72. Zhang, S., Li, C., Zhang, X., Sun, X., Wang, K., Ma, Y.: High Performance Lithium-Ion Hybrid Capacitors Employing Fe_3O_4–Graphene Composite Anode and Activated Carbon Cathode ACS Appl. Mater. Interfaces **20**, 17136 (2017)
73. Hu, X., Deng, Z., Suo, J., Pan, Z.: A high rate, high capacity and long life ($LiMn_2O_4$ + AC)/$Li_4Ti_5O_{12}$ hybrid battery–supercapacitor J. Power Sources **187**, 635 (2009)

Nanoelectronics Devices (Field-Effect Transistors, Electrochromic Devices, Light-Emitting Diodes, Dielectrics, Neurotransmitters)

Percy J. Sephra, Pari Baraneedharan, and Arunachalam Arulraj

Abstract Nanoelectronics is one of the exciting thrust areas which bring conventional electronics and nanotechnology together. These are governed with a motive to make smaller devices ensure their efficiencies remain the same as the conventional one. In recent decades, the field has been significantly emerging owing to its availability of different nanomaterials. Nanoelectronics devices can be constructed in many forms using various nanostructured materials, conducting polymers, and so on. Depending upon the different existing materials on constructing nanoelectronics devices, polymers are one of the chiefly used host matrices for nanomaterials to make the devices in practical ways. These kinds of devices are usually constructed through compositing with nanomaterials, films in the form of coating, and in some instances, it can be used on their own. The main features of conducting polymers are its inherent flexibility and its conductive nature, which makes it well-positioned for wearable electronics, transparent electronics, and nanoelectronics devices. This chapter more keenly focuses on conducting polymers and its composites for various nanoelectronics devices such as field-effect transistors (FETs), electrochromic displays, light-emitting diodes (LEDs), dielectrics, and neurotransmitters. This chapter will also provide insights into each aspect of the conducting polymers applications with its future trends and opportunities.

Keywords Conducting polymers · FETS · Electrochromic devices · LEDS · Dielectric and neurotransmitters

P. J. Sephra (✉)
Department of Chemistry, University College of Engineering-BIT Campus, Tiruchirappalli, India
e-mail: percysephra@gmail.com

P. Baraneedharan
Department of Materials Science, Central University of Tamil Nadu, Thiruvarur, India

A. Arulraj
Graphene and Advanced 2D Materials Research Group (GAMRG), School of Science and Technology, Sunway University, Subang Jaya, Malaysia
e-mail: arulnanotech@gmail.com

Advanced Ceramics and Nanotechnology Laboratory, Department of Materials Engineering, University of Concepción, Concepción, Chile

© Springer Nature Switzerland AG 2021
S. Shahabuddin et al. (eds.), *Advances in Hybrid Conducting Polymer Technology*, Engineering Materials, https://doi.org/10.1007/978-3-030-62090-5_4

1 Introduction

Nanotechnology, a promising advancement platform covers an extensive range of potential applications from electronics, solar cells, engineering, defense, nanobiological systems, nanomedicines, etc. This field of technology requires contribution from multidisciplinary researchers starting from synthesis, process nanostructures, to design and fabricate nanodevices. These nanomaterials' unique physical properties depend upon the desired size, morphology, chemical compositions, and microstructure. The inherent dependence on size and shape is exhibited as unique physicochemical properties differing from bulk materials. Based on the dimensions, nanomaterials are characterized into different shapes such as nanorods, nanospheres, nanosheets, nanoparticles which plays a vital role in the prototyping of devices. Besides these mentioned characteristics, the modern generation devices are more keen on development on devices with flexible characteristics. Generally, the flexibility will be achieved either compositing/doping materials with polymers or direct instance of conducting polymers [1–5].

1.1 Conducting Polymers

Conducting polymers are one of the indispensable materials with highly π-conjugated polymeric chain possessing optimistic optoelectronic and other properties thereby finds its application in various areas including development of nanoelectronics devices such as wearable electronics, field emitting transistors, electrochemical power sources, displays, etc. [6–8]. The conducting polymers properties can be modified by the doping process, which involves oxidation/reduction of conducting polymers. During redox reactions, the charge transfer mechanism alters the polymeric chains results in outstanding electrical properties [9, 10]. Its tunable and exclusive properties have gained an interest in both academic and industrial research. In the year, 2000 Noble Prize in Chemistry has been awarded for the eminent scientist for their discovery of polyacetylene (conducting polymer). The conductivity of polyacetylene (PAc) exposed 10^8 fold increments as compared with other conducting polymers [11]. Followed by these various other conducting polymers have been developed and studied further for developing nanoelectronics devices. So far, the most commonly studied and used conducting polymers in nanoelectronics devices are polyaniline (PANI), polythiophene (PTh), polypyrrole (PPy), polyacetylene (PAc), poly(3,4-ethylene dioxythiophene) (PEDOT) [12–14]. The schematic illustration of these polymers structure is shown in Fig. 1, and their conductivity is given in Table 1.

Aforementioned the unique property of the conducting polymers is its flexibility, which can impose for wearable nanoelectronics devices apart from this as electronic devices; it should possess high electrical conductivity as well as thermal stability. These properties (conductivity and stability) are significantly enhanced

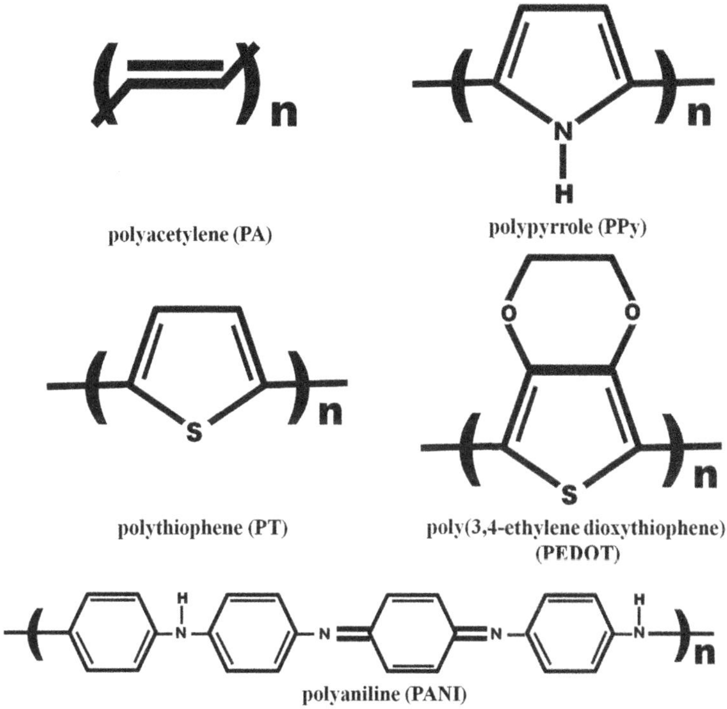

Fig. 1 Schematic illustration of chemical structures of the different conducting polymers

Table 1 Conductivity of various conducting polymers with its corresponding oxidant [15]

Polymers	Oxidant	Conductivity (S m^{-1})
Polyacetylene	I_2, Br_2, AsF_5, Na, Li	10^6
Polypyrrole	BF_4^-, ClO_4^-	5×10^4–75×10^4
Polythiophene	BF_4^-, ClO_4^-, Toyslate, $FeCl_4^-$	10^5
Poly(3-alkyl thiophene)	BF_4^-, ClO_4^-, $FeCl_4^-$	10^5–10^6
Polyphenylsulfide	AsF_5	5×10^4
Polyphenylene-vinylene	AsF_5	10^6
Polyphenylene-vinylene	AsF_5	27×10^4
Polyphenylene	AsF_5, Li, K	10^5
Poly isothi-anaphtene	BF_4^-, ClO_4^-	5×10^3
Polyazulene	BF_4^-, ClO_4^-	100
Polyfuran	BF_4^-, ClO_4^-	10^4
Polyaniline	HCl	2×10^4

using doping or compositing with nanomaterials [16]. In such a case, it can be termed polymeric or polymer nanocomposites (PNCs), where the polymers will act as a matrix to reinforce the nanomaterials. Intensive investigations over the formation of PNCs with varying types of nanofillers ranging from metals, semiconductors, nanostructures including nanotubes, nanowires (conducting metals), metal oxide nanostructures, chalcogenides, carbonaceous materials like carbon nanofibers (CNF), carbon nanotubes (CNT), graphene has been carried out and reported [17–20]. Various methods have been used for the PNCs synthesis, including self-assembly, seed-mediated approach, electrospinning, interfacial polymerization which are the template-free approach and template-assisted physical method including hard and soft template methods [21, 22]. Depending upon the nanofillers, the PNCs can be imparted with an enhancement in structural as well as functional property. The synergistic performance of the PNCs will be significantly enhanced at much lower loadings as compared with higher mass loadings [12].

Moreover, the uniform dispersion of nanofillers over polymer matrix plays a crucial role in the formation of strong bonding (covalent bond) which has effectiveness over flexibility (interfacial stress) and improvement in conductivity supports in device fabrication. Similarly, PNCs will exhibit transition behavior from insulator to a conductor through controlling the loading level of fillers which refers to the percolation threshold. The threshold (percolation threshold) relies strongly on the aspect ratio of nanofillers [12].

Several synthesis strategies have been useful in forming conjugated polymeric structures. Polymerization methodologies including chemical polymerization, electro polymerization [13], plasma polymerization, solid-state method, photochemical method. Among the different synthetic strategies, chemical polymerization is one of the commonly utilized methods for the production of conducting polymers in bulk form. As mentioned, polyaniline (PANI) is commonly studied polymeric material because of its chemical, environmental stability, facile synthesis, excellent yield through polymerization process, simple doping by protonic acids, ease availability of aniline monomer, etc. The presence of –NH– group in the polymeric chain (PANI) supports its physicochemical properties and protonation and deprotonation.

Chemical oxidative polymerization, a two-electron exchange process is the most favorable method for the preparation of PANI which requires strong acids like sulphuric acid and hydrochloric acid (HCl) as a medium, ammonium persulfate as oxidant. For oxidation of monomer (i.e., aniline), one mole of oxidant (ammonium persulphate) has been required and during its reaction, oxidant will support in taking away an electron from nitrogen atom of $-NH_2$ group of aniline. Depending on the conditions of polymerization PANI exists in varying oxidation states. Degree of polymerization is strongly affected by the concentration of oxidant, the concentration of aniline, types of acid used, polymerization temperature, electronegativity, etc. [23].

Electropolymerization method is another method used to synthesis conducting polymers in aqueous protonic acid by applying current/voltage at the working electrode. Electrochemical synthesis is carried out in a potentiostat method in a fixed potential range or galvanostatic method by applying a stable current [13, 24]. Similar approaches have been used for synthesizing other types of conducting polymer and composites of conducting polymer with nanomaterials. The synthesized conducting polymeric materials are effectively used for fabricating different nanoelectronics devices. In the following sections, the applications of the conducting polymers, particularly in the construction of the nanoelectronics devices are discussed in a detailed manner.

2 Applications of Conducting Polymers (Nanoelectronics Devices)

Conducting polymers are used in the fabrication of microelectronic devices since late 1990. A polymer processed chip fabricated using PANI drastically reduces the cost of transistors after its launch by Philips laboratories [25]. For practical applications, the conducting polymers specifically of π-conjugated polymers elicit the possibility of enhancing its properties. Recently, conducting polymers is mostly used for the construction of nanoelectronics devices owing to its specific function as wettability, optoelectronic properties lead to applications in specialized systems such as displays, LEDs, FETs, and others.

2.1 Field-Effect Transistors (FETs)

Field-effect Transistors (FETs) are one of the significant nanoelectronics devices that contribute as ON/OFF switches, amplifiers, an integral part in electrical circuits, etc. Most of these FETs existing in the technological market are made of using silicon (Si). Silicon-based FETs are conquering the market owing to its high mobilities, but the modern generation is looking for smart and wearable electronics where the silicon technology fails to exhibit. Here comes into play of polymers and conducting polymers that are highly flexible and by tuning its properties, it can perform higher or equivalent to the existing Si materials mainly in the field of flexible displays, sensors, LEDs, electronic papers, etc. [26–29]. Conducting polymers-based FETs is promising in modern electronics, which results in reducing the usage of silicon in the electronics market. The general architecture of FETs comprised of three electrodes viz. gate, source, drain as shown in Fig. 2.

Wrighton and co-workers conducted pioneering studies on conducting polymers for FETs through demonstrating the magnitude of current between conducting polymer (thin film) and two microelectrodes. The authors also addressed that the transistor (FET) action arises between the ion corporation (expulsion) and electrolyte due to oxidation and/or reduction of conducting polymers caused by its conductivity modulation [30–32]. Conducting polymer-based FETs, these conducting polymers serve as an insulating layer, thereby preventing the flow of current from source to drain in transistor at OFF state. In order to turn ON, the gate electrode will be biased

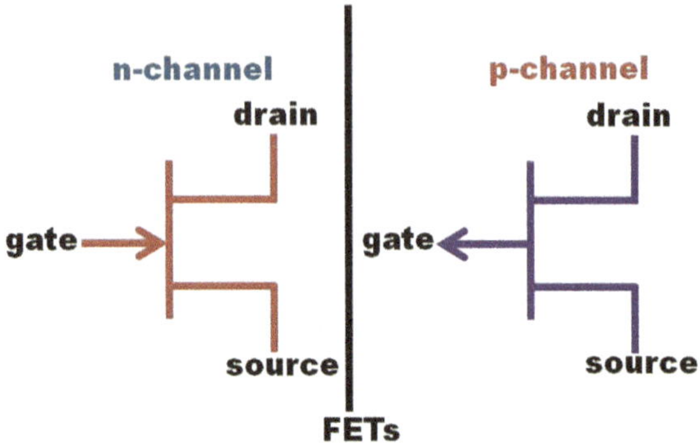

Fig. 2 Schematic representations of field effect transistors (n and p channel)

and when a positive bias is given to the gate, an electric field will be generated and attracts the negative charges in the conducting polymers (i.e., between dielectric interface and conducting polymer layer). Similarly, when a negative bias is applied all the positive charges will be accumulated at the conducting polymer. The accumulated charges in turn form a channel results in the flow of electrons to the drain through source electrodes [29].

In FET, poly(3-hexylthiophene) (P3HT) conducting polymer has been used most often due to its higher mobility value (0.2 cm^2 versus 21) as compared with others [29]. The higher mobility of P3HT attributed to its degree of order of polymeric chains (conjugated polymeric structure with π-π interchain stacking). These π-stacked polymer chains strongly influence the carrier mobility and its electronic transportation. The π-stacking in the P3HT chains highly relies on processing conditions (solvent, thickness, film fabrication method) and molecular parameters (molecular weight, regioregularity) [33–37]. The same has been applicable for most of the polymeric chain reaction with slight modifications. Hence from a practical view, device fabrication is an essential tool for determining its performance characteristics.

Recently, the photocatalytic oxidation method has been demonstrated for fabricating full polymer-based FETs with cost-effective, large-area solution processing techniques. In the device fabrication, the source, drain, and gate electrodes are patterned over the PET and PMMA substrates. The fabrication step involves the deposition of a persulfate aqueous solution over the patterned PET substrates masked of quartz glass (Fig. 3a). The PET substrates were irradiated under ultraviolet (UV) light for few minutes followed by the formation of hydrophilic and hydrophobic regions over the layered PET substrates. Conducting polymer (PEDOT/PSS) was used to form source and drain electrodes over the PET substrates either by dip coating or spin coating process. PEDOT/PSS will get deposited only on the hydrophilic area due to its wettability thereby forming source/drain arrays (Fig. 3b). Finally, P3HT has been coated over the patterned PET substrates followed by PMMA as a dielectric layer on top (Fig. 3c, d). The conducting polymer-based FETs exhibit excellent electrical properties along with its superior flexibility thus claiming the photo-oxidative technique to be one of the most straightforward and less processing time approaches with large scale processing [38].

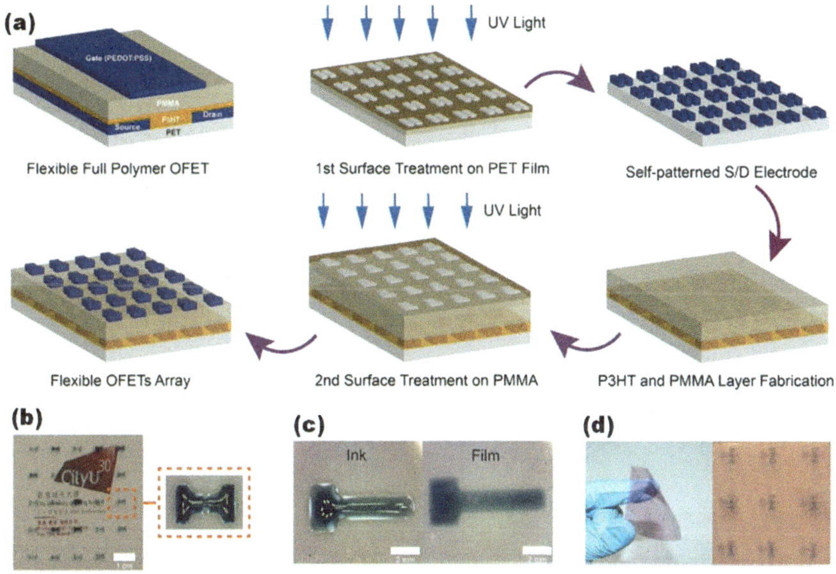

Fig. 3 FET fabrication process. **a** Schematic diagram depicting a polymer FET device; **b** An optical image of the PEDOT/PSS solution self-aligned on the patterned S/D regions; **c** PEDOT/PSS (1:1) solution self-aligned on the patterned gate regions and **d** flexible polymer FETs based on PET substrates [38]

Knopfmacher et al. reported a flexible polymeric organic field-effect transistors (OFETs) with high yield for marine environmental sensors. The developed polymeric OFETs exhibit stable electrical conductivity performance in an aqueous environment (seawater) apart from stable ambient conditions. Moreover, with the simple functionalization of polymeric OFET, the mercury (Hg) contamination in the seawater was detected and supported in controlling the marine environment. The electrical characteristics of the polymeric OFET are shown in Fig. 4 and it also posses remarkably good stability over prolonged time in harshest aqueous (seawater) environments in which an OFET sensor can be operated [39].

Thus, FETs fabricated using conducting polymers exhibiting good performance as nanoelectronics devices. Besides its performance, the challenge associate with conducting polymer-based FETs are its thermal stability. Most of conducting polymer FETs can be used only for ambient conditions or under inert conditions. Hence research has been devoted more towards developing conducting polymeric FETs by compositing with nanomaterials, dielectric materials, buffer layers, and so on [40–42]. Therefore the researchers who are interested in working on conducting polymers for FETs can address the existing challenges.

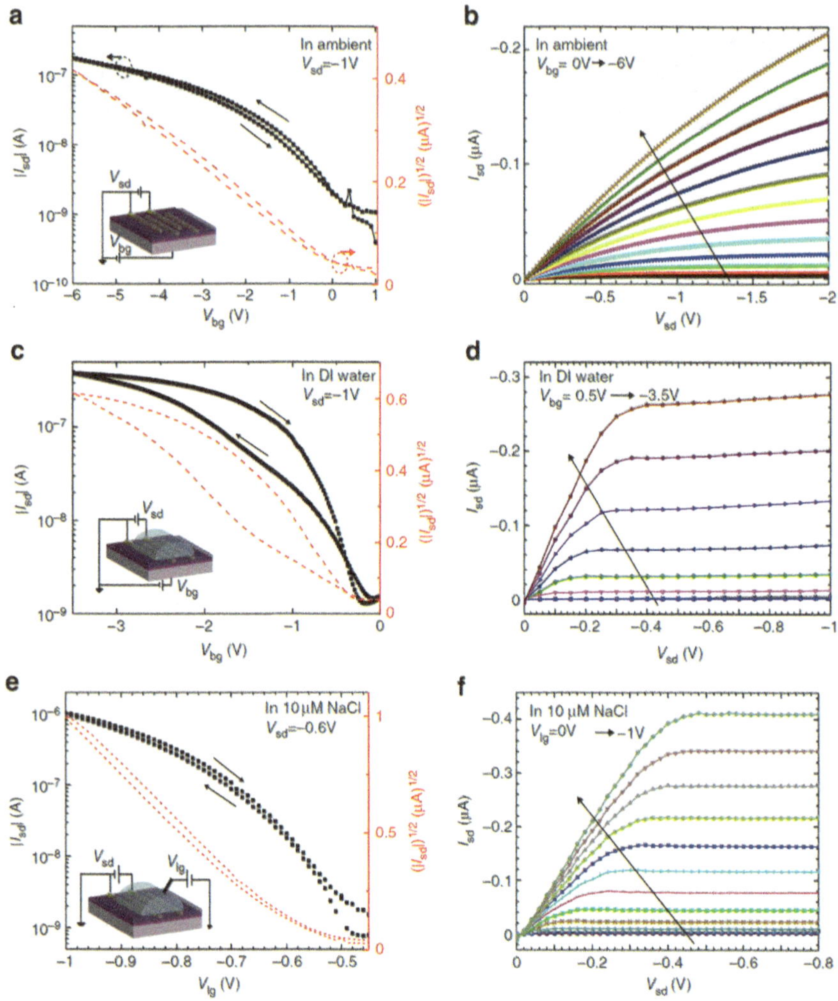

Fig. 4 Electrical characteristics of the polymeric OFETs under different environmental conditions [39]

2.2 *Electrochromic Devices (ECDs)*

Chromism refers to the phenomena of reversible colour tone changes induced by stimuli and "electrochromism" refers to a cause induced due to electrical energy. Electrochromic devices (ECDs) are one of the emerging potential fields in constructing nanoelectronics devices like e-paper, colour displays for windows, etc. To date, there was no successful product on ECDs; even though the work on ECDs started long back in the 1980s still now it is one of the hot topics in the field of research [43]. Electrochromic materials exhibit an apparent color change from colourless which is essential in constructing the imaging devices. The change in colour tone was due to the change in the electronic transfer of material and electrodes. Based on its principles, electrochromic materials gains their interests in the construction of nanoelectronics devices (reflective displays). EC materials are also classified under some unique groups like electrophoresis, liquid crystal devices, etc. Tungsten oxide (WO_x) has been recognized as EC materials since long back and it was one of the benchmarking materials in the field ECDs [44]. Recently, various inorganic, organic, and conducting polymeric materials are emerging as EC materials owing to its multicolour approach, lower operating voltage, etc.

In this regard, the utilization of conducting polymers as EC materials are the focal prospects in developing different nanoelectronics devices. Some of the polymers like poly(3-methyl thiophene), poly(bi-thiophene) delivers colour efficiency of 240 and 100 cm^2 C^{-1}. Similar to this, some polymers like poly(3,4-ethylene dioxythiophene) (PEDOT) are emerging up recently as EC materials. PEDOT possesses a group of ethylenedioxy with attached to thiophene ring results in enhancing the switching stability (redox) and conductivity (due to high electronic density). Additionally, these polymeric networks (copolymers) induce variation color effects in the displays including red, green, blue. Based on it certain copolymers expose the higher value of coloration efficiency (1000 cm^2 C^{-1}) [8]. The remarkable progress of conducting polymers as ECs includes its quick response, variation in color, etc. [45–48] (Table 2).

On the other hand, conducting polymeric materials is debating towards the realization of constructing fully transparent devices (crystal displays, wearable electronics, etc.) due to its narrow bandgap and the conductivity. Aforementioned, conducting polymers shows color variation at its redox and neutral states which have been disadvantageous in developing or realizing full-color representation clearly [50].

Electrochromic devices with high resolution are constructed using conducting polymers (PEDOT and polypyrrole) by approaching the vapor phase polymerization with an exposed to UV-light. By approaching this technique, the researcher demonstrates the electrochromic contrast of the PEDOT and PPy (conducting polymer). The contrast of the conducting polymers enhanced with enriching its optical properties. Typical construction of electrochromic devices involves pattering of PEDOT for exposure of 900 s over the ITO-coated PET substrates (Fig. 5a). Prior to polymerization, the oxidant will be exposed to the UV-light resulting in high resolution

Table 2 Electrochromic characteristics of some polymers [49]

Polymer	λ (nm) π-π*	E_g (eV)	Colour	
			Undoped	Doped
Poly (pyrrole)	380	3.2	Yellow	Blue-violet
Poly(2,2′-bithiophene)	480	2.6	Red	Deep-blue
Poly(dithienothiophene)	490	2.5	Red	Blue
Poly(3-methylthiophene)	530	2.3	Purple	Pale-blue
Poly(isothionaphtene)	825	1.5	Deep-blue	Pale-yellow-green

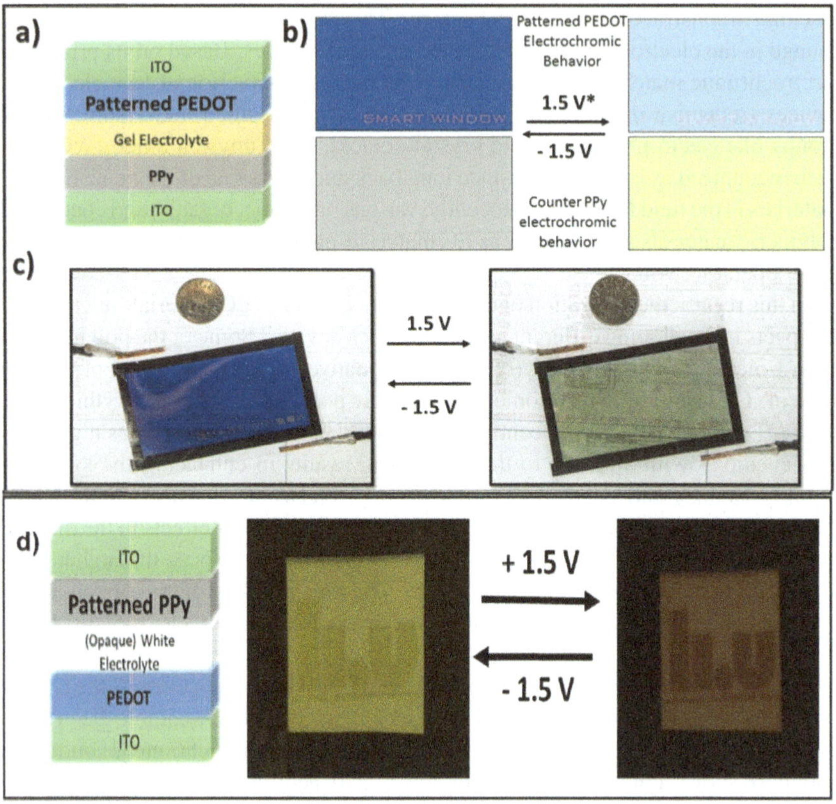

Fig. 5 Prototype patterned electrochromic devices: **a** device architecture; **b** individual electrochromic behavior of the two CP layers; **c** the 'watermarked' smart window prototype in both states and **d** electrochromic device design incorporating a UV-light patterned PPy layer with PPy as the counter polymer [51]

of pattering (100 μm) the devices. Upon applying bias voltage optical properties will be dissimilar from one another due to its different redox states. A prototype ECDs developed with transparent gel electrolyte aid sandwiched between two polymers (PEDOT) (Fig. 5b). PEDOT based ECDs have been carried out by applying a biased voltage of +1.5 V. Figure 5c shows that the conducting polymer-based ECDs exposing transparent characteristics of the fabricated device by depicting the images in the background. Similar device construction has been developed with a prototype design for PPy (conducting polymer) employing white electrolyte and transparent electrolyte for better comparative purposes (Fig. 5d). PPy based ECDs showed that the images are not clear but still visible (than opaque) in an oxidized state and disappeared at reduced state. Compare with PEDOT, visible characteristics of PPy can be enhanced to transparent from opaqueness. The precise mechanisms behind the difference between the two polymeric characteristics are still under investigation. In a combination of both conducting polymer (PEDOT and PPy) with transparent electrolytes, there is a higher possibility for achieving a smart window device including a pattern visible with moderate contrast [51].

2.3 LEDs

Conducting polymer light-emitting diodes (CP-LEDs) are intensively investigated in recent years owing to its potential as emissive elements which can be applied for general lighting (white-light source), flat-panel displays, etc. [52, 53]. The primary interest in conducting polymers for LEDs emerged after the demonstration of electroluminescence (EL) of small molecules operating at low voltages by Kodak [54, 55]. Soon after this demonstration, a research team from Cambridge reported polymer-based LEDs in the early '90s that impacted the field [56, 57]. Enormous progress has been made in enhancing properties like color, luminescence efficiency, reliability, device characteristics, etc. Typical polymer-based LEDs structure, a layer of the polymer will be sandwiched between two electrodes (conducting substrates and metal cathode). The schematic representation of polymer LEDs is given in Fig. 6a. The conducting polymer functions dually in LEDs, one as a light-emission layer and other as a hole transporting layer (HTL). Conducting polymeric LEDs at forward-biased electrons are injected from the cathode (metal contact) and holes injected from anodes. Upon applying electric potential, electrons and holes move towards the opposite direction at the electrodes, and the recombination of generated electron–hole pair occurs within the region of the emissive polymer layer. By releasing recombination energy, it results in the emission of light.

Looking at the structure of CP-LEDs, conducting substrate (i.e., indium doped tin oxide (ITO)) will be near to highly occupied molecular orbital (HOMO) level of polymer which has high work function relatively thereby used as an electrode for hole-injection and electron injection metals such as aluminum (Al), barium (Ba), etc. will be used due to its low-work function. The energy level diagram for CP-LED operational is represented in Fig. 6b. EL quantum efficiency of CP-LED was characterized either by internal or external quantum efficiency. Internal quantum

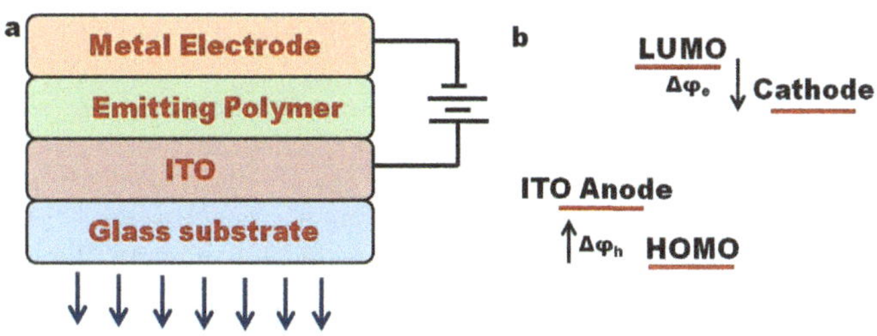

Fig. 6 **a** Basic structure of polymeric LEDs and **b** its corresponding energy level

efficiency (η_{int}) defined as the number of photons (produced within the device) to the number of electrons (flows in an external circuit (γ)) given by:

$$\eta_{int} = \gamma r_{st} q$$

r_{st} is a fraction of excitons, and q is radiative decay efficiency. Conducting polymers emits luminescence below its energy gap between the π bonding and π^* antibonding. For example, the conducting polymer poly(1,4-phenylene)(PPV) has an energy gap of 2.5 eV thereby produces yellow-green luminescence [58].

Niu et al. demonstrated a white polymer LED using PEDOT: PSS, the conducting polymer was thermally cross-linked over the tris(4-carbazole)triphenylamine (TCTA). The conducting PEDOT: PSS polymer serves as hole-injection and transport bilayer in LEDs. The fabricated white polymeric LEDs deliver high efficiency of 11 lm W during emission of light and 5.61 lm W during forwarding emission; the forward brightness recorded for the polymeric LEDs is 800 Cd m^2. Moreover the constructed LEDs emits quite stable light (white color) with CIE coordinates ranging from (0.379, 0.367) to (0.328, 0.351), which are very close to standard white lighting (0.333, 0.333) [59].

Similarly, a research group from the United States in collaboration with South Korea, experimented with surfactants in conducting polymer for constructive nanoelectronics devices fabrication (LEDs). The researcher tested using PEDOT: PSS and reported that surfactant plays a role in achieving the viscoelastic property of the conducting polymers. Upon increasing the surfactant concentration into PEDOT: PSS, the polymeric chain is tuned from brittle into viscoelastic. In order to construct the nanoelectronics devices, the strong modulus (E') versus loss modulus (E'') of the polymer has been adjusted by tuning the polymer's viscoelasticity. A 2D complex micropatterns were developed on PEDOT: PSS films by hand pressing the PDMS (Fig. 7). The arrays of LEDs were achieved with spin coating poly(9,9-dioctylfluorenyl-2,7-diyl) (PFO) with a thickness of 100 nm as light-emitting materials. The customized displays shown in Fig. 7 exhibit stretchability of 0.1 ε by itself and the elastomer, which possesses 0.57 ε. The surfactant loaded conducting polymers enables the characteristics of viscoelasticity, stable conductivity (metallic) during stretching and folding (suitable for wearable nanoelectronics devices), repetitive healing, and customized designing of nanoelectronics devices [60].

2.4 Dielectrics

In the recent years, there has been an urge to develop modern flexible devices with good dielectric properties, good mechanical flexibility, and low-cost capacitive storage systems. Dielectric materials with high dielectric constant (ε), a low dielectric tangent loss (tan δ), low processing temperature, and high dielectric strength (E_b) have been a great deal of interest from various applications [61]. High dielectric

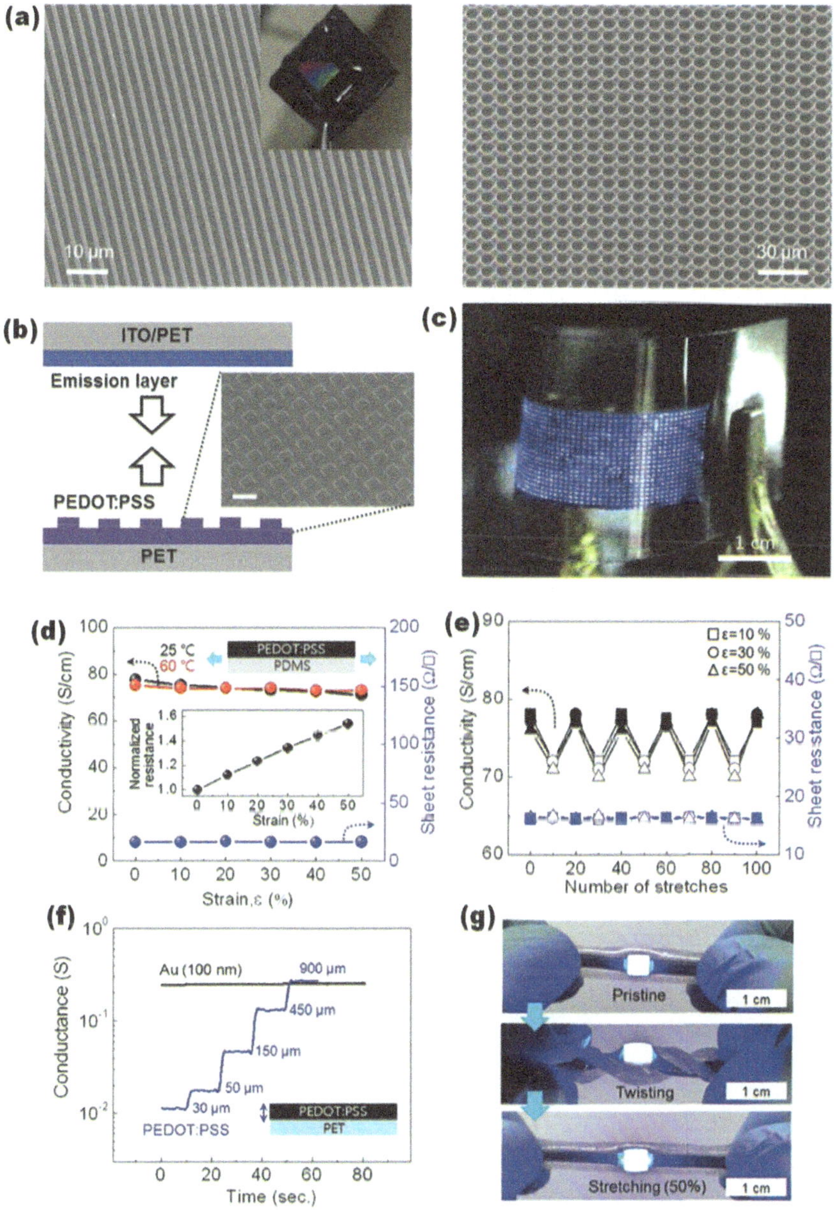

Fig. 7 PEDOT: PSS in customized electronics. **a** Line-and-space pattern; **b** hole pattern; **c** customized OLED using a square-shaped PEDOT:PSS pattern over PDMS substrates; **d** conductivity; **e** changes in conductivity and sheet resistance; **f** change of conductance and **g** demonstration of a deformable LED fabricated on a PEDOT:PSS circuit [60]

permittivity polymer-based composite materials have received attention for electrical stress control applications, actuators, and organic field-effect transistors. Dielectric materials can be made using organic and inorganic materials, metal oxides and ceramics are used as inorganic dielectrics. They exhibit high dielectric constant (ε) by requires high processing temperature, exhibit low E_b and are brittle. Whereas in the case of organic dielectrics, polymers are flexible and exhibit high E_b, which can be processed at low temperatures, but it possesses low dielectric constant. Hence it necessitates developing a composite of organic–inorganic dielectric material to commemorate the needs of various technological applications. Introducing high dielectric constant ceramics over 50 volume percentage into polymer matrix would enhance the dielectric constant, but deteriorates the mechanical properties of the composites, thus making it unsuitable for applications [62]. The dielectric constant of polymer composites is enriched by packing with fillers like carbon-based nanomaterials, conducting polymers, metal nanoparticles, and tackling the issues. Hence, intensive research on the use of conductive fillers through controlled dispersion and self-passivation of metal particles into the dielectric matrix have been carried out. There is a need to control the electrical conducting phenomenon for useful dielectric layers, to overcome the underlying problem that the dielectric layers should be insulating and problems about the high content of conductors in dielectric composites.

Poly(vinyl pyrrolidone) (PVP) encapsulate with one dimensional (1D) nanorods structured polyaniline (PANI) as a filler to fabricate polymer nanocomposites. PVP wrapped PANI stabilizes through strong bonding between PANI (hydrogen bonds) and PVP (carbonyl groups). Moreover, the PVP acts as a barrier layer rendering the dispersion of PANI in PVDF thereby results in declination of dielectric loss as well as leakage current.

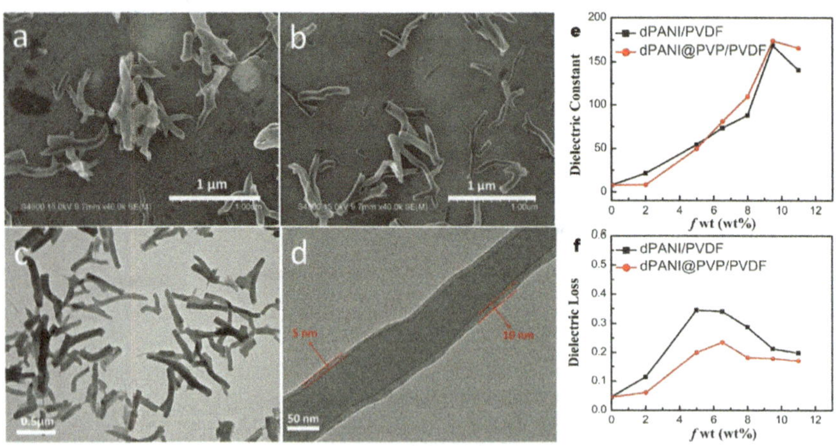

Fig. 8 FE-SEM micrographs: (**a** and **b**) sPANI and dPANI nanorods (**c** and **d**) TEM and HRTEM images of dPANI@PVP nanorods; **e** dielectric constant and **f** dielectric loss of dPANI/PVDF and dPANI@PVP/PVDF nanocomposite films at 100 Hz [63]

Figure 8 shows the dielectric measurements (dielectric loss and constant) carried out at 100 Hz for the PANI@PVP composites. The research group carried out a series of experiments by varying the weight percentage (wt%) of PVP in the polymer nanocomposites. Among the different wt%, 5.5 wt% of PVP in PANI@PVP/PVDF composites yields lower dielectric constant than others, and upon increasing the concentration further, the dielectric constant increases rapidly. As mentioned the PVP facilitates the formation/distribution of 1D nanorods shaped PANI in the PVDF matrix. The increase in the concentration of PVP results in enhancing the micro-capacities in the conducting polymer nanocomposites which led to an increase in dielectric constant beyond 5.5 wt%.

Similarly, the dielectric loss of the conducting polymer nanocomposites (PANI@PVP/PVDF) has been carried out at 100 Hz. It is found that at higher concentration (9.5 wt% of PVP), the polymeric nanocomposites were 0.17 whereas for PANI/PVDF it was obtained to be 0.22. The dielectric loss and the leakage current of the polymeric nanocomposites attributed mainly to the insulating property of PVP in turns cut off the conductive paths caused due to the formation of nanorods structured PANI [63].

Recently, nanocomposite of two-dimensional (2D) materials like graphene oxide (GO), carbon nanotubes (CNTs) with conducting polymers are gaining much interest in the research community. For example, homogeneous dispersion of GO and CNTs with polyimide (PI) conducting polymer via the solvent-exchange method. In this in-situ polymerization and thermal imidization have been carried out for conducting polymer-2D nanomaterials dielectric composites. The conducting polymer (PI) will act as fillers attributing to the improvement in flexibility of the polymeric nanocomposites. GO with strong π-π interaction implies that the formation of CNT-GO composites boosts the conductivity of the composites (Fig. 9a). GO triggers a problem by invariably reducing the electrical conductivity of CNT and agglomeration of GO in organic solvents. Polyimide (PI) is used in this work, a well-known high-performance polymer with excellent mechanical properties and thermal stabilities. A

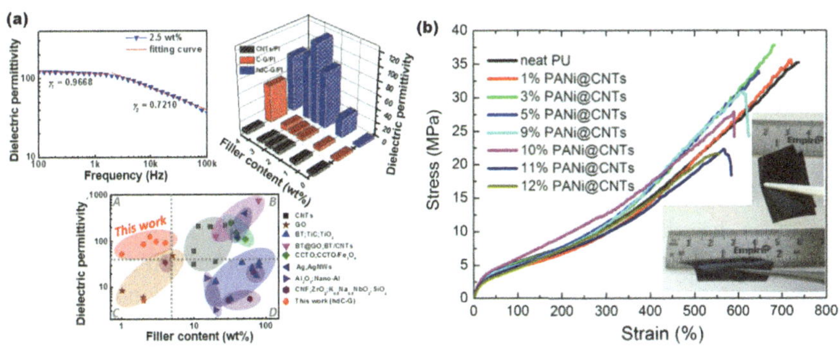

Fig. 9 **a** Dielectric permittivity studies of conducting polymers and **b** stress–strain curves of PANI@CNTs composites [64, 65]

small addition of fillers (hdC-G), the hdC-G/PI composites could achieve high dielectric permittivity, enhanced tensile strength, and thermal stabilities without sacrificing the flexibility of the materials [64].

Wenhui Xu et al. reported the synthesis of core–shell nanostructured using PANI. PANI will be shell covering the core nanomaterials (CNTs), providing good dispersion for the fabrication of flexible and foldable nanoelectronics devices. The core–shell nanostructured conducting polymers exhibit promising dielectric permittivity, the permittivity value recorded for the composites at 9 and 10 wt% (87.7 and 171.4 (@ 100 Hz) respectively (Fig. 9b). The dielectric loss recorded for the samples is obtained to less than 2.0 indicating good conductivity. The significant increase is due to the formation of many micro-capacitors near the percolation threshold [65].

2.5 *Electrochemical Detection of Neurotransmitters Using Conducting Polymers*

Neurotransmitters, an endogenous chemical messenger released from neurons, influence our bodies' diverse physical and psychological functions. These neurotransmitters carry messages and communicate to other cells by crossing a tiny gap called a synapse and this process is known as neurotransmission. Based on the neurotransmitter binding to receptor sites the actions can be both excitatory and inhibitory. These neurotransmitters take hold of a significant role in having proper brain-body coordination and functioning [66]. However, it is complicated and challenging to analyze them to understand the disorders in the brain. In this regard, electrochemical biosensors have taken their role as a sensitive and quantitative technique; it creates an electrical signal in response to the analyte and records them. There are three biogenic amine neurotransmitters namely catecholamine (dopamine, epinephrine and norepinephrine), histamine and serotonin. These simple molecules play an imperative role in regulating the central and peripheral nervous systems of the human body. Epinephrine is used for information transfer in the central nervous system and also used to boost the supply of oxygen and glucose to the brain. Norepinephrine neurotransmitter is present at central nervous systems and abnormalities associated with several psychological problems, Alzheimer's and Parkinson's disease. Dopamines play a notable role in controlling the flow of information necessary for problem solving, attention, memory and contribute to overall mental well being. Serotonin plays essential role in human emotions and its deformity leads to mental retardation, sleep disorder and depression. The primary function in sleep–wake-cycle is contributed by Histamine neurotransmitter and it is also involved in regulation quite a few physiological functioning. Electrochemical biosensors are a kind of electrochemical cell comprises of three electrodes namely working electrode (active material coated on the surface), reference electrode (Ag/AgCl—silver metal coated with layer of silver chloride) and counter electrode (platinum wire) placed in suitable electrolyte solution [67].

2.5.1 Why Conducting Polymers in Neurotransmitters Detection?

Among various other materials explored for electrochemical biosensor design, conducting polymers has become well accepted due to its exceptional electrical and optical properties. The significant contribution of polymers in electrochemical sensor applications brought in from the discovery of conducting polymers (CPs). They are made up of monomers capable of attaining positive or negative charges based on the redox reactions that they undergo. This charge acquiring contribute to electrical conductivity which is primary for any sensor device and its design. In continuation, there are other two classes of polymers namely redox polymers and ionically conducting polymers which are also explored by researchers in electrochemical sensor design for detecting neurotransmitters. Having a localized electron redox site, redox polymers are less conductive than CPs whereas, movement of ions contributes to conductivity in ionically conducting polymers. However, their utility in the electrochemical detection is maximum is hindered due to their low ionic conductivity at room temperature and increase in resistance over a while under operational conditions. Thus considering the above-believed concern, the utmost utilization of CPs in the design of the electrochemical sensor has been explored by large research community.

Interestingly they are capable enough to achieve excellent sensitivity and stability besides having exceptional real time performance. Recent advances in electrochemical sensors are achieved with the utilization of CPs namely polypyrrole (PPy), polyaniline (PANI), polythiophene (PT) and polyacetylene (PA). Their natural laboratory processing approach and scalability make them ideal for electrochemical sensor design. Moreover, the ease of altering their surface area and fine-tuning electron transport property by suitable dopant addition facilitate with better sensing performance. During the preparation process of conducting polymers, if oxidation occurs, it is facilitated by p-doping and results in the formation of positive polarons and the reverse happens when the reduction process is carried out. This charged polaron possesses self-localized separate charges and dramatically helps in electrochemical detection of neurotransmitters—binding happens based on the charges present [68]. Thus immobilization of active material based on the analyte charges stands as key factor in deciding the sensor performance. Hence this immobilization process should be done efficiently with no damage to the sensor probe or material. Generally, covalent attachments and physical adsorption are preferred compared to other available techniques. Conversely several reaction environment parameters like pH, solvents used and temperature should be controlled precisely for competent electrochemical design and detection. As an optional approach utilization of Nafion consisting of tetrafluoroethylene main chain with side chains terminated with sulphonic acid groups aids superior selectivity towards neurotransmitters lowers the interference effect by ascorbic and uric acid [69].

2.5.2 Measurement Strategy in Neurotransmitters Detection

The detection or measurement technique dramatically depends on the redox potentials of CPs used in the device design. If the polymer used becomes electroactive at low potentials, electrochemical impedance spectroscopy remains the most elegant sensitive tool. Further, the detection can be either by direct redox potential or with the utilization of redox indicators such as potassium ferricyanide/potassium ferrocyanide $K_3[Fe(CN)_6]/K_4[Fe(CN)_6]$ or ruthenium complexes. In another way, if the CPs based fabricated electrode immersed in a solution containing analyte molecules and if the electrode-analyte interaction needs to recorded cyclic voltammetry (CV) functions as an efficient detection platform. This CV can very well picturize the change in electrical properties owing to their binding or interactions. Thus it is always to have a suitable measurement tool desired to our designed sensor for a more significant act. On the whole, to increase the practical usability and to contend with clinical confronts all the above said design strategies should carefully be considered and incorporated in sensor design [70].

2.5.3 Surface Modification and Detection Approaches

One of the critical downsides in the electrochemical sensor is the poor electron transfer between the electrode and analyte owing to tunneling distance. Hence, CPs are surface modified extensively with carbon nanotubes (CNTs), graphene, and carbon black to enhance the sensitivity of electrochemical neurotransmitter sensors. The presence of outstanding physicochemical properties, stability, and electrocatalytic effect makes them ideal for this sensor design. The high surface area possessed by the carbon systems makes them ideal for using them as fillers in the polymer matrix [70]. Fayemi et al. developed electrochemical sensors by using PANI/Metal oxide composite and could achieve sensitive detection with no interference from ascorbic and uric acid with detection limit 0.176×10^{-7} M [71]. PANI nanocomposites film doped with metal oxide supported on CNT's prepared and explored for electrochemical sensing by Tsele et al., for epinephrine neurotransmitters. The use of different metal oxide nanostructures gave exceptional behavior compared to stand-alone performance. They were capable of having >0.20 μM limits of detection (LOD) using their modified electrodes with excellent electron transport properties with no interference from other species. These PANI—Metal oxide—CNTs based nanocomposites were also evaluated for real-time pharmaceutical samples detection and produced satisfactory results [72]. Yan et al. achieved selective dopamine detection in interfering species ascorbic acid using the AgCl@PANI core–shell nanocomposites system. The nanocomposites showed excellent inhibitive effect with ascorbic acid in the reaction environment, giving selective detection of dopamine. It is also mentioned that the C=O functional group's presence easily forms hydrogen bonds with the hydroxyl group of ascorbic acid and prevents its oxidation during the measurement steps [73]. Feng et al. reported the synthesis of PANI-based nanotubes with the utilization of self-degradable templates and explored them as electrode material for dopamine. Interestingly the fabricated sensor exhibited superior electrocatalytic performance near the neutral pH which necessarily implies its real-time practical efficacy [74]. PPy—Gold nanoparticles based highly sensitive serotonin sensor developed by Tertis et al., and this three-dimensional material architecture were produced by in-situ electrochemical deposition. The synergistic effect between gold and PPy contributed to the excellent electrocatalytic activity of the sensor. The diagnostic outcomes showed 320 times better capability compared to bare electrode systems [75]. A ternary composite system comprising CNT/PPy/Silver nanoparticles developed and investigated for serotonin electrochemical detection. The detection approach has been applied to serum samples and the results demonstrated the absence of interference-effect. Further, the sensor exhibited LOD of 0.15 μM with 2.2% reproducibility and 1.7% of repeatability values [76]. PEDOT doped with CNTs for the selective detection of dopamine in the presence of ascorbic acid developed by Xu et al., using electrodeposition technique. The doped PEDOT system exhibited high electrocatalytic activity towards the redox reactions of dopamine with LOD of 20 nM. Interestingly the designed sensor possesses excellent stability, reusability and free from ascorbic acid interference with higher concentrations of 1000 times higher than dopamine [77]. Qian et al. produced polymeric film-based sensors for the detection of dopamine and catechol. The polymeric film sensor made up of PEDOT: PSS combining cyclodextrin. The fabricated polymeric film-based systems produced good electrocatalytic activity towards the analytes with LOD of 9.5 nM and 0.027 μM for dopamine and catechol without the use of noble metals in its design [78]. The molecular imprinting based materials are the threshold for specific receptor detection; the Molecularly Imprinted Polymers (MIPs) holds futuristic ease of synthesis, higher stability, lower production cost and also does the functions of enzymes, creation of three-dimensional cavities of specific size and shape for biomolecules recognition. MIPs possess detecting a specific type of biomolecules along with a few other applications like antibody mimics, molecular recognition, drug release, and catalysis [79]. Wei et al. reported the development of MIP based detection systems for epinephrine and Norepinephrine. The electrode material comprised MIPs anchored with dual-color quantum dots and synthesized using a template-based process. The developed sensor exhibited superior detection in the range of 0.08–20 μM with LOD of 9 and 12 nM (Fig. 10) [80].

Multimode quantum dots encoded MIPs were methodically explored for the quantitative detection of dopamine. The change in color coding concerning specific dopamine adsorption recorded and the sensor exhibited LOD of 2 $\mu g\ L^{-1}$ and 0.5 $\mu g\ L^{-1}$ [81]. Based on the previous reports, the synchronization between the electrode and analyte binding is considered as a more predominant one in determining the efficiency of the sensor performance.

Thus the significance of neurotransmitters for diagnosis strategy are addressed in a detailed manner. It also highlighted the importance of polymers in the design strategy of electrochemical sensors and certain immobilization concepts. A new class of materials MIPs was introduced and a brief presentation on its application in neurotransmitters were also presented. Though several materials being explored for the electrochemical detection at the laboratory level, the direct implementations

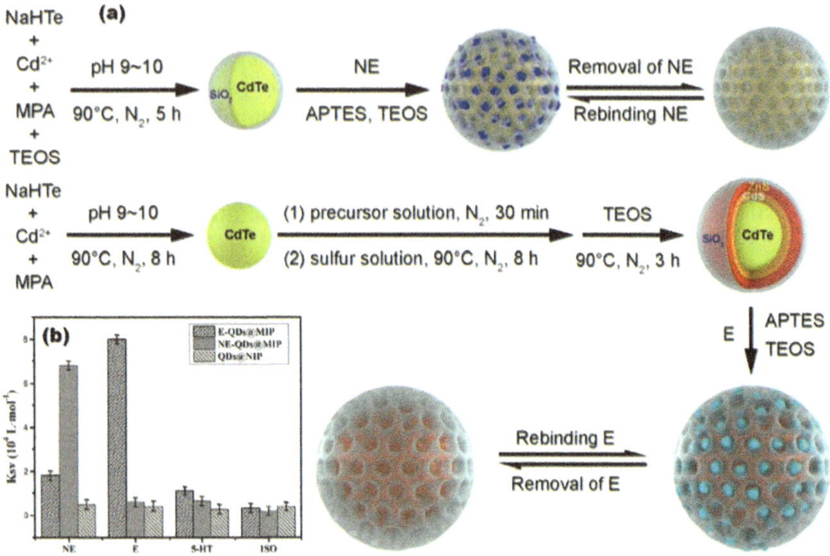

Fig. 10 **a** Schematic illustration and **b** selectivity of dual-color QDs@MIPs [80]

to the human is still under trail—which open ups and put down the path for future advancements.

3 Conclusion

In summary, conducting polymers, its kinds, and its practical usage in the nanoelectronics devices have been briefed. The application of conducting polymers are scattered and in this chapter a keen focus has been provided for few emerging applications. Field-effective transistor, how the conducting polymers will be applied, patterning of the devices its effects are discussed. Electrochromic one of the emergent applications in the fields of technological market, the fundamental aspects of chromism and its construction in displays, crystal devices etc. are briefed. The conceptualization of LEDs and dielectric are explained in a detailed manner towards the construction of nanoelectronics devices. Finally, the applications of conducting polymers in electrochemical detection of neurotransmitters are briefed in a detailed manner. Thus the future electronic devices will be heavily reliant on creating efficient nanoelectronic devices using conducting polymers.

Acknowledgments The author Dr. Arunachalam Arulraj duly acknowledges the FONDECYT Postdoctoral Fellowship (Project No.: 3200076), University of Concepcion, Concepcion, Chile, Santiago for the financial assistance.

References

1. Das, T.K., Prutsy, S.: Review on conducting polymers and their applications. Polym-Plast. Technol. **51**, 1487 (2012)
2. Ramos, P., Cruz, M.A.E., Tovani, C.B., Ciancaglini, P.: Biomedical applications of nanotechnology. Biophys. Rev. **9**, 79 (2017)
3. M. Kesavan, A. Arulraj, K. Rajendran, P. Anbarasu, P. Anandha Ganesh, D. Jeyakumar, M. Ramesh: Performance of dye-sensitized solar cells employing polymer gel as an electrolyte and the influence of nano-porous materials as fillers. Mater. Res. Exp., **5**(11)5305 (2018)
4. Lele, A.: Role of Nanotechnology in Defence. Strat. Anal. **33**, 229 (2009)
5. Arulraj, A., Ramesh, M., Subramanian, B., Senguttuvan, G.: In-situ temperature and thickness control grown 2D-MoS2 via pulsed laser ablation for photovoltaic devices. Sol. Energy **174**, 286 (2018)
6. Andersson, B.P., Nilsson, D., Svensson, P., Chen, M., Malmström, A., Remonen, T.: Active matrix displays based on all-organic electrochemical smart pixels printed on paper. Adv. Mater. **20**, 1460 (2010)
7. Argun, A.A., Aubert, P., Thompson, B.C., Schwendeman, I., Gaupp, C.L., Hwang, J., Pinto, N.J., Tanner, D.B., Macdiarmid, A.G., Reynolds, J.R.: Multicolored electrochromism in polymers: structures and devices. Chem. Mater. **16**, 4401 (2004)
8. Gaupp, C.L., Welsh, D.M., Rauh, R.D., Reynolds, J.R.: Composite coloration efficiency measurements of electrochromic polymers based on 3,4-alkylenedioxythiophenes. Chem. Mater. **14**, 3964 (2002)
9. Facchetti, B.A., Yoon, M., Marks, T.J.: Gate dielectrics for organic field-effect transistors: new opportunities for organic electronics. Adv. Mater. **17**, 1705 (2005)
10. Inzelt, G., Pineri, M., Schultze, J.W., Vorotyntsev, M.A.: Electron and proton conducting polymers: recent developments and prospects. Electrochem. Acta **45**, 2403 (2000)
11. Herrmann, S., Ritchie, C., Streba, C.: Polyoxometalate—conductive polymer composites for energy conversion, energy storage and nanostructured sensors. Dalton Trans. **44**, 7092 (2015)
12. Yang, H., Wei, L., Guan, J., Guo, Y., Wang, X., Yan, X., Zhang, S., Wei, Z., Guo, J.: Polymer nanocomposites for energy storage, energy saving, and anticorrosion. Mater. Chem. A **3**, 14929 (2015)
13. Kardas, G., Solmaz, R.: Electrochemical synthesis and characterization of a new conducting polymer: polyrhodanine. Appl. Surf. Sci. **253**, 3402 (2007)
14. Zhan, G., Yu, Y., Lu, L., Wang, E., Wujcik, S., We, J.: Conductive polymer nanocomposites: a critical review of modern advanced devices. Mater. Chem. C **5**, 1569 (2017)
15. Kumar, R.C.: Sharma. Eur. Polymer J. **34**, 1053 (1998)
16. Inzelt, G.: Recent advances in the field of conducting polymers. J. Solid State Electrochem **21**, 1965 (2017)
17. Kesavan, M., Arulraj, A., Sannasi, V., Rajendran, K., Anbarasu, P., Jeyakumar, D., Ramesh, M.: Mater. Res. Innov. **24**, 1 (2020)
18. Ma, R., Wang, Y., Qi, H., Shi, C., Wei, G., Xiao, L., Huang, Z., Liu, S., Yu, H., Teng, C., Liu, H., Murugadoss, V.: Nanocomposite sponges of sodium alginate/graphene oxide/polyvinyl alcohol as potential wound dressing: in vitro and in vivo evaluation. Compos. B **167**, 396 (2019)
19. Fernández, J., Bonastre, J., Molina, J., Cases, F.: Electrochemical study on an activated carbon cloth modified by cyclic voltammetry with polypyrrole/anthraquinone sulfonate and reduced graphene oxide as electrode for energy storage. Eur. Polymer J. **103**, 179 (2018)
20. Zhanga, Y., Ana, Y., Wua, L., Chena, H., Lia, Z., Doua, H., Murugadoss, V., Fanb, J., Zhanga, X., Mai, X., Guob, Z.: Metal-free energy storage systems: combining batteries with capacitors based on a methylene blue functionalized graphene cathode. J. Mater. Chem. A **7**, 19668 (2019)
21. Yin, Z., Zheng, Q.: Controlled synthesis and energy applications of one-dimensional conducting polymer nanostructures: an overview. Adv. Energy Mater. **2**, 179 (2012)
22. Pan, L., Qiu, H., Dou, C., Li, Y., Pu, L., Xu, J., Shi, Y.: Int. J. Mat. Sci. **11**, 2636 (2010)
23. Toshima, N., Hara, S.: Direct synthesis of conducting polymers from simple monomers. Prog. Polym. Sci. **20**, 155 (1995)

24. Bargon, J., Waltman, R.J.: Electrochemical synthesis of electrically conducting polymers from aromatic compounds. IBM J. Res. Develop. **27** (1983)
25. Kincal, A., Kumar, A.D., Child, J.R.: Reynolds Synth. Met. **92**, 53 (1998)
26. Rogers, J.A., Bao, Z., Raju, V.R.: Nonphotolithographic fabrication of organic transistors with micron feature sizes. Appl. Phys. Lett. **72**, 2716 (1998)
27. Sirringhaus, H., Kawase, T., Friend, R.H., Shimoda, T., Inbasekaran, M., Wu, W.: High-resolution inkjet printing of all-polymer transistor circuits. Science **290**, 2123 (2000)
28. Park, J., Lee, S., Lee, H.H.: High-mobility polymer thin-film transistors fabricated by solvent-assisted drop-casting. Org. Electron. **7**, 256 (2006)
29. Wang, J., Swensen, D., Moses, A.J.H.: Increased mobility from regioregular poly(3-hexylthiophene) field-effect transistors. J. Appl. Phys. **93**, 6137 (2003)
30. Ofer, D., Crooks, R.M., Wrighton, M.S.: Potential dependence of the conductivity of highly oxidized polythiophenes, polypyrroles, and polyaniline: finite windows of high conductivity. J. Am. Chem. Soc. **112**, 7869 (1990)
31. Mccoy, C.H., Wrighton, M.S.: Potential-dependent conductivity of conducting polymers yields opportunities for molecule-based devices: a microelectrochemical push-pull amplifier based on two different conducting polymer transistors. Chem. Mater. **5**, 914 (1993)
32. Paul, E.W., Ricco, A.J., Wrighton, M.S.: Resistance of polyaniline films as a function of electrochemical potential and the fabrication of polyaniline-based microelectronic devices. J. Phys. Chem. **89**, 1441 (1985)
33. Verilhac, J.M., LeBlevennec, G., Djurado, D., Rieutord, F., Chouiki, M., Travers, J.P.: Synth. Met. **156**, 815 (2000)
34. Zen, J., Pflaum, S., Hirschmann, W., Zhuang, F., Jaiser, U.: Asawapirom. Adv. Funct. Mater. **14**, 757 (2004)
35. Park, Y.D., Kim, D.H., Jang, Y., Cho, J.H., Hwang, M., Lee, H.S.: Effect of side chain length on molecular ordering and field-effect mobility in poly(3-alkylthiophene) transistors. Org. Electron. **7**, 514 (2006)
36. Sirringhaus, P.J., Brown, R.H., Friend, M.M., Nielsen, K., Bechgaard, B.M.W.: Langeveld-Voss. Synth. Met. **111**, 129 (2000)
37. Surin, M., Leclere, Ph., Lazzaroni, R., Yuen, J.D., Wang, G., Moses, D.: Relationship between the microscopic morphology and the charge transport properties in poly(3-hexylthiophene) field-effect transistors. J. Appl. Phys. **100**, 033712 (2006)
38. Yan, Y., Huang, L.B., Zhou, Y., Han, S.T., Zhou, L., Zhuang, J., Xu, Z.X., Roy, V.A.L.: Self-aligned, full solution process polymer field-effect transistor on flexible substrates. Sci. Rep. **5**, 15770 (2015)
39. Knopfmacher, O., Hammock, M.L., Appleton, A.L., Schwartz, G., Mei, J., Lei, T., Pei, J., Bao, Z.: Highly stable organic polymer field-effect transistor sensor for selective detection in the marine environment. Nat. Comm. **5**, 2954 (2014)
40. Lee, J.G., Seol, Y.G., Lee, N.E.: Polymer thin film transistor with electroplated source and drain electrodes on a flexible substrate. Thin Solid Films **515**, 805 (2006)
41. Seol, Y.G., Lee, J.G., Lee, N.E., Lee, S.S., Ahn, J.: Electrical characteristics of poly(3-hexylthiophene) organic thin film transistor with electroplated metal gate electrodes on polyimide. Thin Solid Films **515**, 5065 (2007)
42. Xue, F., Liu, Z., Su, Y., Varahramyan, K.: Inkjet printed silver source/drain electrodes for low-cost polymer thin film transistors. Microelectron. Eng. **83**, 298 (2006)
43. Araki, S., Nakamura, K., Kobayashi, K., Tsuboi, A., Kobayashi, N.: Electrochemical optical-modulation device with reversible transformation between transparent, mirror, and black. Adv. Mater. **24**, OP122 (2012)
44. Deb, S.K.: Optical and photoelectric properties and colour centres in thin films of tungsten oxide. Philos. Mag. **27**, 801 (1973)
45. Bulloch, R.H., Kerszulis, J.A., Dyer, A.L., Reynolds, J.R.: Mapping the broad CMY subtractive primary color gamut using a dual-active electrochromic device. ACS Appl. Mater. Interfaces **6**, 6623 (2014)

46. Liu, D.Y., Chilton, A.D., Shi, P., Craig, M.R., Miles, S.D., Dyer, A.L., Ballarotto, V.W., Reynolds, J.R.: In situ spectroscopic analysis of sub-second switching polymer electrochromes. Adv. Funct. Mater. **21**, 4535 (2011)
47. Beaujuge, P.M., Reynolds, J.R.: Color control in π-conjugated organic polymers for use in electrochromic devices. Chem. Rev. **110**, 268 (2010)
48. Amb, C.M., Dyer, A.L., Reynolds, J.R.: Navigating the color palette of solution-processable electrochromic polymers. Chem. Mater. **23**, 397 (2011)
49. Mastragostino, M.: Electrochromic devices. In: Scrosati, B. (ed.) Applications of Electroactive Polymers. Chapman & Hall, London.
50. Kobayashi, N.: Electrochromic Displays. Handbook of Visual Display Technology. Springer Publishers (2015)
51. Brooke, R., Edberg, J., Iandolo, D., Berggren, M., Crispin, X., Engquist, I.: Controlling the electrochromic properties of conductive polymers using UV-light. J. Mater. Chem. C **6**, 4663 (2018)
52. Pope, M., Kallmann, H., Magnante, P.J.: Chem. Phys. **38**, 2042 (1963)
53. Helfrich, W., Schneider, W.G.: Recombination radiation in anthracene crystals. Phys. Rev. Lett. **14**, 229 (1965)
54. Tang, C.W., Vanslyke, S.A.: Organic electroluminescent diodes. Appl. Phys. Lett. **51**, 913 (1987)
55. Adachi, C., Tsutsui, T., Saito, S.: Organic electroluminescent device having a hole conductor as an emitting layer. Appl. Phys. Lett. **55**, 1489 (1989)
56. Burroughes, D.D.C., Bradley, A.R., Brown, R.N., Marks, K., Mackay, R.H., Friend, P.L., Burns, A.B.: Holmes. Nature **347**, 539 (1990)
57. Braun, D., Heeger, A.J.: Visible light emission from semiconducting polymer diodes. Appl. Phys. Lett. **58**, 1982 (1991)
58. Grüner, J., Cacialli, F., Friend, R.H., Emission enhancement in single-layer conjugated polymer microcavities. J. Appl. Phys. **80**, 207 (1996)
59. Niu, B.Y.H., Liu, M.S., Ka, J.W., Bardeker, J., Zin, M.T., Schofield, R., Chi, Y., Jen, A.K.Y.: Crosslinkable hole-transport layer on conducting polymer for high-efficiency white polymer light-emitting diodes. Adv. Mater. **19**, 300 (2007)
60. Oh, Y., Kim, S., Baik, H.K., Jeong, U.: Conducting polymer dough for deformable electronics. Adv. Mater. **28**, 4455 (2016)
61. Huang, X., Jiang, P.: Adv. Mater. **1** (2014)
62. . Zhang, L., Wang, W., Wang, X., Bass, P., Cheng, Z., Zhang, L., Wang, W., Wang, X., Bass, P., Cheng, Z.: Appl. Phy. Lett. **103**, 232903 (2014)
63. Yu, S., Qin, F., Wang, G.: Improving the dielectric properties of poly(vinylidene fluoride) composites by using poly(vinyl pyrrolidone)-encapsulated polyaniline nanorods. J. Mater. Chem. C. **4**, 1504 (2016)
64. Liao, X., Ye, W., Chen, L., Jiang, S., Wang, G., Zhang, L., Hou, H.: Flexible hdC-G reinforced polyimide composites with high dielectric permittivity. Compos. A Appl. Sci. Manuf. **101**, 50 (2017)
65. Xu, W., Ding, Y., Yu, Y., Jiang, S., Chen, L., Hou, H.: Highly foldable PANi@CNTs/PU dielectric composites toward thin-film capacitor application. Mater. Lett. **192**, 25 (2017)
66. Nambiar, S., Yeow, J.T.W.: Biosens. Bioelectron. **26**, 1826 (2011)
67. Moon, J.M., Thapliyal, N., Hussain, K.K., Goyal, R.N., Shim, Y.B.: Conducting polymer-based electrochemical biosensors for neurotransmitters: a review. Biosens. Bioelectron. **102**, 540 (2018)
68. Aydemir, N., Malmstrom, J., Sejdic, J.T.: Phy. Chem. Chem. Phy. **1**, 1 (2013)
69. Ardakani, Z.T., Hosu, O., Cristea, C., Ardakani, M.M., Marrazza, G.: Localization of sliding movements using soft tactile sensing systems with three-axis accelerometers. Sensor **19**, 2036 (2019)
70. Ou, Y., Buchanan, A.M., Witt, C.E., Hashemi, P.: Frontiers in electrochemical sensors for neurotransmitter detection: towards measuring neurotransmitters as chemical diagnostics for brain disorders. Anal. Methods **11**, 2738 (2019)

71. Fayemi, O.E., Adekunle, A.S., Kumaraswamy, B.E., Ebenso, E.E., J. Electroanan. Chem. **818**, 236 (2018)
72. Tsele, T.P., Adekunle, A.S., Fayemi, O.E., Ebenso, E.E.: Electrochemical detection of epinephrine using polyaniline nanocomposite films doped with TiO_2 and RuO_2 nanoparticles on multi-walled carbon nanotube. Electro. Chem. Acta **243**, 331 (2017)
73. Yan, W., Feng, X., Chen, X., Li, X., Zhu, J.J.: BioElectrochem. **72**, 21 (2008)
74. Feng, X., Zhang, Y., Yan, Z., Chen, N., Ma, Y., Liu, X., Yang, X., Hou, W.: J. Mater. Chem. A **1**, 9780 (2013)
75. Tertis, M., Cernat, A., Lacatis, D., Florea, A., Bogdan, D., Sucia, M., Sandulescu, R., Cristea, C.: Highly selective electrochemical detection of serotonin on polypyrrole and gold nanoparticles-based 3D architecture. Electrochem. Commun. **75**, 43 (2017)
76. Cesarino, I., Galesco, H.V., Machado, S.A.S.: Determination of serotonin on platinum electrode modified with carbon nanotubes/polypyrrole/silver nanoparticles nanohybrid. Mater. Sci. Eng. C **40**, 49 (2014)
77. Xu, G., Li, B., Cui, X.T., Ling, L., Luo, X.: Electrodeposited conducting polymer PEDOT doped with pure carbon nanotubes for the detection of dopamine in the presence of ascorbic acid. Sens. Actuat. B Chem. **188**, 405 (2013)
78. Qian, Y., Ma, C., Zhang, S., Gao, J., Liu, M., Xie, K., Wang, S., Sun, K., Song, H.: High performance electrochemical electrode based on polymeric composite film for sensing of dopamine and catechol. Sens. Actuat. B Chem. **255**, 1655 (2018)
79. Aleandar, S., Baraneedharan, P., Shriya, B., Ramaprabhu, S.: Highly sensitive and selective non enzymatic electrochemical glucose sensors based on graphene oxide-molecular imprinted polymer. Mater. Sci. Eng. C **78**, 124 (2017)
80. Wei, F., Xu, G., Wu, Y., Wang, X., Yang, J., Liu, L., Zhou, P., Hu, Q.: Molecularly imprinted polymers on dual-color quantum dots for simultaneous detection of norepinephrine and epinephrine. Sens. Actuat. B Chem. **229**, 38 (2016)
81. Zhao, X., Cui, Y., He, Y., Wang, S., Wang, J.: Sens. Actuat. B Chem. **304**, 1 (2020)

Energy Harvesting Devices

Shehroz Razzaq, Ali Asghar, and Sami Iqbal

Abstract This chapter provides detail introductory information on conducting polymers based energy harvesting devices such as thermoelectric, piezoelectric, and solar cells. Energy harvesting is the capture and conversion of small amounts of readily available energy in the environment into usable electrical energy. Electric energy can be regulated for direct use or accumulated and stored for future use. This provides an alternative energy source for applications in areas where there is no grid power, where wind turbines or solar panels are inefficient. Apart from free solar energy, no small energy source can provide a large amount of energy. However, the energy captured is sufficient for most wireless applications, remote sensing, human implants, radio frequency identification, and other applications with a lower power spectrum. Even if the energy obtained is too low to power the device, it can still extend battery life. Energy harvesting is also known as energy scavenging or micro energy harvesting. Most low-power electronic products, such as remote sensors and embedded devices, are powered by batteries. However, even long-life batteries have a limited life and must be replaced every few years. When there are hundreds of sensors in remote areas, replacement costs can be high. On the other hand, energy harvesting technology provides unlimited service life for low-power devices and eliminates the need to replace batteries at a high cost, impractical or dangerous conditions. Most energy harvesting applications are designed to be self-sustaining, cost-effective, and require little or no service for many years.

Keywords Energy harvesting devices · Thermoelectric · Piezoelectric · Solar cells · Conductive polymers · Thermoelectric generators (TEG)

Thermoelectric Energy Harvesting Devices

S. Razzaq (✉) · A. Asghar · S. Iqbal
Joint International Research Laboratory of Information Display and Visualization, School of Electronic Science and Engineering, Nanjing 210096, P.R. China
e-mail: Shahroz.razzaq506@yahoo.com

© Springer Nature Switzerland AG 2021
S. Shahabuddin et al. (eds.), *Advances in Hybrid Conducting Polymer Technology*, Engineering Materials, https://doi.org/10.1007/978-3-030-62090-5_5

1 Introduction

The emergence of energy conversion technologies are contributing to global warming and large amounts of carbon dioxide into the environment due to heat emissions and pollution. As a result, thermoelectric power generation converts thermal gradients into useful electrical energy, which can better serve the field to provide clean and green options. The device used for thermoelectric power generation is called a thermoelectric generator (TEG). The primary prerequisite is to provide a large number of p-type and n-type thermal elements. These devices alternately connect in the family and are thermally connected in parallel mode, providing sufficient power for small electronic devices and low-power electrical devices. TEG has many advantages, such as (i) noise-free operation, no mechanical components, (ii) extended service life, (iii) environmentally friendly without any harmful by-products (iv) easy to transport and commission in remote areas. However, due to the inefficiency of conversion (usually less than 6%) and high cost, these applications still do not find enough commercialisation potential. Efforts are being focused on finding alternative materials for TEG applications. Most materials are made up of inorganic materials involving multiple expensive manufacturing processes, expensive raw materials (using less abundant metals such as Te and Ge), and are toxic and fragile. In the past ten years, the demand for new energy materials has been expanded with the rapid development of economy and environmental protection. Thermoelectric technology is thought to produce electrical energy from temperature differences and is useful for energy harvesting. Organic thermoelectric materials cannot be used at high temperatures. Most of the waste heat comes from temperatures below 150 °C. As a result, organic materials can be used to recover electricity from waste heat below 150 °C.

Thermoelectric materials play a decisive role in renewable energy technologies and can solve global energy shortage problems. They have proved to be the right solution for high-end technological employment such as rockets and space applications. Thermoelectric properties of the device depend mainly on the type of material used and their characteristics, such as the Seebeck coefficient, conductivity, thermal conductivity, and thermal stability. Compared to other organic materials, traditional inorganic materials are essential because of improved thermoelectric responses. There are emerging materials, such as carbon nanomaterials, electronic conductive polymers. The strategy of improving the thermoelectric properties of materials and insight into semiconductor physics are highlighted. Methods such as nanostructures, nanocomposites, and stimulants improve the thermoelectric response by simultaneously adjusting the different properties of the material. Recent trends in thermoelectric research suggest that high-performance thermoelectric materials could be adapted for energy production and energy sustainability shortly.

2 Thermoelectric Background

Table 1 presents the thermoelectric progress timeline.

Table 1 Historical background of thermoelectric

Dates	History
1821–1909	Seebeck effect was first observed in 1821. In 1834, the Peltier effect was discovered, and in 1851, Gustav Magnus defined the potential difference between two thermocouple junctions was proportional to the temperature difference of the application. In 1856, the Thomson effect was demonstrated experimentally [1, 2].
1950–1959	Ioffe stressed that the thermal conductivity of semiconductors is a function of atomic weight. In general, high atomic weights have low thermal conductivity (although low thermal conductivity is also reported in low-density materials). Horowitz presented a pattern that showed low-temperature difference and lattice balance. Goldsmid studied changes in conductivity, crystalline structure and electron mobility; Goldsmid suggests that the ratio of electron mobility to thermal conductivity is a function of atomic weight. Ioffe hypothesized that alloying semiconductors with homogeneous materials could reduce thermal conductivity without affecting its conductivity. As a result, modified semiconductor alloys are used in thermoelectric applications. In this regard, Ioffe recommends the figure of merit (ZT) as a measure to quantify the thermoelectric properties of materials [1, 3, 4].
1960–1969	Chasmar and Stratton introduce hardware parameters (adjust parameters) based on carrier mobility, effective mass, doping, temperature, and thermal conductivity [3]. Using the Ioffe model, it can demonstrate the effect of band slots on ZT. Large band gaps result in high thermal conductivity and low wearer mobility. Littman and Davidson argue that there is no ZT on the upper limit of irreversible thermodynamics.
1970–2015	Dresselhaus and Hicks point out that thermoelectric materials such as super crystals, highly anisotropy superpositions, should increase ZT. However, the presence of different directions/interfaces reduces thermal conductivity due to acoustic diffusion. Sofa and Mahan further revised their arguments so that the alternating super-rat barrier layers had limited thermal conductivity and tunnel probability in quantum wells. They also suggest that the quantum mixing in the well changes the density of states (DOS) from 2D to 3D, and predicted an increase in ZT, but below that predicted by Dresselhaus Hicks [5–7]. Bulusu and Walker propose an unbalanced green functional method for determining the thermoelectric properties of silicon nanofilms, nanowires, and super Si/Ge/Si crystals. The dispersion of each band is used to simulate quantum effects. The effects of limitation and dispersion on the electrical conductivity and Seebeck coefficients in superlattices were studied [4].

3 Thermoelectric Performance of the Conducting Polymers

3.1 Electrical Parameters and Conductivity

In general, polymers can be doped by electrochemical or chemical methods. The use of stimulants by Bredas et al. [8] has made it clear that "stimulant" is the wrong term for polymers. In fact, in the case of conducting polymers (CPs), "doping" is an oxidation reaction in which non-conductive polymers are converted into ion complexes consisting of polymeric cations/anions and anti-ions, i.e., the reduction/oxidation form of oxidation/reducing agents. This reaction is highly popular with π-bonding unsaturated polymers, which have π-electrons that are easily de-localized without interrupting the Sigma bond, responsible for keeping the polymer intact. The detailed mechanisms of doping, i.e., the generation of charge carriers (including electro-chemical/chemicals), include (i) monomer oxidation/reduction, electron elimination/acceptance of the backbone π system, resulting in free radicals and less rotation/positive charge. The reduction/oxidation of the doping is converted into negative secretions, neutralizing the positive and negative charges introduced in π-electronic systems; (ii) the local resonance of the charge (i.e., the lattice deformation) and the base (with ½ spin) result in the connection of the charging bit to the base, which is called the pole; (iii) the formation of this pole, which can be the base or ions of the underlying or base ions. In the gap, a new local electron state is created, a single unpaired electron occupies the low energy state, and (iv) removal/addition of electrons from the pole produces a new spin defect, called bipolarity, which is a pair of basic ions connected to polar distortion, which, depending on the chemical structure of the polymer, can affect three to four monomers.

3.2 Seebeck Coefficient

The Seebeck coefficient is the ratio of open-circuit voltage obtained between the hot and cold ends of the material to the temperature difference between them. Seebeck coefficient can arise from three different contributions, namely electronic, phonon, and electron-phonon. In metallic systems with many free electrons, the electronic contribution to the Seebeck coefficient dominates. Thus, the migration of thermally excited electrons of the hot-end towards the cold-end creates the voltage difference, thus originating from the electronic contribution. In this case, the Seebeck coefficient is linearly proportional to the temperature. Phonon contribution to the Seebeck coefficient known as "Phonon drag" is usually observed at low temperature. Phonon contribution to the Seebeck coefficient known as "Phonon drag" is usually observed at low temperature (<200 K) where phonon mean free path is large. When mean free paths for both the electron and phonon become comparable, electron-phonon scattering becomes significant. Highly conducting polymers with good crystalline structure at low temperature may exhibit significant electron-phonon scattering. The heavily doped polymers such as polyacetylene, polyaniline, and polypyrrole exhibit a small positive Seebeck coefficient (<14 μV/K near room temperature) with a linear decrease value with temperature. However, deviation from linearity can sometimes be caused due to electron-phonon interactions. In lightly doped CPs, the Seebeck coefficient has a larger magnitude than heavily doped ones. For such CPs, the Seebeck coefficient either decreases or increases non-linearly with temperature. Sometimes the non-linear decrease of the Seebeck coefficient follows $T^{1/2}$ dependence and can be explained using Mott's variable range hopping transport between localized states. The nearest neighbour hopping within the localized states gives rise to 1/T dependence of the Seebeck coefficient. Whereas, in the case of heavily doped CPs, the Seebeck coefficient shows linear dependence with temperature.

3.3 Thermal Conductivity

Unlike inorganic materials, where the electronic part of thermal conductivity is coupled with the electrical conductivity, this synergy in CPs is often invalid (contrary to Wiedemann-law) because of stronger charge-lattice coupling. Besides, due to low electrical conductivity (σ) compared to inorganic solids, the electronic contribution to thermal conductivity (k) of material is minimal. Therefore, in CPs majority of the heat is transported by the phonons (i.e., quantized lattice vibrations) rather than by charge carriers (ke), i.e., kl > ke. The k in CPs depends on their molecular weight/shape and the difference in their chain structure. Figure 1a shows the four main structures of

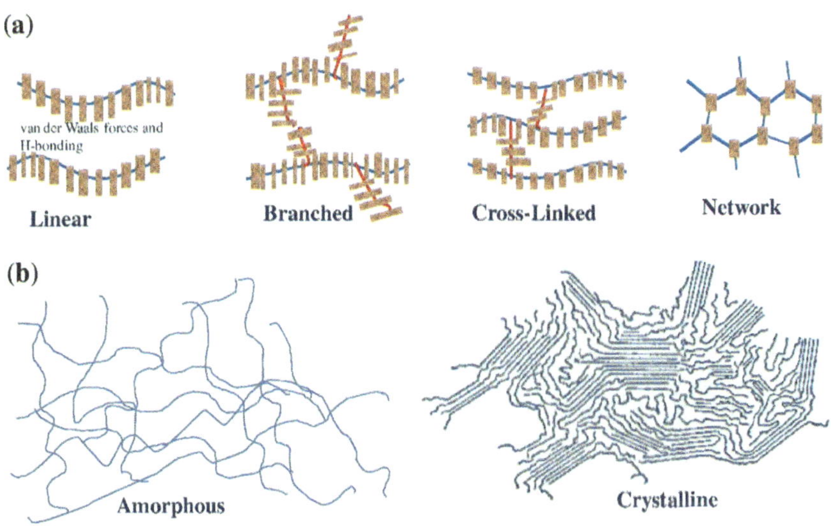

Fig. 1 Shows **a** polymers molecular structures; **b** polymer chains showing amorphous and crystalline structures [9]

polymer molecular structure: (i) linear polymers, (ii) branching polymers, (iii) cross polymers; and (iv) network polymers. The polymer chain structure is so different that amorphous and semicrystalline polymer structures are produced in different ways (Fig. 1b). These different structures lead to heat transfer and the spread of anisotropy. Such anisotropy is more when the chains are ordered for a particular CP.

4 Classification of Thermoelectric Materials

Thermoelectric materials are mainly divided into intermetallic metals, stone tools, Clarets, Half-Heulser (HH), oxides, and of rare earth joules, Zintl phase materials, silicon, nickel, and its super-iron structure. All of these materials are degraded semiconductors, each with a complex ribbon structure. Carbon nanomaterials, such as carbon nanotubes (CNTs) and graphene, as well as electronic conductive polymers such as polyaniline (PANI) and poly 3,4-ethylenedioxythiophene (PEDOT), have also been shown to be useful in thermoelectric devices. In recent years, the nano synthesis of various carbon nanomaterials and electronic conductive polymers has become a potential thermoelectric system for developing the next generation of flexible thermoelectric devices.

5 Conducting Polymer Thermo-Electric Material

In 1970, Alan Heeger, Alan MacDiarmid, and Hideki Shirakawa were the first to discover conductive polymers. They won the Nobel Prize in 2000 for their discovery. Later, several conductive polymers, such as polyaniline (PANI), polypyrrole (PPY), poly (3.4-ethyl dioxythiophene) doped with poly(styrenesulfonate) (PEDOT: PSS), polycarbazoles (PC) and its corresponding derivatives were found. As compared to conventional polymers, synthetic polymers have low thermal conductivity when synthesized with inorganic materials, which is conducive to high ZT. Also, conductive polymers are low-density, low-cost, easy-to-synthesize, and versatile processing. Hence, the recent study of thermoelectric (TE) applications has attracted considerable attention.

5.1 P-type Conducting Polymers PEDOT: PSS

PEDOT is a successful conductive polymer due to its high conductivity, low density, excellent environmental stability, and ease of synthesis when suitable for stimulants. One of the main problems that are limiting the PEDOT applications is solubility in water and common solvents. This can be resolved by using PSS emulsification. Then forming aqueous solutions such as PEDOT: PSS water solution (PH1000), which has been mass-produced commercially. There are different ways to improve the performance of conductive polymers such as doping and de-doping, and crystallization and alignment.

5.1.1 Doping and De-doping

Generally, doping and de-doping have a substantial effect on the wearer's mobility, wear-resistant density, and oxidation levels, which affect the conductivity and conductivity polymers of the Seebeck coefficient. A range of chemicals can be used as stimulants such as dimethyl sulfoxide (DMSO), tetrahydrofuran (THF), and potassium hydroxide (KOH). The conductivity of conductive polymers can sometimes be increased by several orders of magnitude by proper doping, mainly because doping helps to adjust the conductive polymer chain and improve carrier transport.

5.1.2 Crystallinity and Alignment

Carriers can move both along the conducting polymer chain and interchain; however, the carriers' mobility along the chain is higher than for hopping events. Therefore, the electrical conductivity and TE properties of conducting polymers can also be improved by enhancing the crystallinity and chain alignment. For example, single-crystal PEDOT nanowires with high crystallinity and high electrical conductivity ($\sim$8000 S/cm) were fabricated via direct printing combined with vapor phase polymerization process by Cho et al. [10]. This high electrical conductivity is mainly because of suitable crystalline structures, which enhances the charge-carrier mobility in PEDOT nanowires

6 N-type Conducting Polymer Thermoelectric Materials

So far, most of the reported conductive polymers are p-type materials, and the corresponding ZT values have been significantly increased (RT15) to 0.42. For TE devices, p and n conductive polymers are required. However, most n-type conductive polymers are unstable in the air, limiting the application of conductive polymers in TEG. The causes of instability of n-type doping agents and conductive polymers in the air are mainly due to the reduction of polymer chains and O_2 oxide ions (e.g., alkaline metal ions). As a result, most studies have focused on p-type conductive polymers.

It is encouraging to focus on n-type conductive polymers, and several n-type conductive polymers have been reported. For example, in 2012, Sun et al. [11] synthesized n-type poly [Nax (Ni-ett)], poly [Kx (Ni-ett)], and p-type poly [Cux (Cu-ett)], (1,1,2,2-ethenetetrathiolate (ett)) materials. The electrical conductivity, Seebeck coefficient and ZT values at 440K are $\sim$60 S/cm, -151.7 μV/K, and 0.2 for the poly[Kx(Ni-ett)], respectively. Although poly[Kx (Ni-ett)] show high TE

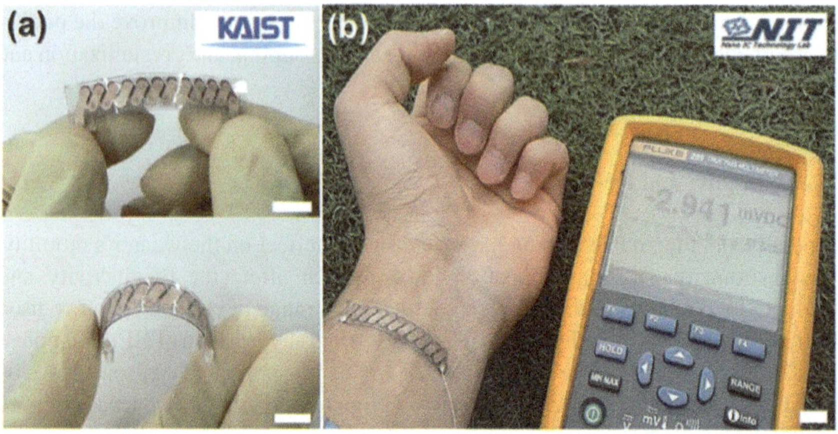

Fig. 2 Demonstration of band-type flexible TE generator for harvesting thermal energy from human skin: (**a**) photos of band-type flexible TE generator and (**b**) electricity generation measured on human skin at an air temperature of 15 °C. Scale bar, 1 cm [12]

performance, it is insoluble and limits its application. Subsequently, many solutions are treated as n-type conductive polymers. Kim et al. [12] prepared a flexible TEG by successively screen printing Bi_2Te_3 and Sb_2Te_3 pastes on the glass fabric before annealed under 530 °C or 500 °C in N2 atmosphere, respectively (Fig. 2). An open-circuit voltage, output power per unit area, and output power per unit weight of 90 mV, 3.8 mW/cm^2, and 28 mW/g, at $\Delta T = 50K$ was achieved for the generator with 8 thermocouples, respectively. The as-prepared band-type flexible TEG can generate an open-circuit voltage and output power of 2.9 mV and 3W, respectively, at an environmental temperature of 15 °C when worn on human skin.

7 Flexible Thermoelectric Power Generators—TEGs

In 1999, Kishi et al. [13] developed a TEG wristwatch. The TEG consists of 104 Bi-Te compound plates with elements size is 80μm × 80μm × 600μm. Maximum voltage when wearing the watch for the first time. Then the tension drops until the wrist and air balance for 30 min. The results show that the residual heat (voltage is 300mv) can be used to power portable electronic devices using TEG. Therefore, more and more research is focused on portable, flexible TEG. Many methods have used to manufacture portable, flexible TEG, such as the integration of commercial thermoelectric reactors into textiles, using only p or n-type materials, p and p-type materials, and fabrics that produce TE energy. In Fig. 3, a flexible TEG was fabricated by deposited Bi_2Te_3 and Sb_2Te_3 thin films (thickness: 500 nm) on Kapton HN polyimide foil with a total size of 70 mm × 30 mm by RF magnetron co-sputtering technique [14]. The device's internal resistance with 100 thermocouples was 380 kΩ,

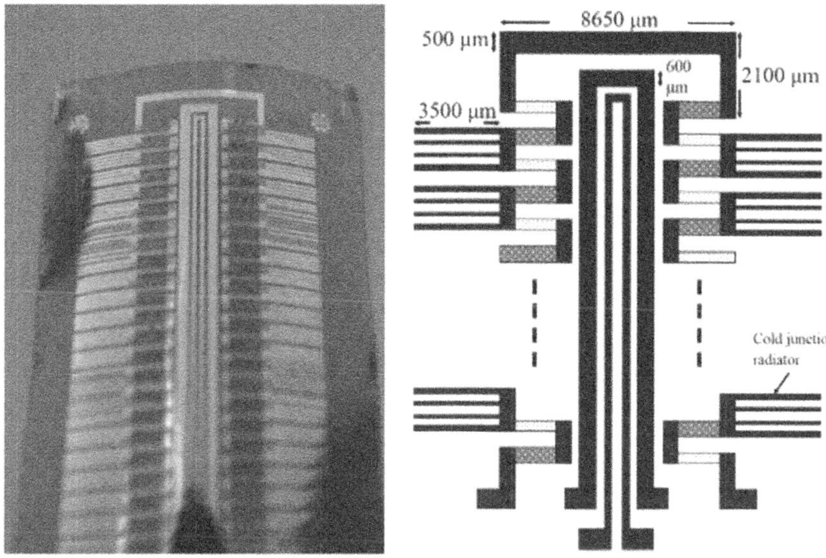

Fig. 3 **a** Photograph of fabricated flexible μTEG on Kapton HN. **b** Schematic of a flexible thermoelectric generator [14]

which is much higher than that of the calculated internal resistance (43.5 kΩ), mainly because of the contact resistance and the overlapping of gold contacts active regions. An open-circuit voltage and maximum output power of 430mV and 32nW, respectively, was achieved for the as-prepared TEG at a temperature difference of 40K. This section provides some key findings to guide and focus on the flexible application of TEG. It also summarizes the fabrication methods, materials, TE properties, size of the flexible TEGs covered in this review for reference purposes. Over the past decade, the ZT values and TE characteristics (output voltage, output power, output power density, flexibility, etc.) of portable TEG conductive polymers, inorganic/conductor polymer nano complexes, etc. have improved significantly. However, flexible TEG depends on many challenges in the future before it can be widely used. The TE properties of conductive polymers depend on their chemical and microstructure. The use of stimulants and de-benzene, post-treatment, crystallization and alignment are effective ways to increase their ZT value. However, technical conditions and processes still need to be optimized. To improve the ZT value of nano-synthetic inorganic/conductive polymers, it is necessary to match the fermi level of inorganic nanostructure and a conductive polymer. Besides, the improved energy filtering function on the interface should be improved. So far, due to the low stability of n-type conductive polymers in the air, most conductive polymers and their corresponding TE composites are P-type materials, seriously affecting the development of portable TEG. Although poly[KX(Ni-ett)] is the best yield for all n-type conductive polymers, its insoluble properties greatly limit their application. Therefore, there is an urgent need to develop an n-type stable conductor polymer and its corresponding TE composite materials. For portable applications, the temperature of the generator is lower than the ambient temperature due to the thermal resistance of human skin and air. This significantly reduces the output voltage, power, and conversion efficiency of portable TEGs.Therefore, increasing the TEG thermal resistance and reducing the contact thermal resistance is important for maintaining the high-temperature gradient in the TEG. Besides, many flexible TEGs are not truly portable due to their reduced wear resistance, waterproof, air and humidity. Maintaining the permeability of the fabric and giving the fabric TE energy generation function is the future research direction of flexible TEG. Due to the ZT value of p-type legs and n-type legs used for wearable TEGs are typically not equal. Therefore the geometric cross-sectional areas of the p-type and n-type legs should be optimized. Also, the skin is not smooth, which is the challenge of improving TEG conversion efficiency. Therefore, more work needs to be done to optimize portable TEG structures such as device geometry, size, structure, layout, etc. Inorganic materials are required for applications of high-temperature systems. One of the main problems with commercial equipment manufactured from bulk inorganic materials is low mechanical flexibility. So far, inorganic materials have been made in the form of films, such as inorganic films deposited on flexible organic substrates, CNT-based films, covering films, and other inorganic film materials, based on 2D material film thermoelectricity, is a popular choice for TEG Flexible. However, at this stage, this study is still in its infancy. Finally, although different systems are proposed to describe the TE characteristics of commercial bulk TE, no established criteria for measuring flexible TE characteristics have been developed.

8 Piezoelectric Energy Harvesting Devices

The increasing demand for fully self-powered electronics have led to an increase in the study of power acquisition equipment over the past decade. With advances in wireless and low-power electronics, sensors are being developed and can be placed almost anywhere. However, because these sensors are wireless, they need their power supply, which in most cases, is a traditional electrochemical battery. Once these limited power supplies are powered off, the sensor must be acquired, and the battery replaced. The task of replacing batteries can be tedious and can become very expensive when the sensor is placed in a remote location. These problems can be potentially mitigated by using power acquisition equipment. The goal of the power collector is to capture the energy that is usually lost around the system and convert it into available energy, powered by gas equipment consumption. By leveraging these untapped energy sources, electronics that do not rely on limited power sources, such as batteries can be developed. One energy source that is usually lost is the vibration of the environment around most machines and biological systems. This energy source is ideal for using piezoelectric materials, which convert mechanical strain energy into electrical energy and vice versa. The historical background is shown in Table 2.

Table 2 Historical background of piezoelectric

Dates	History
1800	In the mid-eighteenth century Carl Linaeus and Franz Aepinus first time observed the piezoelectric effect. The piezoelectric effect is a kind of molecular phenomenon, that can be observed at the macro level as a change in electric potential created when a piezoelectric substance is deformed.
1900	In the early 1900s, the converse of this was used in the first submarine sonars. Scientists created piezoelectric transducers from quartz crystals sandwiched between steel plates. A current passed through one transducer emitted a high-frequency ping that would bounce off objects and then be picked up by another transducer sensitive enough to convert the echo precisely. Later it was found that alternatively, the crystals would vibrate when a current was passed through them. Depending on the waveform of the voltage, the sound could be generated.
1984	One early study into power harvesting by Hausler and Stein investigated the ability to generate energy from the expansion and contraction of the rib cage during breathing. A prototype of the power harvesting system was constructed using polyvinylidene fluoride (PVDF) film and was implemented in vivo on a mongrel dog. The prototype was demonstrated to produce a peak voltage of 18 V, which corresponded to a power of about 17 mW.
1996	Umeda et al. [15] quantified the amount of energy that could be produced when a steel ball impacted a piezoelectric plate. The authors used an equivalent circuit model to predict energy while modifying numerous system parameters to find the best combination. It was determined that a significant amount of energy was returned to the steel ball in the form of kinetic energy as it bounced off the plate, making the system ineffective.
2001	Another investigation into the ability to use piezoelectric materials for power harvesting from the motion of humans and animals, was performed by Ramsey and Clark [16]. They studied the ability to power an in vivo micro electro mechanical system (MEMS) application. The research used a thin square plate driven by blood pressure to provide power and was shown to be capable of powering the electronics if they were used intermittently. Another form of excitation commonly used is the ambient vibration of mechanical structures.
2004	Sodano et al. [17] formulated a model of a power harvesting system that consisted of a cantilever beam with piezoelectric patches attached. The model was developed such that any combination of boundary conditions and location of piezoelectric material could be accommodated, but was verified on a cantilever beam experiencing a base excitation from the clamped condition. The model was found to accurately estimate the energy generated and was also used to demonstrate the damping effect of power harvesting. With the research into power harvesting devices growing, it was determined that the amount of energy generated by piezoelectric materials was not sufficient to power most electronic devices. Thus, for power harvesting technology to make its way into the commercial market, methods of accumulating and storing the harvested energy until a sufficient amount can be recovered to power the portable electronics, are the key to a successful power harvesting system [18].

9 A New Hybrid Piezoelectric Electromagnetic Micro-vibration Energy Harvester

As an alternative to batteries in ultra-low-power microelectronics, independent piezo-electric vibration energy collectors, or hybrid batteries that couple the two conversion mechanisms together, have attracted widespread attention. All previously reported mixed energy harvesters have shown significant improvements in energy conversion efficiency. However with this, the problem of working frequency and volume also arises. The new hybrid vibration energy collector has a low operating frequency, small volume, and large output energy.

10 The Efficiency of Piezoelectric Device

The effectiveness of Macro Fiber Composite (MFC), Quick Pack (QP), and Lead Zirconate Titanate (PZT) as power capture equipment needs to be compared first. This is done by determining the efficiency of each device used in the experiment. Using data from laser vibrators, force sensors, and piezoelectric voltage outputs, the following data are numerically calculated to determine the average efficiency.

$$\eta = \frac{P_{out}}{P_{in}} \times 100\% = \sum_{n=2}^{m} \frac{\frac{(V_n - V_{n-1})/R}{((F_n - F_{n-1})(d_n - d_{n-1}))(t_n - t_{n-1})}}{m} \times 100\% \tag{1}$$

where is the efficiency, V is the voltage drop across the load resistance R, F is the force applied to the board base, d is the displacement of the board, t is the time increment between the data points, n is the data point index, m is the total number of measured data points. The efficiency of each press unit is calculated at the first resonant frequency (PZT is 50 Hz, MFC is 108 Hz, QP is 32 Hz), its tweeting frequency is 0 to 500 Hz, and the random signal is 0 to 500 Hz. As mentioned earlier, the automotive compressor randomly vibrates. Therefore, the efficiency corresponding to random vibration is most likely to represent that the piezoelectric unit is affected by environmental vibration. The resulting efficiency is shown in Table 3.

It is noted that efficiency does not represent the efficiency of the actuator itself, as the experimental configuration and other factors may vary. However, these efficiencies do provide a comparison between the three actuators tested. For each signal, three measurements were made to show consistency. When excited with all three signals, the PZT board is relatively consistent and higher than the other two devices. The resonant efficiency of the PZT is slightly lower because the resonant frequency used is the frequency with the largest voltage output, rather than the frequency with the optimal pressure and voltage output characteristics. Also, QP can be seen performing poorly in resonance, which is thought to be due to the device's construction with four separate piezoelectric patches, which results in an area without piezoelectric material

Table 3 The efficiency of the PZT, MFC, and QP with three different inputs

Signal	PZT η (%)	MFC η (%)	QP η (%)
Resonant	4.54	1.7871	0.4662
	4.51	1.7211	0.6094
	4.2312	1.7377	0.946
Chirp 0–500 Hz	3.102	0.2927	1.6505
	3.0725	0.3033	1.2611
	3.0293	0.3368	1.492
Random 0–500 Hz	6.57	1.2103	3.097
	6.954	1.3013	2.9664
	6.8562	1.4663	3.1551

in the middle of the beam. This leads to a decrease in the area's stiffness, causing most stains to be concentrated in the area. However, since the excitation method used does not transfer the energy of all applications to the piezoelectric unit, all efficiency is relatively low. From Table 3, it can be seen that the MFC is not performing well with all the incentive signals used. Through these tests, it is found that MFC is poorly performed as an energy collection medium. The electrical output of the MFC contains a huge voltage component, but the current is very low. Even with a large voltage and a low current, the power can still be considered the same, but for MFC, the power generated is less than 10 times, which shows the current output of each device at the first natural frequency. It is believed that the increase reduces the performance of the MFC in impedance caused by the use of digital inter-electrodes. Another way to consider this scenario is to consider each part of the piezoelectric fiber between the digital electrodes.

11 Classification of Piezoelectric Materials

Piezoelectric ceramics materials are classified into piezoelectric crystals and piezoceramic tube transducer. Piezoelectric crystals generally refer to single piezoelectric crystals, which refer to crystals grown in the long-range order according to the crystal's lattice. This crystal has no symmetry center. Such as crystal, lithium gallate, lithium niobate, titanium niobate, and lithium transistor so on. Piezoelectric ceramics generally refer to piezoelectric polycrystals. Piezoelectric ceramics are polycrystals obtained by mixing, forming, high-temperature sintering with a raw material of necessary component, and irregularly collecting fine crystal grains obtained by a solid phase reaction and sintering process between the powder particles. Piezoelectric ceramics are called piezoelectric and ferroelectric ceramics.

Piezoelectric ceramic are information-functional ceramic materials that can convert mechanical energy and electrical energy into each other. For the piezoelectric effect, in addition to piezoelectricity, piezoelectric ceramics have piezoelectric properties. It also has dielectric properties and elasticity. It has been widely used in medical imaging, acoustic sensors, an acoustic transducer, ultrasonic motors, and the like. Piezoelectric ceramics are made using the materials under mechanical stress, causing the relative displacement of the center of the positive and negative charges to be polarized, resulting in the opposite side of the material with the opposite sign of the trapped charge. The piezoelectric effect has sensitive characteristics. In general, piezoelectric ceramics are ceramic materials that generate voltage by external stimulation. Both piezoelectric ceramic and electrostrictive ceramics are dielectric. The dielectric has two effects under an electric field, namely the inverse piezoelectric effect and electrostrictive effect. The inverse piezoelectric effect refers to the dielectric.

11.1 Materials and Fabrication

For the implementation of piezoelectric gloves, five piezoelectric sensors are used as the base material, and each is integrated on the finger. The piezoelectric sensor has a diameter of 2 cm and has the characteristics of converting machinery into electrical energy, which is determined by the piezoelectric effect. It is also obvious that piezoelectric sensors are considered AC voltage (AC) sources. Some initial measurements of piezoelectric sensors (sensors for each finger) are made using a digital meter. At first, it was noted that each sensor produced about 1.2 V. When the force exerted by the finger is maximized, 1.8 V is produced. In terms of sensor connectivity, choose parallel connections as the best solution. Besides, the bridge rectifier DB105 is used to convert AC voltage to a continuous voltage source (AC to DC) and capacitor 220μF 6.3 V for the temporary storage of the energy generated by piezoelectric sensors. Nickel metal hydride (NiMh) batteries 1.2 V is also used to store the energy produced. This battery is widely used in portable medical applications such as:

- Infusion pumps
- Portable X-ray systems
- Respirators
- Defibulators
- Surgical tools
- Patient monitoring devices
- Motorized beds and carts
- Muscle stimulators.

12 Application of Piezoelectric Transducers

With the social and economic development and the continued depletion of fossil energy, declining environmental quality, combined with our sustainable development strategy, how to efficiently collect and facilitate clean energy has been the focus of energy research. As an important national infrastructure, highways bear important functions such as human and cargo transportation. Millions of vehicles vibrate on the road, which is significant, and the vibration of vehicles causes damage to the road and leads to a waste of energy. How to collect vibrational energy and convert it into energy that can be used effectively is a challenge that needs to be solved. At present, based on the conversion mechanism, the energy collection of the environment can be divided into three types:

- electromagnetic
- piezoelectric
- electrostatic.

Piezoelectric energy harvesting technology uses the characteristics of electromechanical coupling of piezoelectric materials to convert mechanical energy into electrical energy directly. Compared with electromagnetic and electrostatic power generation, piezoelectric power generation has the original small size and does not need an external power drive. Based on these advantages, piezoelectric technology is developing rapidly, but most piezoelectric parts are used to collect fixed mechanical equipment vibration energy, tidal energy and wind power, applications. Piezoelectric technology is still in the research stage on the road. At home and abroad, the piezoelectric road has achieved certain results in theoretical research.

13 Developments, Challenges, and Future of Piezoelectric Devices

The days when piezoelectric energy harvesting was considered unreliable and low-power output were long gone. With the development of ultra-low-power electronic devices, energy harvesters have also significantly improved. In just a few years, piezoelectric harvesters have moved from harvest capacity to equipment shown at the 2008 harvest. Microstrain, a company recently acquired by the Lord Corporation (Williston VT, USA), further integrates piezoelectric energy harvesters in applications as niche as the condition monitoring of rotor blades in helicopters, powering the harvester through the rich vibration environment in which it operates. Piezoelectric technology is still in the research stage. At home and abroad, the piezoelectric road has achieved specific results in theoretical research.

14 Piezoelectric Markets

15 Piezoelectric Materials

More than 200 piezoelectric materials are available for energy harvesting, and each application is selected for the appropriate application. Although iron titanate is the first piezoelectric ceramic to be discovered, ceramic lead silicate, also known as PZT, remains the most commonly used material for piezoelectric harvesting. Alternatives are receiving some attention, and often the efficiency and temperature performance of the material itself are not primary considerations, but rather factors such as flexibility, lightweight, and even toxicity. For example, sodium nitrate exhibits properties very similar to PZT, without lead, or has polymer piezoelectric materials such as polyvinylidene difluoride (PVDF), where piezoelectric is due to the application of electric fields when interwoven long-chain molecules attract and repel each other.

16 Innovative Architectures of Piezoelectric Devices

One of the challenges of Piezoelectric is that, although they are highly efficient at optimal resonance, they only vary slightly with the optimum resonance frequency, resulting in a significant reduction in energy production - the bell curve is very steep. Some people try to compensate by using magnets to limit vibrations in the device's main resonance, which increases volume and cost and fails to enable true broadband capacity. Another solution provided is to pre-position piezoelectric materials electronically. The charge of the material acts as a damper, requiring the material to do more work on it. As a result, the power gain is up to 20 times compared to the optimal resistive load. Innovative architectures of piezoelectric harvesters have also enabled higher power outputs, as demonstrated by the NIA (National Institute of Aerospace) in the USA. Focusing on basic scientific principles that demonstrated how the "33" (longitudinal) excitation mode on piezoelectric harvesters is characterized by 3 times higher energy conversion efficiency than the "31" (transverse) excitation mode, the NIA researchers described the design and construction of a hybrid piezoelectric energy harvesting transducer that can harvest 4 times more energy than a traditional "31" harvester.

17 Polymers Applications in Solar Cells

17.1 Organic Solar Cells

One possible alternative to the crystalline-silicon solar cells is cells made from thin-films of conjugated and photoactive polymers, which can be molded onto flexible substrates over a large surface area by using wet processing techniques. The strong absorption of photoactive polymers and their low-cost deposition on the flexible substrates make them favorable semiconductors for solar cells. The cells made with a polymer and double electrodes (Fig. 4a) are inefficient because the built-in electric field cannot separate the photo-generated excitons, which appears from the differences in the electrode work-function. The splitting of excitons at the interface of two semiconductors with offset energy levels can increase cells' efficiency (Fig. 4b). In different conjugated polymers, the diffusion length of exciton has been measured from 4–20nm. The conjugated polymers suffer from the short diffusion length of excitons. To address this problem, two conjugated polymers with offset energy levels are mixed so that all excitons formed near an interface, as shown in Fig. 4c [19]. This cell structure, known as bulk heterojunction (BHJ), provided a route to all photo-generated excitons in the thin-film that could be split into free carriers. The bi-continuous blend design is beneficial to transport the free carriers towards the electrode. The top active-layer consists of two inter-penetrating sub-networks with the donor, respectively acceptor layer, and is sandwiched between two charge carrier contacts [20]. The BHJ devices are based on two inter-penetrating sub-networks, the electron acceptor and conjugated polymer, randomly distributed

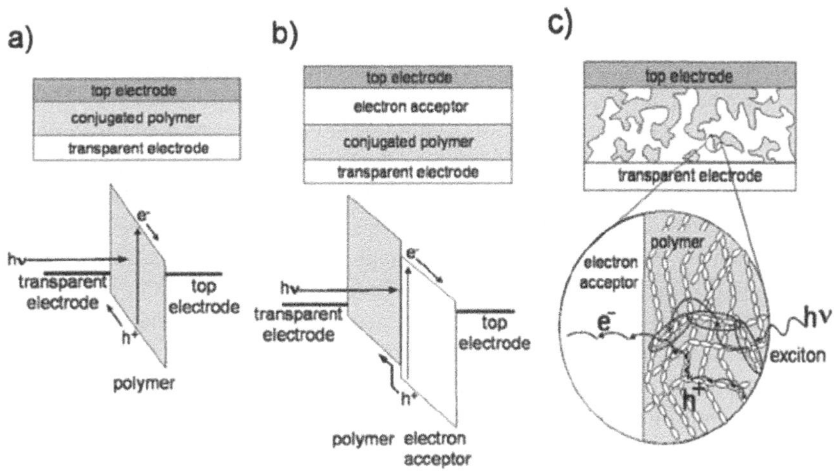

Fig. 4 Conjugated polymer-based thin film device architecture [20]

throughout the thin-film. In polymer cadmium-selenide and polymer PCBM (phenyl-C61-butyric acid methyl ester) Nanorod devices, the electron acceptors are randomly distributed, leading to electron trapping on isolated acceptors unless a large mass fraction of acceptors is employed. Besides, the randomly distributed interface between the two semiconductors can lead to incomplete PL quenching in the conjugated polymer in regions of the polymer that are more than an exciton diffusion length away from an acceptor. In the perfect condition, each exciton will be found on the conjugated polymer within the diffusion-length of the electron acceptor, even though quantitatively model has been pointed out that some electromagnetic light-emission will have occurred in the polymer. The primary purpose is to understand the correlation between the installed morphologies and observed photoexcitation properties. The limited performance of organic solar cells is due to poor charge mobility and exciton separation due to high recombination rates of exciton and low charge transportation in polymers. The blended and layered heterojunctions can be used to overcome this challenging task partially. Such morphological structure offers several exciton dissociation sites and distinct charge path-ways; therefore, limited exciton recombination rate allows for thicker and more absorber, polymer thin films. However, the lifetime of an excited state cannot be correlated with the best device performance. The morphology differences, charge path-ways that enable charge transportation efficiently frequently outweigh the influence of the charge transfer. The best improvement of nano-scale morphology still lacks due to the complications of the structure changes from the simple nanostructure, which can be transformed into a photovoltaic device. The examples of conducting polymers are shown in Fig. 5. The poly phenylenevinylene (PPV) belongs to the group of polymer's family. It serves as proto-typical conjugated polymers for the application and the basic understanding of whole electronic procedures in the conjugated polymer. The objective is to attain the electro-luminescence that spans the visible and infrared regions by a suitable chemical structure change. The processing capability of poly-phenylene vinylene can be improved on the polymer's backbone, and the flexible chains have been introduced, resulting in poly-phenylene vinylene derivates (MEH-PPV and others). Recently, poly(3-hexylthiophene) (P3HT) is a donor material, is receiving considerable attention because of the high absorption coefficient close to the maximum photon-flux in the solar spectrum and lower band-gap, as compared to other materials such as Poly-para-(phenylene-vinylene). Poly(3-hexylthiophene) is a commonly well-known material in the field of organic materials (electronic) as a high mobility material. Since the thickness of the active layer of photovoltaic (PV) cells is a trade-off between the absorption of electromagnetic light and extraction of the charge carrier, the charge carrier mobility is considered essential. The thermal treatment strongly affects the efficiency of poly(3-hexylthiophene) based solar cells, usually attributed to the ordering of the poly(3-hexylthiophene) due to its regioregularity structure. Annealing can recrystallize the poly(3-hexylthiophene), enhance the absorption spectrum of solar cells, and charge mobility inside the P3HT. An example of OPV is shown in Fig. 6.

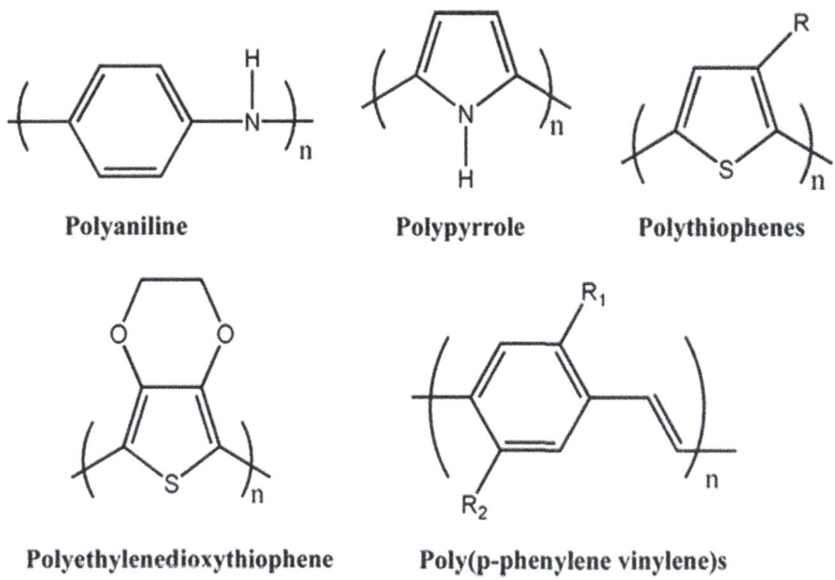

Fig. 5 Examples of conducting polymers

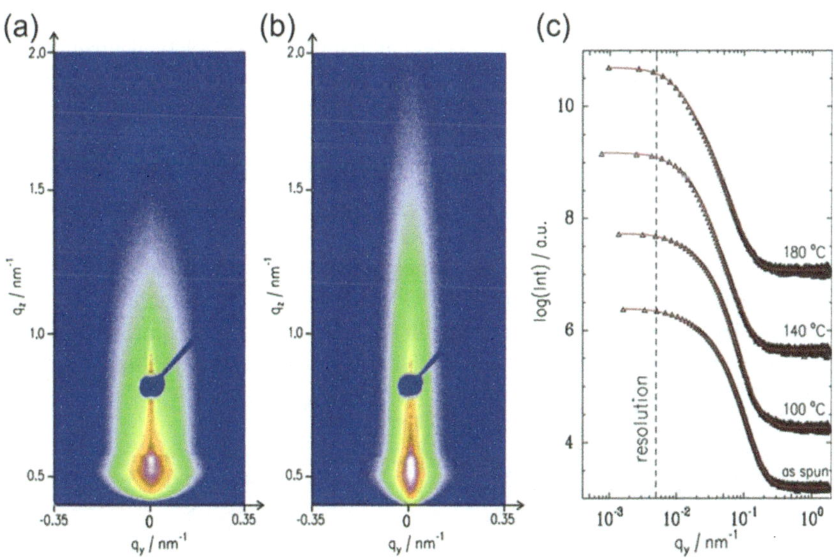

Fig. 6 Example of GISAXS data measured at an OPV film **a** 2D GISAXS patterns of as-spun and annealed P3HT:F8TBT films. The color code represents the scattering intensity. A beam stop shields the specular reflection. **b** Double logarithmic plot of the out-of-plane cuts of the GISAXS data taken at the critical q-value of F8TBT [21]

17.2 Polymer-Based Hybrid Cells

Solar energy is the most abundant, free, inexhaustible and clean energy resource on the planet. The solar radiation can be captured through PV cells and transformed into electrical power. It has been recorded that solar energy that reaches the earth's surface delivers ten-thousands times more energy than what we consume. It has enormous potential to meet future energy consumption requirements. Despite significant developments in the solar industry over the past several decades, the production of highly efficient and cost-effective solar cells remains a challenging task. The conventional silicon-based solar systems are suffering from high fabrication costs and limited adaptation to the novel requirements. Solar cells based on organic materials are recently receiving considerable attention due to low fabrication cost, easy processing, and flexible substrate. Figure 7 the sketch of hybrid solar cells, a combination of DSSC (dye-sensitized solar cells) and polymer solar cells. They consist of p-type and n-type conjugated polymers. The dye-sensitized solar cells consist of nanostructured titanium dioxide (TiO_2) on a compact transparent semiconductor oxide, and in between two electrodes, the multiple component redox electrolytes has been filled. Albeit there are some cost-effective organic photovoltaic (OPV), the existence of fluid electrolytes creates a sealing issue, which also initiates the search for another hole conducting material. In hybrid solar cells, the electrolyte can be replaced by a p-type conjugated polymer, which acts as a hole conducting and electron blocking the device's layer. Combining a conjugated polymer and inorganic offers a practical advantage of organic material such as easy processing, mechanically flexible, synthetically tailor property, and high electrons mobility. The efficiency of these hybrids solar cells mostly shows their dependence on the morphology of the nanostructure in-organic material, titania, as the volume to surface ratio can be determined by morphology, therefore the available surface for interface reactions. Furthermore, the morphology affects the positive and negative charge carrier routes and electron-hole pair recombination probability near the surface. Whence, the tailoring of desire morphology yield an efficient performance of hybrid PV cells is of considerable significance and considers the key step of layer stack building. The primary requirement for the consecutive phases of the preparation of hybrid OPV is the preservation of tailored

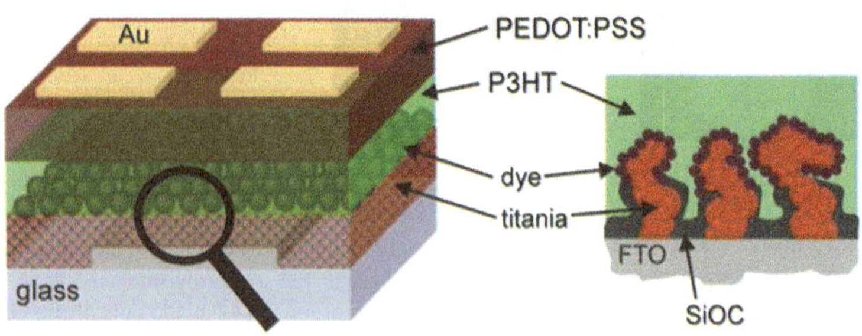

Fig. 7 A sketch of a hybrid cell based on titania and a conducting polymer

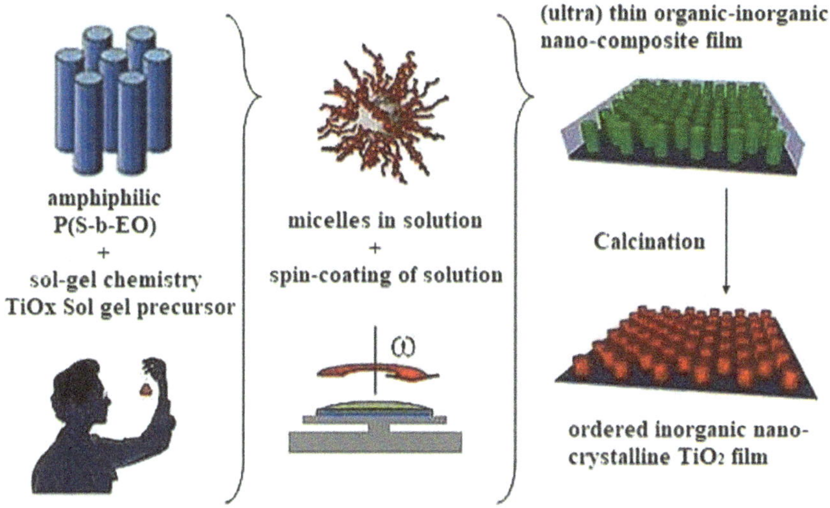

Fig. 8 Schematic of the applied sol–gel route to achieve nanostructured titania films

morphology. To fulfill the requirement arises from this flexible application, it is essential to make the desired morphologies with maximum reproducible and homogenous spread-out across the whole area. Several methods have been reported to synthesize titania at nano-level with distinct morphologies, e.g., nanowires, nanorods, nanoparticles, nanotubes, nano-vesicles, numerous structures of the mesoscale network, and lamella. Various synthesis approaches have been made to make a combination of organic and inorganic material. In contrast, the organic component is responsible for controlling the mesoscale structure such as copolymer block, and titania, which is the inorganic component, offers a particular functionality resulting in the evolution of the functional well-structured hybrid PV systems. The major structure control has been based on the formation of separate micro-stages and the self-organized mechanism of the block copolymers. To attain the titania thin-films with varying morphology, and the thin-film can be reproduced. Implementing an easy formula using an amphiphilic di-block copolymer P(S-b-EO) as an agent and combining with so-called excellent and poor solvent pair induced stage separation combined with sol-gel chemistry (As shown in Fig. 8). The di-block can be dissolved in 1.4 dioxanes, which is considered

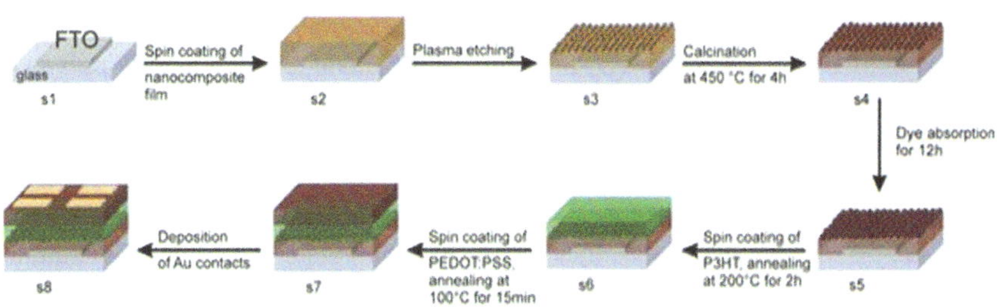

Fig. 9 A sketch of a hybrid cell based on titania and a conducting polymer [22]

the best solvent for both blocks (main PS and minor PEO constitution components). The addition of hydrochloric acid (HCl) acts as a poor solvent for PS results in the cylindrical micelles in the solution. The addition of the sol-gel precursor of titanium tetra-isopropoxide is combined with the surface of the PEO domain and can be incorporated into the PEO domain. Finally, the self-organized mechanism guides the formation of the nanostructured by thin-film preparation through the spin coating. The structure templating process has been investigated recently in detail. It has summarized by the phase diagrams of the morphologies of titania, representing the rich capability of the preparation procedure of tailor-made morphologies. Figure 9 shows the schematic presentation of the phases of Hybrid photovoltaic cells. Multiple layers combine to make a sandwich-type structure between two electrodes in the classic hybrid solar cells. In the presented system, the two electrodes are on the side of the anode of fluorine (F) doped tin oxide (SnO) (FTO) on the top of the glass substrate and another side by the cathode of a thermal evaporated gold layer. The anode of FTO is considered as the bottom electrode, since FTO on glass is the substrate from where the stack buils-up starts. The electromagnetic light will incident on this top electrode, and it will begin to work. The stacks build-up will be started with the nanostructure titania nestling the top surface of the FTO. The novel approach has been employed in the absence of the relatively compact titania layer beneath the nanostructure titania. The above procedure contains the titania nanostructured; therefore, the encapsulation process is supplied by its preparation component.The solution contains the three-block copolymer poly(ethyleneoxidemethacrylate block dimethylsiloxane block ethylene oxide methacrylate) ((PEO)MA-b-PDMS-b-MA(PEO)). They combined with the titania sol-gel chemistry is spin-coated resulted in a thin-polymer nanocomposite thin-film. The thin-film has been made by the three-block copolymer acts as a structure-directing agent which arises from the formation of parted micro stages and the self-organized mechanism of the block copolymers. The sol-gel chemistry supplies the titanium to the (PEO)MA phases, which is the instigator of the in-corporation into this specific stage. Because of PEO MA block-ratio to PDMS, the PDMS phase forms the matrix with embedded PEO MA phases, therefore, embedded titania. Later on, the PDMS matrix will be removed successively by the typical plasma etching method in exposing titania nanostructures. In the last step, the residual polymer nanocomposite thin-film with titania sticking-out is calcining at the high temperatures. The titania converted into crystalline anatase-phase, and the residual PDMS change into SiOC (silicon oxy carbide) ceramic. The conventional approach has been employed in the compact titania layer; the silicon oxy carbide will act as the blockage layer between the transparent anode and conducting polymers across the titania nanostructures. During initial preparation stages, a ruthenium-dye has been anchored to the titania and formed a single layer of the dye molecules of the top surface of the nanostructures as sketched in the cross-section scheme in Fig. 9. The pure organic layer of P3HT has been spin-coated on the in-organic nanostructure in the next step. The topmost sandwiched layer has direct contact with the cathode, a single layer poly(3,4-ethylenedioxythiophene), poly(styrene sulfonate) (PEDOT: PSS) has been spin-coated on the P3HT. The title role of the PEDOT: PSS mixture is used for the blockage of free electrons from reaches the cathode, but it will enable the conduction of the holes towards the cathode.

18 Conclusion

This chapter has addressed the structures and applications of conducting polymers in TEG, Piezoelectric, and solar cells in various contexts. It has emerged that the TEG is a viable solution for energy harvesting, able to supply electrical loads in relatively low-power applications. TEG efficiency is also typically low. Thereby, the advantages of using TEG have to be found in the characteristics of specific applications. There is a significantly high-temperature difference across the TEG system, and other solutions with higher efficiency cannot be applied because of various limitations. These limitations may be the relatively high temperatures for the materials adopted, the strict requirements on the system to be used (regarding the type of operation, emissions of pollutants, the device's position during operation, or noise). In these cases, TEGs may be fully competitive with other solutions. In particular, the use of TEGs is entirely consistent with the provision of green energy through energy harvesting from even small temperature differences. Some low-power applications have been identified on electronic circuits, sensors, waste heat recovery, residential energy harvesting, and automotive systems. In other applications to enhance the efficiency of energy production systems with a higher power, the efficiency increase is still somewhat limited, considering that an investment in TEG integration may be profitable. Nevertheless, there is a growing interest in the potential of thermoelectric applications. For the future, rapid development of TEG solutions can be expected in a broader range of green energy applications. This development depends on improvements in the TEG technology, better information on the TEG characteristics, and new solutions to promote better energy production integration.

References

1. Rowe, D.M.: Handbook of thermoelectrics, micro to nano. 1st Edition. Boca Raton: CRC Press Taylor & Francis Group 6000 Broken Sound Parkway NW (2006)
2. Goldsmid, H.J.: Introduction to Thermoelectricity. In: Hull, R., OSgood, R.M., Parisi, Jr. J., Warlimont, H. (eds.) Springer Series in Material Science. Springer, Berlin, Heidelberg (2010). Available from: http://www.springer.com/series/856
3. Chasmar, R.P., Stratton, R.: The thermoelectric figure of merit and its relation to thermoelectric generators. J. Electr. Control. **7**, 52–72 (1959)
4. Bulusu, A., Walker, D.G.: Review of electronic transport models for thermoelectric materials. Superlattices Microstruct. **44**, 1–36 (2008)
5. Applegarth, A.: Effect of quantum-well structures on the thermoelectric Bgure of merit. Phys. Rev. B. **47**, 1097–107 (1993)
6. Hicks, L., Dresselhaus, M.S.: Thermoelectric figure of merit of a one-dimensional conductor. Phys. Rev. B. **47**, 8–11 (1993)

7. Dresselhaus, M.S., Chen, G., Tang, M.Y., Yang, R., Lee, H., Wang, D., et al.: New directions for low-dimensional thermoelectric materials. Adv. Mater. **19**, 1043–1053 (2007)
8. Bredas, J.L., Street, G.B.: Polarons, bipolarons, and solitons in conducting polymers. Acc. Chem. Res. **18**, 309–315 (1985)
9. Bharti, M., Singh, A., Samanta, S., Aswal, D.K.: Conductive polymers for thermoelectric power generation. Prog. Mater. Sci. **93**, 270–310 (2018)
10. Cho, B., Park, K.S., Baek, J., Oh, H.S., Koo Lee, Y.E., Sung, M.M.: Single-crystal poly(3,4-ethylenedioxythiophene) nanowires with ultrahigh conductivity. Nano Lett. **14**, 3321–3327 (2014)
11. Sun, Y., Sheng, P., Di, C., Jiao, F., Xu, W., Qiu, D., et al.: Organic thermoelectric materials and devices based on p- and n-type poly(metal 1,1,2,2-ethenetetrathiolate)s. Adv. Mater. **24**, 932–937 (2012)
12. Kim, S.J., We, J.H., Cho, B.J.: A wearable thermoelectric generator fabricated on a glass fabric. Energy Environ. Sci. 7, 1959–1965 (2014)
13. Kishi, M., Nemoto, H., Hamao, T., Yamamoto, M., Sudou, S., Mandai, M., et al.: Micro thermoelectric modules and their application to wristwatches as an energy source. In: Eighteenth International Conference on Thermoelectrics. Proceedings, ICT'99 (Cat. No. 99TH8407), pp. 301–307. IEEE (1999)
14. Francioso, L., De Pascali, C., Farella, I., Martucci, C., Cret, P., Siciliano, P., et al.: Flexible thermoelectric generator for ambient assisted living wearable biometric sensors. J. Power Sources [Internet]. Elsevier B.V. **196**, 3239–3243 (2011). Available from: http://dx.doi.org/10.1016/j.jpowsour.2010.11.081
15. Umeda, M., Nakamura, K., Ueha, S.: Analysis of the transformation of mechanical impact energy to electric energy using Piezoelectric vibrator. Jpn. J. Appl. Phys. [Internet]. IOP Publishing **35**, 3267–3273 (1996). Available from: http://dx.doi.org/10.1143/JJAP.35.3267
16. Ramsay, M.J., Clark, W.W.: Piezoelectric energy harvesting for bio-MEMS applications. ProcSPIE [Internet] (2001). Available from: https://doi.org/10.1117/12.429684
17. Sodano, H.A., Inman, D.J., Park, G.: A review of power harvesting from vibration using Piezoelectric materials. Shock Vib. Dig. **36**, 197–205
18. Sodano, H.A., Park, G., Inman, D.J.: Estimation of electric charge output for piezoelectric energy harvesting. Strain **40**, 49–58 (2004)
19. Yu, G., Heeger, A.J.: Charge separation and photovoltaic conversion in polymer composites with internal donor/acceptor heterojunctions. J. Appl. Phys. **78**, 4510–4515 (1995)
20. Coakley, K.M., McGehee, M.D.: Conjugated polymer photovoltaic cells. Chem. Mater. **16**, 4533–4542 (2004)
21. Kaune, G., Müller-buschbaum P.: Gradient-doping of a conductive polymer film with a layer-by-layer approach pss. **54**, 52–54 (2010)
22. Perlich, J., Memesa, M., Diethert, A., Metwalli, E., Wang, W.: Preservation of the morphology of a self-encapsulated thin titania film in a functional multilayer stack : An X-Ray Scattering Study. 799–805 (2009)

Perspectives of Conducting Polymers Towards Heat Transfer Applications

Syed Nadeem Abbas Shah, Syed Shahabuddin, and Mohd Faizul Mohd Sabri

Abstract The heat dissipation requirements of many industrial applications demand efficient, economical and eco-friendly coolants and lubricants with enhanced thermal transport characteristics. In this context, nanofluids have been extensively studied and praised as well. However, most of the nanomaterial additives are expensive, unstable within base fluids and a potential threat towards environmental pollution while dumping is needed. Therefore, the current review unveil the potential of conducting polymers (CPs) to develop stable, cost-effective and eco-friendly nanofluids within the base fluid networks. Only scant work is performed on CPs pertinent to thermal transport applications but a tremendous potential exists to manipulate them. Thus the conventional nanomaterials could be replaced with CPs particularly with polyamine (PANI) for possible deployment in convective heat transport devices. Besides the thermal transport mechanisms associated with CPs have also been discussed with intention to tailor them for specific applications. In this way, the present review could be considered as a starting point to explore further the potential of conducting polymers regarding heat transfer applications predominantly in nanofluids.

Keywords Conducting polymers · Thermal transport · Nanofluids

S. N. A. Shah · M. F. M. Sabri
Department of Mechanical Engineering, University of Malaya, Kuala Lumpur, Malaysia
e-mail: engr.nadeem@uet.edu.pk

S. N. A. Shah
Department of Mechanical Engineering, University of Engineering and Technology (UET), Main Campus, Lahore 54890, Pakistan

S. Shahabuddin (✉)
Department of Science, School of Technology, Pandit Deendayal Petroleum University, Raisan Village, Gandhinagar, Gujarat 382007, India
e-mail: syedshahab.hyd@gmail.com

© Springer Nature Switzerland AG 2021

S. Shahabuddin et al. (eds.), *Advances in Hybrid Conducting Polymer Technology*, Engineering Materials, https://doi.org/10.1007/978-3-030-62090-5_6

1 Introduction

The recent developments in science and technology have paved the new pathways for energy efficient systems. In this context, a large number of multifunctional nanomaterials based nanofluids have been utilized as energy conversion and transport mediums [1–7]. But the standing issue remained there in terms of weak colloidal, chemical and thermal stability, poor economy, environment protection and un-controlled apparent dynamic viscosity enhancement. In the meantime a number of CPs such as polyaniline (PANI), polypyrrole, polythiophene and derivatives are getting much attention due to their tuneable properties in a variety of applications [8, 9]. The reason behind the usage of CPs is their chemical stability, cost effectiveness and eco-friendly nature. For instance, the poly (3, 4-ethylenedioxythiophene) (PEDOT) has shown tremendous performance in thermoelectric generators applications [10]. On the other hand the coupling of PEDOT with the polystyrene sulfonate (PSS) has shown prestigious conducting properties along with outstanding stretch ability [11]. In addition, there are many other applications of CPs, for example, biomedical, tissue engineering and photocatalytic [12–14]. Among all CPs, polyaniline (PANI) has been extensively studied for energy storage applications since the 1980s due to its eco-friendly nature, high specific capacity and promising conductivity [15]. There has been reported the potential of CPs to make nanocomposites with graphene analogous materials such as WS_2, BN, MoS_2 and MXenes for various applications [8]. This characteristic of CPs is attracting research community for further exploration including heat transfer. However, very limited work has been carried out pertinent to heat transfer applications using CPs, particularly in convective heat transport applications by employing nanofluids.

2 Thermal Transport of Conducting Polymers

In heat transport applications, the thermal conductivity of the materials plays a pivotal role. The thermal conductivity is the material property which depends on the movement of energy carriers. The fundamental energy carriers in energy transport mechanisms are electron, phonon, photon and molecules [16]. Generally, phonons act as energy carrier in polymers. But the presence of amorphous nature make polymers insulators with very small thermal conductivity. The major cause of reduced thermal conductivity is the phonon scattering due to crystal defects caused by the low degree of crystallinity [17]. However, the intrinsic conducting polymers (CPs) might show different thermal behaviour due to presence of free moving electrons. Likewise to the metals, the free electrons in CPs can also transfer energy from one end to another end of the materials resulting in higher thermal conductivity [18]. Nevertheless, both the electrons and phonons have been regarded as contributing factors for thermal conductivity in CPs [19]. As most of the CPs are rich in heterogeneous (amorphous/crystalline) nature which could be appealing to investigate owing to fact that amorphous materials thermal conductivity increase with temperature (above plateau region at ~10 K) when compared with purely crystalline. It was concluded that the extended acoustic phonons conduct heat in crystalline regions, while optical phonon conduct in amorphous region as a result of hopping with acoustic phonons [20]. In short, the thermal conductivity can be tuned by controlling the charge mobility in certain direction of the material.

Here it is worth mentioning that the polymeric materials thermal transport could be tuned by employing chemical routes such as doping with materials with high thermal transport properties. This approach requires lot of optimisation based on the applications whether one is interested to deploy solid materials or dispersions based. Because in case of dispersion of CPs with improved thermal transport also demands long term colloidal stability which limits the amount of dopant.

Furthermore, the thermal transport behaviour of conductive polymers have been investigated by employing both experimental and theoretical approaches. It has been noticed that the solitons, polarons and bipolarons are acting as energy carriers in case of conducting polymers [21]. The propagation of these energy carriers is influenced by various factors such as temperature, level of doping, interactions among polymer chains etc. It is widely accepted fact that in case of solid materials the charge carriers and phonons serve as heat conduction mechanisms. But the problem arises when these charge carriers and phonon got scattered as a result of crystal defects in the materials. Therefore a well ordered polymeric structure might be the game changer to conduct maximum heat energy along the axial direction. In this context, literature has also shown that the polymeric materials with ordered structures found to have extremely high phonon lattice thermal conductivity [22]. Thus the well alignment of the CPs fibres is indispensable for very high thermal conductivity. However the chain length of the polymer fibres should well match with the mean free path of the energy carriers. For example if the chain length of the synthesized material is shorter than the mean free path of the energy carrier (charge and phonon) then scattering effects may diminishes the heat conduction. The detailed illustration of strategy to tune thermal transport in PEDOT poly (3,4 ethylenedioxythiophene) CPs is shown in Fig. 1. It was shown that for the ideal PEDOT crystal the lattice thermal conductivity appear around 41.7 and 61.2 W/m K in chain backbone direction corresponding to intrinsic and lightly doped with p-toluenesulfonate. Such improvement in lattice thermal conductivity is also the indication of strong covalent interaction in chain direction. On the other hand the intrinsic thermal conductivity of PEDOT of ~41.7 W/m K reduced to 0.97 W/m K through doping with poly (styrenesulphonate) (PSS). This reduction was ascribed to the low level of crystallinity developed via PSS.

The kinetic theory of gases shows a relationship between lattice thermal conductivity (k), specific heat capacity (c_V), phonon group velocity (V) and relaxation time (τ) of phonons arising after scattering process as given by Eq. (1). As given in the literature that the increase in temperature cause more interactions among phonons which reduces the mean free path followed by reduced relaxation time. Consequently, a reduction in thermal conductivity is expected in case of conducting polymers.

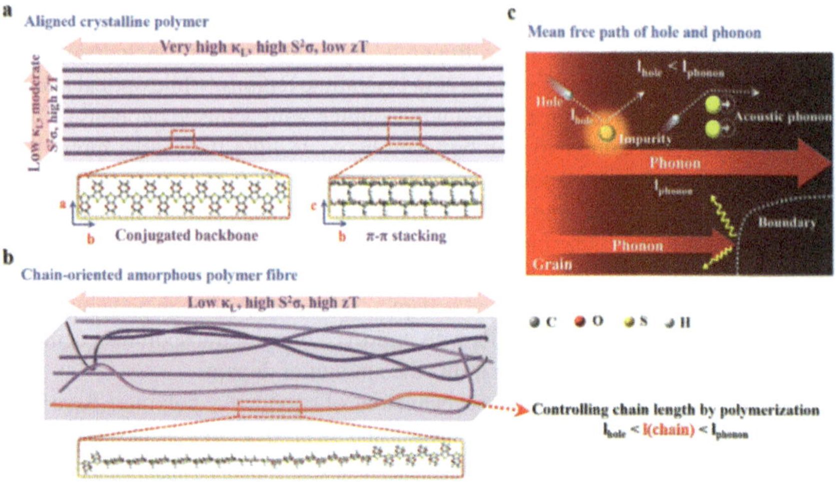

Fig. 1 Schematic illustration of thermal and electrical transport in crystalline polymers and chain-oriented polymeric fibers. **a** Anisotropic thermoelectric transport properties of crystalline polymers in the conjugated backbone and π−π stacking directions, respectively. The chemical structures of PEDOT are also displayed. **b** Suppression of phonon transport in chain-oriented amorphous polymer fibers. The polymer chains extend along the long axial direction of the fiber. The polymer chains are tailored to be shorter than the mean free path of phonons while longer than that of charge carriers by controlling the degree of polymerization. **c** Separated length scales of thermal and charge transport in the same crystallite. In general, the mean free path of phonons is much longer than that of charge carriers. If the grain size is smaller than the mean free path of phonons, the phonons will be predominantly scattered by the grain boundary [17]

$$k = c_v \times V^2 \times \tau \tag{1}$$

3 Integration of PANI in Convective Heat Transfer Research

Regarding heat transfer capabilities of CPs based nanofluids, particularly polyaniline (PANI) nanofibers have been synthesised and evaluated for thermal conductivity improvements. The mechanisms for heat conduction in PANI nanofibers (~7.5 nm) has been presented with major contribution from the phonon-only thermal conductivity. It was speculated that the possible disorder in the structure with temperature rise as a result of doping which might decrease in the phonon-only thermal conductivity owing to Umklapp process. Here the aniline monomers were transformed into PANI using Interfacial polymerization in the presence of ammonium persulphate (APS) oxidant and HCl dopant with different molar concentrations (0.02, 0.04, and 0.06).

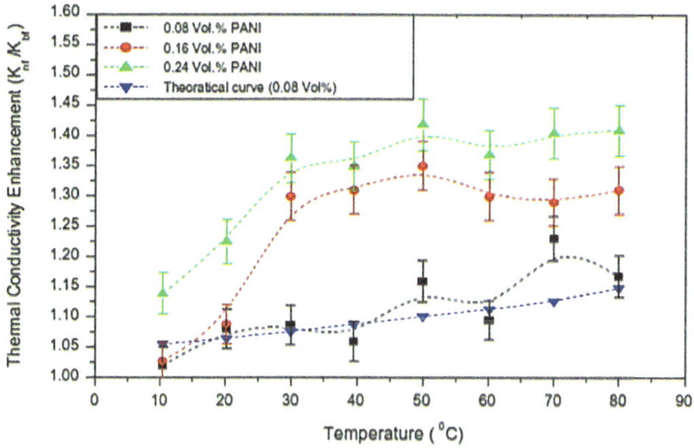

Fig. 2 Knf/Kbf (thermal conductivity ratio) versus temperature [24]

Hence the synthesis plays a major role to control the thermal energy of CPs nanostructure [23]. Nevertheless, further experimental work is still needed to reach at plausible conclusion. In 2012, Wan et al. reported the successful utilization of the synthesized PANI nanofibers (~80 nm dia. and 2 μm length) in water based nanofluids. As evident from Fig. 2, addition of PANI considerably improves the thermal conductivity of DI-water to appreciable extent, which is appealing considering CPs based nanofluids. The maximum ~40% enhancement was recorded at 0.24 vol% PANI loadings corresponding to temperature of 80 °C. Apparently, improvement in thermal conductivity was noted due to the uniform morphology and sufficient crystallinity. However, particle shape, temperature and Brownian motion effects were correlated with thermal conductivity through appropriate model(s). The oxidative polymerization route was adopted for synthesis using iodate as oxidant while HCl (12 M) as dopant [24]. Further in 2014, spherical shape PANI nanopartcles with size in arrange of 50–60 nm were synthesized using reverse micellar polymerization with dodecylbenzenesulfonic acid (DBSA) for possible deployment in deep eutectic ionic liquid (DES) based nanofluids for heat transfer applications. The maximum reported enhancement in thermal conductivity was ~6.47% at 50 °C corresponding to 0.2 wt% loadings in DES and water combination as base fluid. However, pure DES solvent appeared to be ineffective for the reported case [25]. Similarly the researchers in [25] reported the synthesis of PANI using DBSA dopant through oxidative polymerization of aniline in a size range of 15–50 nm. The PANI/water nanofluids were studied and ~5.4% enhancement in thermal conductivity observed [26]. Later in 2015, Gurav and co-workers presented the heat transfer performance evaluation on synthesized PANI nanofibers (50–60 nm) based water nanofluids. As shown in Fig. 3, the integrations of PANI/water nanofluids in a copper tube revealed substantial improvement (33–63%) in heat transfer coefficient corresponding to particles concentration in range of 0.2–1.2 wt% using constant heat flux and variable flow rates. In addition, the one dimensional nature (1-D) of fibrous structure can be dispersed easily in the

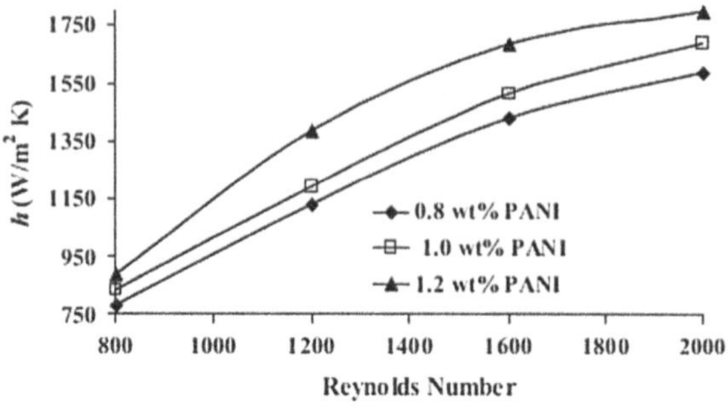

Fig. 3 Heat transfer coefficient as a function of Reynolds number at 0.8, 1.0 and 1.2 wt% of colloidal PANI nanoparticle in reference fluid [27]

base fluid with complementary enhanced specific surface area (SSA). This in turns give rise to improved thermo-physical properties of PANI/water nanofluids. Overall, the small size of particles is the key to tailor the properties owing to enlarged SSA and quantum effects. Here the ultrasound assisted miniemulsion polymerization was used for synthesis using sodium dodecyl sulphate (SDS) surfactant and APS oxidant along with HCl (1 M) as dopant [27].

So far only scant attention has been paid towards CPs integration in the area of convective heat transport. Some of the research reports have reported the intriguing results in improvements of thermo-physical properties and heat transfer coefficient. As in 2018, Bhanvase et al. explored further PANI based water nanofluids in a vertical helical coiled heat exchangers. The PANI nanofibers (around 100 nm) were synthesized through ultrasound assisted emulsion polymerization in the presence of APS oxidant, HCl (1 M) dopant and sodium lauryl sulphate surfactant. The results presented in Fig. 3 indicated an improvement in local heat transfer coefficient in a range 10–70% with particle loadings from 0.1 to 0.5 vol%. However, the thermo-physical properties were only calculated using the empirical relations reported in the literature [28]. Further, Bhanvase and co-workers synthesized PANI and CuO-PANI nanocomposites (size less than 100 nm) via in-situ emulsion polymerization for utilization in water based nanofluids. The synthesized CuO through hydrothermal mean obtained in spherical morphology with size less than 10 nm. The straight tube heat exchanger has been considered for performance evaluation of nanofluids. The results display an improvement by >12% and ~38% in heat transfer coefficient corresponding to 0.5 volume % of PANI and PANI-CuO nanocomposite loadings, respectively. SLS surfactant was employed to achieve better colloidal dispersion [29] (Fig. 4).

Subsequently, a novel synthesis technique (gamma radiolysis) has been employed for Cu-PANI nanocomposites formation with a nominal size ~21 nm. A wide temperature range (10–90 °C) thermal conductivity measurement (Fig. 5) show anomalous enhancement ~107% and 159% corresponding to water based nanofluid temperature of 10 °C and 90 °C, respectively, for 1 vol% particle loadings[30]. Lanjewar et al. also studied CuO-PANI based water nanofluids for performance evaluation of vertical helical coiled tube heat exchanger. The nanocomposites were synthesized by varying the concentration of CuO (1, 3, 5 wt%) in PANI matrix using ultrasound

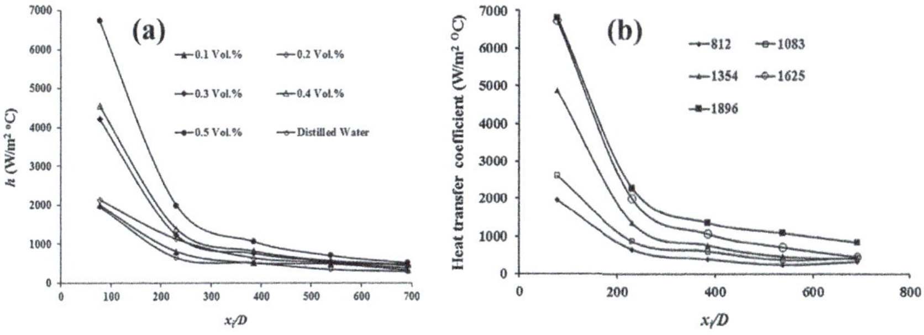

Fig. 4 **a** Effect of PANI nanoparticle concentration in nanofluid on heat transfer coefficient (Re = 1625) and **b** Effect of Reynolds Number of nanofluid on heat transfer coefficient (PANI vol% = 0.5) [28]

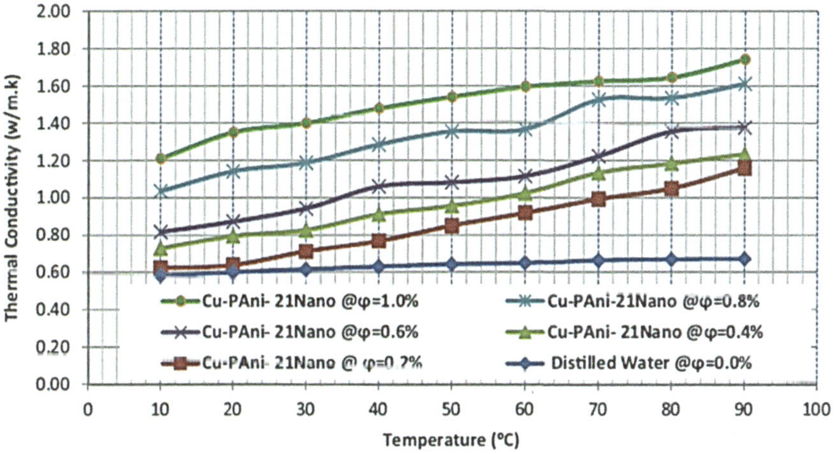

Fig. 5 The thermal conductivity of copper-polyaniline nanocomposites-water nanofluid as a function of temperature at various values of particle volume fraction [30]

assisted polymerization. The CuO showed rod shape morphology with size less than 10 nm while a new technique known as hydrodynamic cavitation has been integrated to achieve uniform nanofluids suspension with minimum agglomeration. The findings show that the thermal conductivity of nanofluids increase with concentration of nanocomposites and temperature. The maximum thermal conductivity ratio found to be ~2.1 at 40 °C corresponding to 0.4 vol% nanofluids concentration and 1 wt% CuO doping. Surprisingly, when the CuO doping level increased to 3 and 5 wt%, the maximum thermal conductivity ratio appeared to be 2.59 (at 40 °C and 0.4 vol%) and 3.87 (at 48.8 °C and 0.5 vol %), respectively. However, the thermal conductivity data of pure CuO and PANI was missing to reach on plausible conclusion as the apparent enhancement in thermal conductivity is astonishing [31].

Moving on, with the increasing interest to develop efficient, stable and environment-friendly nanofluids for various applications. More recently, George et al. have reported the potential of adding PANI as nano additive to improve the thermal storage and thermal conductivity of the paraffin wax based phase change materials. It was shown that the addition of 1 wt% of PANI had increased the latent heat capacity of paraffin wax by 8.2% along with 48.6% thermal conductivity enhancement [32]. On the other hand, Sofiah and co-workers have shown that the addition of PANI nanotubes significantly increase the thermal conductivity (up to 25.76%) of refined, bleached and deodorized palm olein (RBDL) nanolubricants [33]. Meanwhile, Shah et al. also reported the increase in thermal conductivity of ethylene glycol with the addition of PANI nanotubes as well as PANI/WS_2 nanocomposites [34].

4 Conclusions

The limited literature on conducting polymers thermal transport have revealed that the thermal properties of the CPs can be tailored by controlling their growth during synthesis using appropriate dopant and its concentration. The ease of integration of graphene analogous materials can also be charming area interest to develop nanocomposites with improved thermo-physical properties and stability. The thermal behaviour of CPs is astonishing and in contrast to metallic crystals. As the electronic transport also assist in heat conduction along with phonons contribution. However, the temperature response of CPs has shown inhomogeneous structure depicting great potential to tune thermal transport. Apart from improved thermo-physical properties, long term stability is also the need for convective heat transfer nanofluids. Considering the better stability of the CPs, particularly one dimensional architecture of PANI help ease the dispersion which could be utilized to minimize the agglomeration problem as compared to typical metal/metal oxides based nanofluids. Realizing the cost effectiveness and environment friendliness CPs should be explored further for integration into convective heat transfer research.

References

1. Masoud Hosseini, S., Safaei, M.R., Estellé, P., Hadi Jafarnia, S.: Heat transfer of water-based carbon nanotube nanofluids in the shell and tube cooling heat exchangers of the gasoline product of the residue fluid catalytic cracking unit. J. Therm. Anal. Calorim. **5** (2019).https://doi.org/10.1007/s10973-019-08813-5.
2. Ghanbari, S., Javaherdeh, K.: Thermal performance enhancement in perforated baffled annuli by nanoporous graphene non-Newtonian nanofluid. Appl. Therm. Eng. **167**, 114719 (2020). https://doi.org/10.1016/j.applthermaleng.2019.114719
3. Gupta, M., Singh, V., Kumar, S., Dilbaghi, N.: Experimental analysis of heat transfer behavior of silver, MWCNT and hybrid (silver + MWCNT) nanofluids in a laminar tubular flow. J.

Therm. Anal. Calorim. (2020). https://doi.org/10.1007/s10973-020-09453-w
4. Taheri, A., Kazemi, M., Amini, M., Sardarabadi, M., Kianifar, A.: The performance assessment of nanofluid-based PVTs with and without transparent glass cover: outdoor experimental study with thermodynamics analysis. J. Therm. Anal. Calorim. (2020). https://doi.org/10.1007/s10973-020-09311-9
5. Kalidoss, P., Venkatachalapathy, S., Suresh, S.: Photothermal energy conversion enhancement studies using low concentration nanofluids. J. Sol. Energy Eng. Trans. ASME **141**, 1 (2019). https://doi.org/10.1115/1.4043864
6. Rammohan, A., Ramesh Kumar, C.: Liquid cooling system for a high power light emitting diode of an automotive headlamp and its effect on light intensity. Eur. Phys. J. Spec. Top. **228**, 2495–2509 (2019). https://doi.org/10.1140/epjst/e2019-900066-8
7. Kapıcığolu, A., Esen, H.: Experimental investigation on using Al_2O_3/ethylene glycol water nano-fluid in different types of horizontal ground heat exchangers. Appl. Therm. Eng. **165**, 114559 (2019). https://doi.org/10.1016/j.applthermaleng.2019.114559
8. Dunlop, M.J., Bissessur, R.: Nanocomposites based on graphene analogous materials and conducting polymers: a review. J. Mater. Sci. **55**, 6721–6753 (2020). https://doi.org/10.1007/s10853-020-04479-9
9. Inoue, A., Yuk, H., Lu, B., Zhao, X.: Strong adhesion of wet conducting polymers on diverse substrates. Sci. Adv. **6**, 1–11 (2020). https://doi.org/10.1126/sciadv.aay5394
10. Kim, N., Lienemann, S., Petsagkourakis, I., Alemu Mengistie, D., Kee, S., Ederth, T., Gueskine, V., Leclère, P., Lazzaroni, R., Crispin, X., Tybrandt, K.: Elastic conducting polymer composites in thermoelectric modules. Nat. Commun. **11**, 1–10 (2020). https://doi.org/10.1038/s41467-020-15135-w
11. Wang, Y., Zhu, C., Pfattner, R., Yan, H., Jin, L., Chen, S., Molina-lopez, F., Lissel, F., Liu, J., Rabiah, N.I., Chen, Z., Chung, J.W., Linder, C., Toney, M.F., Murmann, B., Bao, Z.: A highly stretchable, transparent, and conductive polymer. Sci. Adv. **3**, 1–11 (2017)
12. Kenry, B.L.: Recent advances in biodegradable conducting polymers and their biomedical applications. Bio Macromol. (2018). https://doi.org/10.1021/acs.biomac.8b00275
13. Iqbal, N., Khan, A.S., Asif, A., Yar, M., John, W.: Recent concepts in biodegradable polymers for tissue engineering paradigms: a critical review. Int. Mater. Rev. (2018). https://doi.org/10.1080/09506608.2018.1460943
14. Shahabuddin, S., Khanam, R., Khalid, M., Sarih, N.M., Ching, J.J., Mohamad, S., Saidur, R.: Synthesis of 2D boron nitride doped polyaniline hybrid nanocomposites for photocatalytic degradation of carcinogenic dyes from aqueous solution. Arab. J. Chem. **11**, 1000–1016 (2018). https://doi.org/10.1016/j.arabjc.2018.05.004
15. Rauhala, T., Davodi, F., Sainio, J., Sorsa, O., Kallio, T.: On the stability of polyaniline/carbon nanotube composites as binder-free positive electrodes for electrochemical energy storage. Electrochim. Acta. **336**, 135735 (2020). https://doi.org/10.1016/j.electacta.2020.135735
16. Chen, G.: Nanoscale Energy Transport and Conversion: A Parallel Treatment of Electrons, Molecules, Phonons, and Photons. Oxford University Press, New York (2005)
17. Shi, W., Shuai, Z., Wang, D.: Tuning thermal transport in chain-oriented conducting polymers for enhanced thermoelectric efficiency: a computational study. Adv. Funct. Mater. **27**, 1–8 (2017). https://doi.org/10.1002/adfm.201702847
18. Pradhan, S.S., Unnikrishnan, L., Mohanty, S., Nayak, S.K.: Thermally conducting polymer composites with EMI shielding: a review. J. Electron. Mater. **49**, 1749–1764 (2020). https://doi.org/10.1007/s11664-019-07908-x
19. Jin, J., Manoharan, M.P., Wang, Q., Haque, M.A.: In-plane thermal conductivity of nanoscale polyaniline thin films. Appl. Phys. Lett. **95**, 2007–2010 (2009). https://doi.org/10.1063/1.3184786
20. Damker, T., Bryksin, V.V., Bo, H.: Phonon hopping in disordered systems. Phys. B Condens. Matter **264**, 263–265 (1999)
21. Xu, X., Zhou, J., Chen, J.: Thermal transport in conductive polymer-based materials. Adv. Funct. Mater. **30**, 1–18 (2020). https://doi.org/10.1002/adfm.201904704

22. Zhou, H., Cai, Y., Zhang, G., Zhang, Y.W.: Quantum thermal transport in stanene. Phys. Rev. B **94**, 1–6 (2016). https://doi.org/10.1103/PhysRevB.94.045423
23. Nath, C., Kumar, A., Syu, K.Z., Kuo, Y.K.: Heat conduction in conducting polyaniline nanofibers. Appl. Phys. Lett. **103**, 1–5 (2013). https://doi.org/10.1063/1.4821656
24. Wan, M., Yadav, R.R., Yadav, K.L., Yadaw, S.B.: Synthesis and experimental investigation on thermal conductivity of nanofluids containing functionalized Polyaniline nanofibers. Exp. Therm. Fluid Sci. **41**, 158–164 (2012). https://doi.org/10.1016/j.expthermflusci.2012.03.030
25. Siong, C.T., Daik, R., Hamid, M.A.A.: Thermally conductive of nano fluid from surfactant doped polyaniline nanoparticle and deep eutectic ionic liquid. AIP Conf. Proc. **1614**, 381–385 (2014). https://doi.org/10.1063/1.4895227
26. Chew, T.S., Daik, R., Hamid, M.A.A.: Thermal conductivity and specific heat capacity of dodecylbenzenesulfonic acid-doped polyaniline particles-water based nanofluid. Polym. (Basel) **7**, 1221–1231 (2015). https://doi.org/10.3390/polym7071221
27. Gurav, P., Naik, S., Bhanvase, B.A., Pinjari, D.V., Sonawane, S.H., Ashokkumar, M.: Heat transfer intensification using polyaniline based nanofluids: preparation and application. Chem. Eng. Process. Process Intensif. **95**, 195–201 (2015). https://doi.org/10.1016/j.cep.2015.06.010
28. Bhanvase, B.A., Sayankar, S.D., Kapre, A., Fule, P.J., Sonawane, S.H.: Experimental investigation on intensified convective heat transfer coefficient of water based PANI nanofluid in vertical helical coiled heat exchanger. Appl. Therm. Eng. **128**, 134–140 (2018). https://doi.org/10.1016/j.applthermaleng.2017.09.009
29. Bhanvase, B.A., Kamath, S.D., Patil, U.P., Patil, H.A., Pandit, A.B., Sonawane, S.H.: Intensification of heat transfer using PANI nanoparticles and PANI-CuO nanocomposite based nanofluids. Chem. Eng. Process. Process Intensif. **104**, 172–180 (2016). https://doi.org/10.1016/j.cep.2016.03.004
30. Alyan, A., Abdel-Samad, S., Massoud, A., Waly, S.A.: Characterization and thermal conductivity investigation of copper-polyaniline nano composite synthesized by gamma radiolysis method. Heat Mass Transf. **55**, 2409–2417 (2019). https://doi.org/10.1007/s00231-019-02588-z
31. Lanjewar, A., Bhanvase, B., Barai, D., Chawhan, S., Sonawane, S.: Intensified thermal conductivity and convective heat transfer of ultrasonically prepared cuo–polyaniline nanocomposite based nanofluids in helical coil heat exchanger, period. Polytech. Chem. Eng. **64**, 271–282 (2020). https://doi.org/10.3311/PPch.13285
32. George, M., Pandey, A.K., Abd, N., Tyagi, V.V., Shahabuddin, S., Saidur, R.: A novel polyaniline (PANI)/ paraffin wax nano composite phase change material: superior transition heat storage capacity, thermal conductivity and thermal reliability. Sol. Energy **204**, 448–458 (2020). https://doi.org/10.1016/j.solener.2020.04.087
33. Sofiah, A.G.N., Samykano, M., Shahabuddin, S., Kadirgama, K., Pandey, A.K.: An experimental study on characterization and properties of eco-friendly nanolubricant containing polyaniline (PANI) nanotubes blended in RBD palm olein oil. J. Therm. Anal. Calorim. (2020). https://doi.org/10.1007/s10973-020-09891-6
34. Shah, S.N.A., Shahabuddin, S., Sabri, M.F.M., Salleh, M.F.M., Said, S.M., Khedher, K.M.: Thermal conductivity, rheology and stability analysis of 2D Tungsten disulfide doped polyaniline based nanofluids: an experimental investigation. Int. J. Energy Res. (2020). https://doi.org/10.1002/er.5697

Conducting Polymer Based Nanoadsorbents for Removal of Heavy Metal Ions/Dyes from Wastewater

Gagandeep Kour, Richa Kothari, Rifat Azam, Pradeep Kumar Majhi, Sunil Dhar, Deepak Pathania, and V. V. Tyagi

Abstract Decline in the availability of potable water due to introduction of enormous contaminants including heavy metals and dyes paves way for the introduction of new and advanced water treatment technologies that ensures suitability of water for drinking purpose and at the same time eliminate pollutants or contaminants present in water. Adsorption based on nanoadsorbents is promising because of cost effectiveness and ease of operation. Therefore, this chapter intends to provide comprehensive detail about the adsorption process which is considered as the best method for heavy metal and dye removal along with the factors affecting the adsorption process. Different types of nanoadsorbents showing greater efficiencies in terms of heavy metal and dye removal are also discussed in detail. Much consideration is being given to conducting polymers. An insight about the synthesis of certain conducting polymer based nanoadsorbents is provided. The excellent properties of conducting polymers have enabled them to be used in remediation of toxic contaminants. Recommendations for future research in the area of conducting polymer based nanoadsorbents for improving the heavy metal and dye removal potentials are also put forth.

Keywords Adsorption · Conducting polymers · Nanoadsorbents · Heavy metals · Dyes · Wastewater

---

G. Kour · R. Kothari (✉) · S. Dhar · D. Pathania
Department of Environmental Sciences, Central University of Jammu, Rahya-Suchani, Bagla, Samba, J&K 181143, India
e-mail: kothariricha21@gmail.com

R. Azam · P. K. Majhi
Department of Environmental Sciences, Babasaheb Bhimrao Ambedkar University, Lucknow 226025, U.P, India

D. Pathania
Department of Chemistry, Sardar Vallabhbhai Patel University, Mandi 175001, Himachal Pradesh, India

V. V. Tyagi
School of Energy Management, Shri Mata Vaishno Devi University, 182320 Jammu & Kashmir, India

© Springer Nature Switzerland AG 2021

S. Shahabuddin et al. (eds.), *Advances in Hybrid Conducting Polymer Technology*, Engineering Materials, https://doi.org/10.1007/978-3-030-62090-5_7

1 Introduction

Reckless development in an advanced technology implying severe environmental crisis in recent era. Among all recent environmental issues, wastewater generation to the aquatic body is arising a serious global concern [42]. The wastewater from different wastewater sources (point and non-point source) comes out and mixes with the fresh water body. The availability of hygienic water for each individual is a basic requisite of human society, which has now become inaccessible due to substantial increase in the release of toxic chemicals into the water body [8]. In general, the amount of undesirable contaminants or materials present in the water bodies have surpass the permissible limit as suggested by regulating agencies, creating environmental concern for society. Among several contaminations of wastewater, the major concerns are heavy metals as well as the extreme load of dissolved solids. So, the microbial pathogens get proper scope for their growth to cause harmful water borne diseases, which is simply not possible to prevent [31]. The decent approach to work over this kind of worldwide problems, there is a need to build up low cost, eco friendly with highly efficient treatment technologies. Nanotechnology covers up the industrialized process with suitable managing of resources at the nano scale level. In recent years, the advances in nanotechnology have been widely used in the disciplines of science and technology like: section dealing with energy, various processes industry, sensors as well as environmental applications [14]. Literature confirms, nanotechnology based approaches are capable to develop efficient system and approaches that can solve lots of difficult problems, which were never solved by conventional methods. Polymer based nanoadsorbents finds a wide array of applications in modern eras with broad spectrum from electronic to photonic, conducting material, medicine, sensors, as well as pollution minimization [5]. Literature confirms, there are numerous adsorbents which can be involved in the treatment of wastewater or aqueous solutions contaminated with metal ions or dyes. Among the many type of polymers available, Polyaniline (PANI) and its derivative are very much efficient as conducting polymers for the eradication of a range of heavy metal ions and dyes which are contributing to the contamination of wastewater or aqueous solutions [73].

1.1 The Importance of Clean Aquatic Environment

Industrialization is a main hazard to the aquatic environment in day to day life. It discharges toxic chemicals, gases and solid wastes into the environment, which are very effective for aquatic life [57]. Aquatic ecosystem consist of rich biodiversity habitats which ranges from chemoautotrophic to oligotrophic mountains lakes and streams all of which are in danger threats, thus it pose for opportunities for the conservation and restoration. The polluted water bodies have a serious threat to human being and aquatic environment. Several anthropological activities as well as release of greenhouse gases from industries have high impact on water body including flora and fauna. Following factors are responsible for clean up the water body, physical, demographic and socio economic status.

1.2 Pollutant in Water

Water is an important renewable resource which is required by all forms of life on earth. It is surely unique gift from nature which is utilized for agricultural purposes, hydropower generation, and production of livestock, industrial activity and forestry. Some large amount of hazards present in water bodies which are not used for the purpose of drinking, bathing, cooking and other uses. The contamination in the environment is referred as pollution. In many parts of the world, human activities effecting the surroundings due to their extensive urbanization, agricultural practices and industrial activity [70]. The pollution in water bodies occur from two different sources that is point as well as non-point sources. In water bodies industrial, commercial and agricultural activities are the key sources of pollution [22]. The main cause responsible for deteriorating the quality of water is leakage of sewage, increasing population rate, dumping of industrial waste into the water bodies, drilling activities, flood during rainfall, heavy metal, mining activities, animal wastes, deforestation and eroded sediments (Fig. 1). When the rainfall occurs in excessive amount, the herbicides and pesticides were washed with rain water, which causes a serious problem in living organisms. Sustainable development is a good initiative for the improvement of a society without any harm to the environment. It provides a better opportunity to the human being for the conservation and management of environment [29]. Therefore,

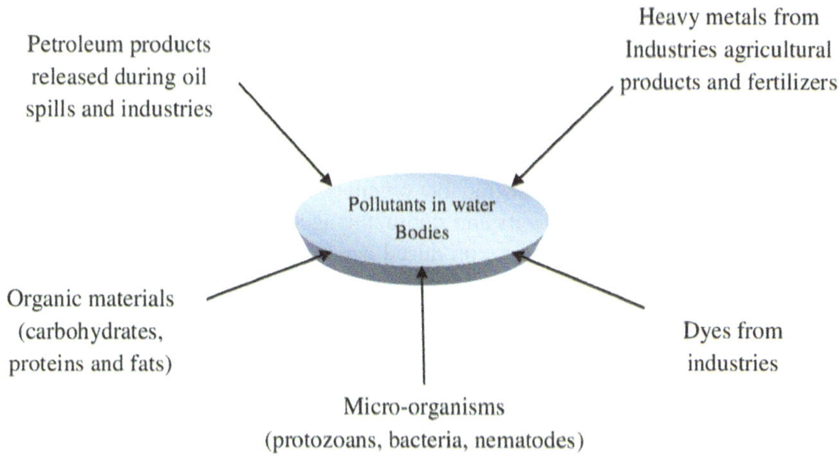

Fig. 1 Common pollutants in water bodies

protective water quality is part of societal agenda because of serious water pollution issues from surface to groundwater and results are global level shortage of water.

1.2.1 Heavy Metals

The major concern with water pollution is presence of heavy metals due to release of influents/effluents from industries. It occurs in the top list of inorganic pollutant, which has the deadly effects on aquatic organisms, as well as plants and human being [15]. Generally, heavy metal come from different sources into the environment such as mining, surface finishing industries, animal wastes and agricultural wastes. In addition, different industries have released various types of heavy metal like Arsenic, lead, mercury, nickel, cadmium, chromium and copper from burning of coal, insecticides, fungicides, mining activities, paper industry etc. [20]. Arsenic comes in the environment from smelting, mining and pesticides. Mercury, cadmium, lead and copper are introduced from pesticides, sewage sludge and mining [20, 60]. Increasing heavy metal ions affect the plant metabolism rate, photosynthesis and cell membrane damage [37]. The metal toxicity affects the human and animal life through food chain and food web and the effects include skin irritation, headache, diarrhea, blood pressure low and vomiting [53].

The arsenic cause cancer in the lungs, liver and gallbladder, as well as the cadmium damage the kidney, lungs and bones through ingestion of water. In case of children, lead cause memory loss, attention effect and mostly make children aggressive. In pregnant women, high level of lead exposure causes miscarriage, whereas in men, it damages the testis [63, 69]. Mercury is unique, because it travels a long distance and it is known as global pollutants. It occurs in the form of methyl and dimethyl mercury, while it inhaled through blood stream [75].

1.2.2 Dyes

Dyes are highly carcinogenic chemical substances found in water body. They are considered as one of the major culprits for wastewater generation, as contains highly toxic substances in it such as: degradable organics, harmful dyes, detergents, inorganic salts, heavy metals, desizers and stabilizing agents. As per estimation, worldwide 30 million tons of textiles are produced per year with a need of 700,000 tons of dyes per year [27, 61]. It has diverse purpose of uses like fabric sprinting, leather coloring, farming investigate, photo electrochemical cells, and hair complexion. The generated effluents from dye industry possesses high range of pollutants such as: dense dark colour, pH, chemical oxygen demand (COD), biochemical oxygen demand (BOD), Suspended solids (SS), metals, temperature and salts [21]. The exercise of so much large quantity of dyes produces large environmental problems throughout the discharge of their effluents. Generally, during the processing steps (pretreatment, dyeing, printing, and finishing) of dye production, the maximum amounts of pollutants are generated [27]. Based on its toxic chemical nature and difficulties in its removal processes, major attentions are being given to solve this issue. But, in present time the question comes to build up the advanced technology that can be easily operated in less processing time as well as in low cost management plan in a sustainable green approach.

1.2.3 Other Pollutants

Apart from heavy metals and dyes, untreated or improperly treated waste water or water also contains a wide variety of contaminants which may pose serious deleterious health effects. These contaminants may include nitrogen and phosphorous from agricultural runoff, hydrocarbons released from various industrial process, organic matter, microbes, fecal matter [1]. Occurrence of the major nutrients (N&P) in amounts exceeding the desired levels leads to eutrophication, which results in lowering of dissolved oxygen and an increase in the biological oxygen demand. Excess of nitrates in water leads to methanoglobinemia. Release of petroleum products in water bodies pose deleterious effects on marine habitats particularly fishes, aquatic birds, indirectly human health through food chain and in broader aspect, the devastation of ecological balance [74]. Organic materials which generally comprises of carbohydrates, proteins or fat enter the water bodies through agricultural, industrial and commercial sources and impose an oxygen demand on the micro-organisms that help in degradation there by decline in DO availability to other aquatic organisms making the water body a dead zone. In addition to these contaminants, water bodies also holds collection of microbes mainly protozoan, nematodes and worms mainly associated with water borne diseases.

1.3 Methods for Removal of Heavy Metal Ions and Pollutants from Wastewater

1.3.1 Adsorption

The process of adsorption is used not only for colour producing compounds, but is also used as protecting threats to the natural environment from water. Adsorption process involves accumulation of gas or a liquid solute on the surface of a liquid or solid adsorbents including Zeolite, Na-bentonite and Kaolinite etc., which ultimately forms a molecular/atomic film known as adsorbate. In most natural physical, chemical and biological systems, it can be operated, and it is broadly used in industrial applications like activated charcoal, water purification and synthetic resin. Among, these methods, adsorption is presently known as a very suitable method for the treatment of wastewater. Figure 2 depicts the generalized adsorption mechanism. It is very well known, economical and effective techniques for the removal of heavy metals and dyes from various industrial discharges as compared to conventional method.

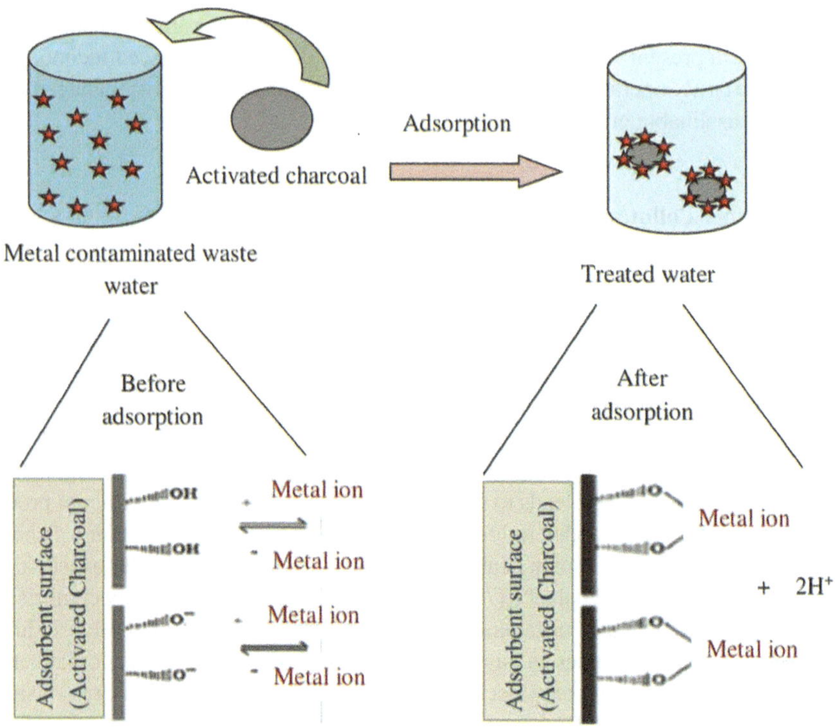

Fig. 2 Generalized adsorption mechanism

Adsorbents are of different types and classified as industrial waste, natural materials, biological and agricultural waste. The following adsorbents such as biological materials, activated carbon and fly ash are commonly applied for chromium (VI) removal. In terms of flexibility and simplicity of design, initial cost, ease of operation and insensitivity to toxic pollutants, it has been found that this method is superior to other methods. In addition, it does not result in the formation of harmful materials. Activated carbon, clay minerals, natural zeolites are the most commonly used adsorbents [69, 71]. Some adsorbents can also be derived from industrial or agricultural waste and are cost effective and easily available in huge amount [48].

1.3.2 Factors Affecting Adsorption

The following factors like pH, temperature, concentration and time, affect the adsorption process for the removal of heavy metal:

1. Effect of adsorbent dosages

The adsorbent amount or adsorbent dosage plays a crucial role in adsorption process, as higher adsorbent reflects greater number of available adsorption sites thereby enabling more number of ions to be adsorbed. As the dosage increases, the ratio of number of adsorption sites to the number of heavy metal ions would increase and there would be plenty of unabsorbed adsorption sites. However, by making use of two or more adsorbents in combination, lower the adsorption rate. The possible reason could be due to the particle aggregation, which would lead to a decrease in total surface area of the adsorbent [68].

2. Effects of pH

pH of the aqueous medium is one of the important factor affecting the adsorption process as it can directly affect the uptake of metal ions or dye by the adsorbent by affecting the metal ionization extent and the surface characteristics of the material on the surface of which adsorption process takes place [28]. pH of the solution control the magnitude of electrostatic charges possessed by the ionized metal or dye molecules thereby affecting the adsorption rate. Generally, in case of dye adsorption, low pH solution decreases the cationic dye removal and increases the anionic dye removal. In contrary to this, the percentage of dye removal will decrease for anionic dye and increase for cationic dye at high a pH value [58].

3. Effects of temperature

When the temperature increases, the thickness of the boundary layer decreases due to risings tendency of the metal ions which results in a decrease in adsorption Also, the magnitude of the desorption decrease as the temperature rises from 40 to 80° C. This is because of weakening of force of attraction between biomass surface and metal ion with the increase in temperature [2]. As the temperature increase the adsorption rate also increase and viscosity factor also goes down [62].

4. Effects of time

In adsorption process, there are different contact times. The contact time affects the adsorption process especially efficiency, economic as well as kinetics adsorption. The adsorption capacities were investigated by researcher with different contact time and ranges. The longer contact time might give a longer time for the adsorbate to interact with the adsorbent surface. But, it was also confirmed by other researchers that when the contact time of adsorption process exceeds more than four hours, the efficiency of the process does not show more enhancement, which indicates to be the saturation limit [23].

5. Effects of concentration

The effect of concentration comes to play when the increased concentration of metal ions in a solution causes an increase in mass transfer of ions from solution to the carbon surface [64]. The removal percentages occur due to the accessibility of overloading biomass and binding sites will increase the concentration of metallic ions. As the concentration increases, the metallic ions also increased due to the gradient force concentration. The process by which the liquid to solid phase transfer will increase the probable striking between the metallic ions and biomass, thus resulting in higher metal ions adsorption [40, 49].

Adsorption process is considered as one of the widely used method for heavy metal as well as dye removal from wastewater. Maintaining the optimal range required and considering the factors responsible for the adsorption process, will enhance the dye and heavy metal removal efficiency. Further, the effectiveness of the adsorption process can be enhanced by the incorporation of nanotechnology. Nanotechnology approaches are capable of eliminating contaminants present in wastewater because of high efficiency and removal potential. New and advanced nanoadsorbents can be applied for this purpose as these materials have very high adsorption capacity, active adsorption sites because of their nanoscale size as compared to their conventional counterparts.

1.4 Nanoadsorbents

Pollution of water is of growing concern since the amount of potable water is decreasing with increasing contamination and overexploitation of the natural resource. The quality of water has degraded drastically within couple of decades and a major proportion of world population is getting short of clean and drinkable water. Contaminants, whether they are organic, inorganic or biological nature, pose significant intimidations to the quality of water resources including surface as well as groundwater. Inorganic pollutants such as lead, cadmium, arsenic, nitrate ions are the major contributors responsible for the contamination of both surface as well as ground water [10]. With the advancement of nanoscale technology, various novel gateways have been opened up in the development and improvement of new and existing water purification facilities and have been considered very effective in providing solutions to various water related problems. Over the last decade, nanotechnology have gained attention because of the uniqueness in physical and chemical properties and have been widely recognized for the treatment of water and wastewater and in environmental remediation.

Among the wide applications of nanotechnology, the use of nanomaterials as adsorbents or catalysts for the removal of toxic and harmful materials from water or wastewater has been paid much attention. The smaller size of nanomaterials increases the chemical activity as well as the capacity to adsorb heavy metals on their surface by providing larger surface area [5]. Adsorption process mainly depends upon adsorption coefficient and partitioning rate in case of heavy metals and organic pollutants, and redox reactions in case of inorganic pollutants [45, 66]. Nanomaterials owing to their small size offer larger surfaces for the adsorption of pollutants on their surface. Nanoadsorbents are the nanoscale materials that are used to adsorb dissolved substances or gases. These have emerged as a potential candidates for the purification water/waste water over the conventional methods because of their efficacy, cost effectiveness, less time consumption and insignificant harm to the environment [4, 11, 19]. Nanoadsorbents have shown higher adsorption rates as compared to conventional activated carbon in case of organic compounds [46] The universally used and most potential nanoadsorbents for the adsorption of heavy metals are carbon nanotubes, polymeric nanoadsorbents (dendrimers), metal oxides (manganese oxide, titanium oxide, zinc oxide, magnesium oxide and ferric oxide), nanocomposites, zeolites etc. [19, 46, 66]. Different researchers suggested various metal oxide based nano-adsorbents to effectively adsorb arsenic (inorganic and organic form) [7].

1.4.1 General Properties of Nanoadsorbents

The applicability of nanoadsorbents mainly depends on the innate surface properties as well as on the external functionalization. Because of their nanosize, nanoadsorbents have larger surface area which provide larger platform for the attachment of functional groups and hence increases the number of active adsorption sites thereby making them effective adsorbers of metal ions and other pollutants. In addition to their substantial metal-binding capacities, nanoadsorbents work rapidly and can be regenerated by chemical processes after being exhausted [36, 52, 72]. The properties of nanomaterials including both physical and chemical are directly linked to the innate properties which include the internal composition, external surface structures, apparent size etc. Presence of large surface area with location of atoms on the surface, active adsorption sites contributing to high adsorptive capacity, high chemical activity; high surface binding energy and absence of internal diffusion resistance are the fundamental innate factors influencing the functioning of nanoadsorbents in solution as well as in solid (substrate) phase. Each of these factors significantly affects the atoms or molecules present on the surface and interface thereby increasing the influence of surface affinity, surface binding energy, surface enrichment and number of active adsorption sites towards particular analytes. Another important property which affect the adsorption capacity of nanomaterials is external functionalization wherein intentional modification in the surface properties of nanomaterial is brought by the addition of various functional groups of interest on the surface of nanomaterials resulting in excellent adsorption properties. In general, nanoadsorbents are characterized by high porosity, superior surface area, high adsorption capability, improved mechanical strength, high surface binding energy, and low temperature modification. Some nanoadsorbents like carbon nanotubes exhibit the characteristics of having superior mechanical and chemical properties, higher conducting capabilities and electronic properties in addition to general properties [34].

1.4.2 Types of Nanoadsorbents

Waste water treatment involve different types of nanomaterials which could effectively and efficiently perform the adsorption processes. Some of the most commonly used nanoadsorbents are carbon based nanoadsorbents, polymer based, organic or inorganic based nanoadsorbents.

1. **Polymer based Nanoadsorbents**

In spite of many advantages offered by nanomaterials, there exists certain limitations. Because of the nanosize, nanomaterials are unstable and tend to form agglomerates, which further decreases their adsorption capacities. This problem can be overcome by modifying the nanomaterials with polymers. Polymeric modifications can be performed in two ways: (1) By coating the particles with polymer; (2) by using the polymer as a medium to synthesize and nurture nanoparticles on it [43, 51]. Polymeric nanoadsorbents such as dendrimers (tailor made cross links) are gaining more and more attention because of their utility in adsorption process due to their reversible process and outstanding properties like larger surface area enhancing the adsorption rate, high mechanical rigidity and porosity, and regeneration of adsorbent at mild condition without decrease in sorption capacities [33]. These have been widely utilized for the removal of heavy metals and organic compounds because of their bi-functional nature. The organic compounds present in water or waste water can be adsorbed by the inner hydrophobic shells while heavy metals can be adsorbed by tailoring of the exterior branches of the dendrimers [46]. Nanopolymer spheres could be considered desirable candidate for separation purpose [34]. An integrated dendrimer-ultrafiltration system was designed for the removal of copper ions present in water and to the very surprise, the technique has shown positive results in terms of removal of the preferred ions i.e., copper ions [18]. The adsorbent can be regenerated after adsorption through a pH shift i.e., by simply bringing the pH to 4 [46].

2. **Carbon based Nanoadsorbents**

Carbon based nanomaterials is a category of nanomaterials, which possess unique physical structures, chemical properties, and electronic modifications making them appropriate for sorption processes. The properties which makes them advantageous over other wastewater treatment processes and water purification facilities are larger surface area, rapid kinetics, high selectivity towards aromatics and an excellent sorption capacity for a broad gamut of contaminants. They have the ability of removing both organic and inorganic pollutants. There is a wide range of carbon nanomaterials and include nanotubes made up of carbon, carbon beads, nanoporous carbon and fibers incorporated with carbon [35]. Among these, extensively used are carbon nanotubes (CNTs) and have gained attention over couple of years. CNTs are basically rolling up of graphene sheets in the form of cylinders of diameter of 1 nm [16, 41]. Based on their structure, nanocarbons can be of two types: Single walled carbon nanotubes (SWCNTs: are single layer of graphene sheet rolled in the form of a cylinder,and Multiwalled carbon nanotubes (MWCNTs: multiple layers of graphene sheet having about 0.34 nm of spacing between the adjacent layers folded in the form of multiple cylinders. Surface modifications of CNT increases its overall adsorption capacity [5, 51]. Iron oxides can be added to the surface of CNTs so as to bring rapid exclusion of chromium from water. Such tailored CNTs are known as composite adsorbents [25, 41]. In spite of having exceptional properties, the applications of CNTs are not widely accepted nor recognized because of their high cost and low volume production [16].

3. **Inorganic and organic based Nanoadsorbents**

The removal of metal ions from water or waste water requires the use of nanoadsorbents that contain metal oxide based nanoparticles like TiO_2, SiO_2, MnO, ZnO and Fe_2O_3; nanoparticles with magnetic properties like Fe_3O_4; metal based nanoparticles; CNTs, Zeolites or graphene oxide (GO) [10, 32, 36, 46]. Both organic and inorganic particles of nanoscale range are considered to be more efficient adsorbents when compared with agricultural residues for the removal of wide variety of contaminants present in water and wastewater. The adsorption rate of these nanoscale particles for various contaminants lie in the range between 0.003 to 1119 mg/g [73]. The adsorption process is pH dependent, in case of heavy metals removal by making use of oxides of Zn, Cu and Fe [33]]. Apart from the widely used nanoadsorbents, metal oxide based nanoadsorbents are frequently applied in the adsorption process targeting heavy metals. The use of magnetic nanoparticles offer much advantages over nonmagnetic nanoparticles. The inorganic pollutants within aqueous medium can be successfully and efficiently separated by the magnetic field of these nanoscale magnetic adsorbents [30, 36]. The exceptional properties infatuated by metal oxide based nanoadsorbents which make them a potential competitor for the other physical and chemical conventional processes are their low solubility, low cost, strong mechanical structure, chemical stability and resistance against organic chemicals, alkalis, oxidants and acids and also these are having minimal environmental impact [33].

4. **Other Nanoadsorbents**

Apart from the above said nanoadsorbents, the list of nano sized materials with good adsorption capacity is ever expanding. Other nano sized adsorbents with promising results are nanoclays, nanofibre, xerogels and aerogels [34]. These can also be efficiently applied in the removal of varied variety of noxious waste present in water and waste water. Possession of properties like high surface-to-volume ratio and presence of more active sites adds nanoclays and nanofibres in the category of effective nanoadsorbers.

1.5 Conducting Polymers

The discovery of conducting polymers was made by Alan Heegar, Dr. Hideki Shirakawa and Alan Macdiarmid [59] and before its inventions the polymers were used as insulators in electronic industrial sectors. The conducting polymers are the organic polymers that have potential for electricity conduction (example: Polyacetylene) and the rate of conductivity depends on the number of electrons present in that specific material and their mobility. A general assumption confirms that the outer electron is free to carry and flow charge due to electron bumping factor [38]. These types of compounds generally have metallic conductivity or semiconductor. The conducting polymers shows a lot of advantages and beyond those the "processability" is considered as one of the major advantages mostly by dispersion. Generally, the polymers are thermoplastic (nonthermo-formable) in nature, but in case of insulting polymers these are organic and are potentially efficient to offer higher electrical conductivity [17]. The advanced dispersion and organic synthesis method are used to fine tune their electrical properties (Table 1).

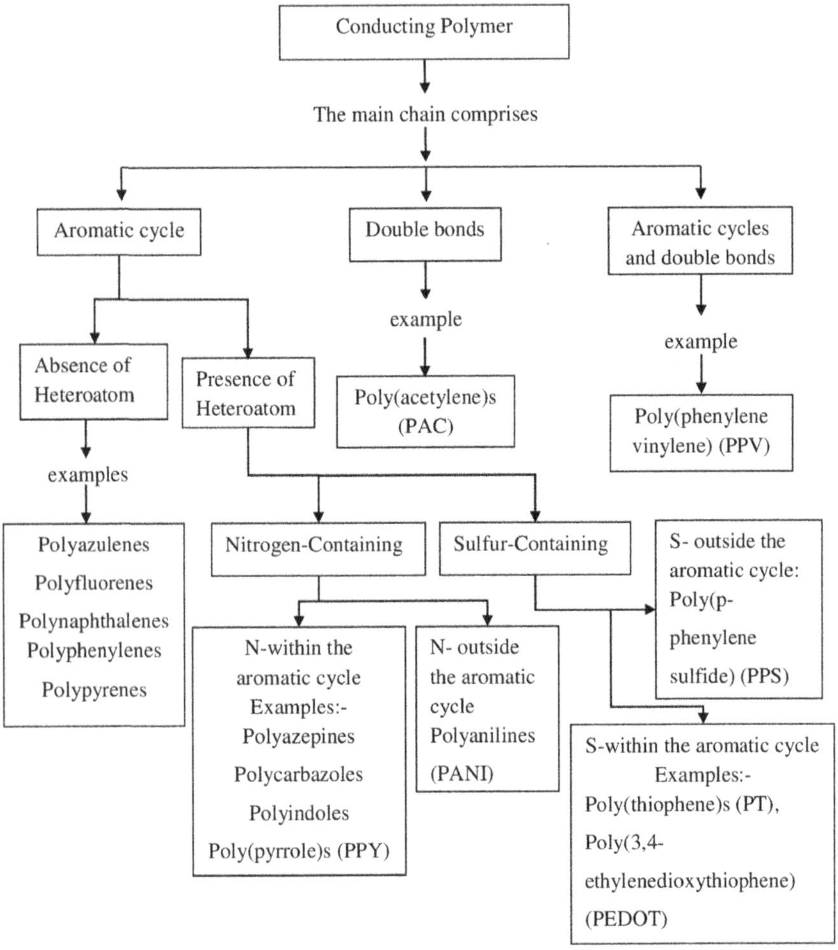

Fig. 3 Types of conducting polymer [50]

Table 1 General properties of conducting polymers

Chemical properties	Possibility of ion transportation, Catalytic properties, Oxidation-reduction behavior, Electrochemical effect, Junction effect and Photoactivity
Mechanical properties	Flexible, Lightweight The surface properties are Non-metallic

1.5.1 Types of Conducting Polymers

The literature confirms that the polymers made up of carbon, as its backbone chain, is called organic polymer and as same when does not contain carbon is called inorganic polymer. But, in the case of conducting polymers the classification is done based on its structure and the attachment of nitrogen and sulfur (described in Fig. 3).

1.5.2 Properties and Applications

The uses of conducting polymers have little large scale several applications (shown in Fig. 4) based on its requirement in specific field due to their poor processability. The conducting polymer have promises for antistatic materials but could be exposed into commercial displays, it has a number of disadvantages due to the manufacturing costs, toxicity, poor solubility in solvents, material inconsistencies, and inability to directly melt process [56]. As per recent literature they are also favorable tools for organic solar cells, and organic light emitting diodes, etc.

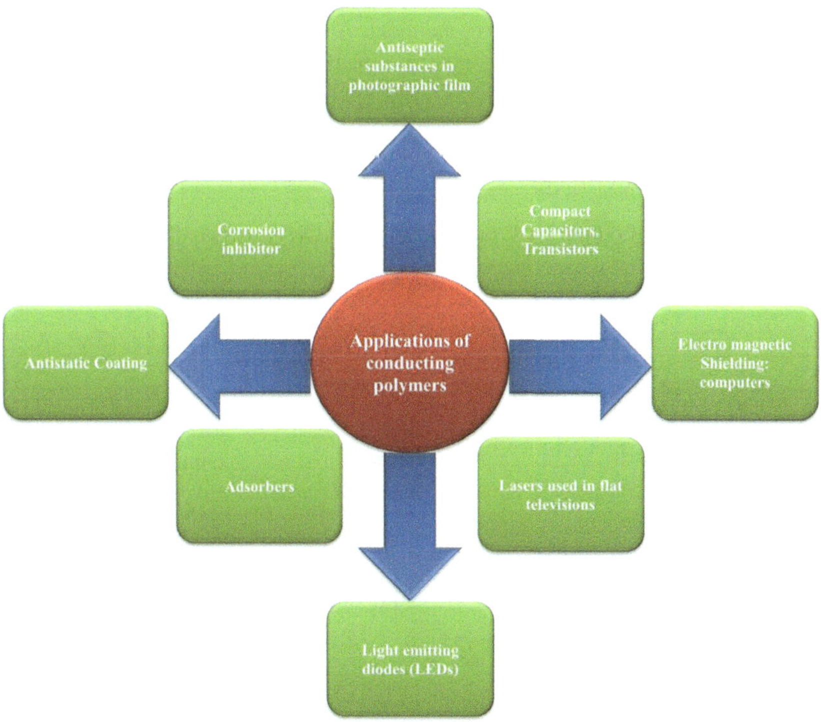

Fig. 4 Different uses of conducting polymers [56, 65]

Current studies on conducting polymers are gaining attraction due to its rapid application with processable things with efficient physical and electrical properties in lower cost. The nanostructured conducting polymers are accomplished as considerably superior capacitance values than their non-nanostructured counterparts. Various researchers have reported applicability of nanostructured conducting polymers as effective adsorbers of heavy metal ions or dye because of their better adsorption capabilities, larger surface area [73].

2 Synthesis of Certain Conducting Polymer Based Nanoadsorbents

Conducting polymers are a class of compounds which consists of monomer units ensuring electrical conductance of the polymer under conditions of doping because of the presence of conjugated chemical bonds in it. There are a number of techniques that control the conjugation bond of conducting polymer. The key requirement is the conservation of conjugate nature of monomers in the synthesis process of conducting polymers [3]. Synthesis of conducting polymers generally consists of two classes-chemical and electrochemical. A chemical polymerization process requires the dissolution of a monomer, a dopant, and an oxidant in a solution at certain optimal temperature. Figure 5 shows a generalized chemical polymerization

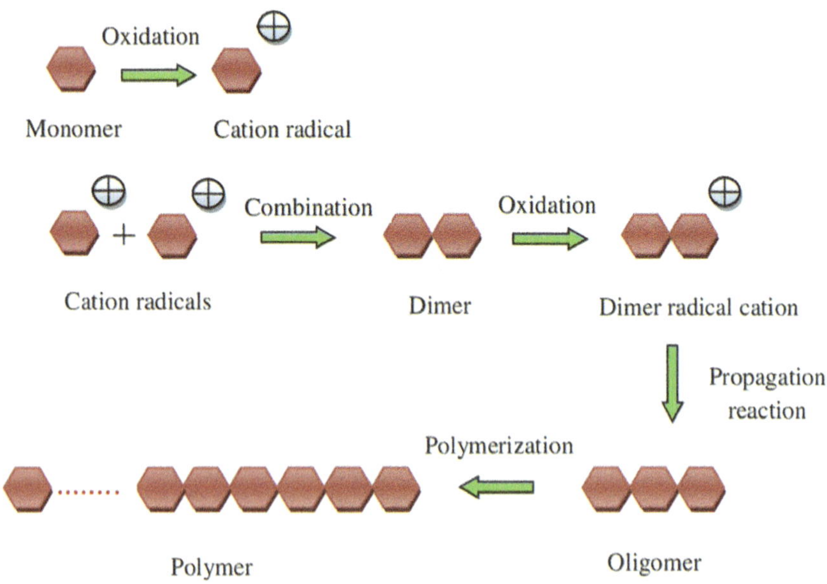

Fig. 5 Generalized chemical polymerization process

process. Although there exist certain uncertainties regarding the mechanism of polymerization, still various researchers agree with the cation-radical mechanism. The mechanism involves oxidation of monomer to form cation radicals having numerous resonance forms. Coupling of two radical cations leads to the formation of dimer which further oxidizes into a dimer radical cation. The reaction propagates resulting in the formation of oligomers followed by polymers and continues until termination of the chain takes place. The entire process of polymerization takes few minutes to some days for completion depending upon the reaction conditions. The resulting product is a mixture of low and high purity conducting polymers which need to be filtered, washed and dried in order to yield the polymer with highest purity. The electrochemical polymerization process is quite similar to chemical polymerization. The only difference that exists among them is in the generation of radical cation which is being done by applying a potential difference across the ends of an electrolyte solution containing a mix of monomer and a dopant.

2.1 Polypyrrole

During the last decade, the rate of studies being conducted on the applicability and benefits derived from conducting polymer - polypyrrole (PPy) have shown a significant increase. The reason behind the increase is probably due to advantageous morphological structures and structural modifications exhibited by PPy [9]. The attention seeking properties of PPy include their uncomplicated pathways of synthesis, enhanced mechanical and conductivity properties and excellent environmental stability [47]. Structurally, pure PPy contains benzene ring (shown in Fig. 6) and is an insulator which can be made to work as a semiconductor by doping with an oxidant.

PPy can be synthesized by two methods viz., chemical and electrochemical method. Synthesis of PPy by chemical polymerization is brought about by either chemical oxidants or by the deposition of chemical vapour in aqueous and non-aqueous solvents and is considered to be a simple and fast process with no requirement of special instrument [47]. Chemical polymerization occurs using a surfactant and an oxidant. The need for surfactant is fulfilled by Sodium dodecylbenzene sulfonate (DBSNa) and Iron(III) chloride hexahydrate ($FeCl_3.6H_2O$) acts as an oxidant. The

Fig. 6 Chemical structure of polypyrrole

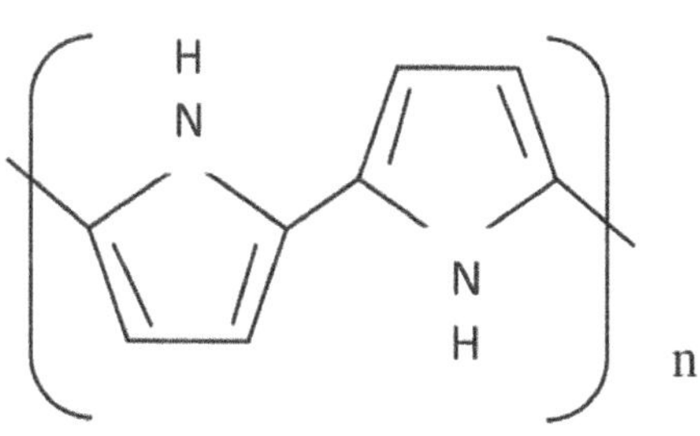

synthesis of Polypyrrole as per [9] involves the dissolution of 3.48 g of DBSNa surfactant and 27.03 g of $FeCl_3.6H_2O$ oxidant in 100 cc of distilled water separately and then mixing of the two solutions in a reaction flask of 500 ml capacity followed by rapid stirring for a period of 15 min at 0 °C and again at 28 °C temperature. In the resulting mixture, drop wise addition of 0.15 molar solution of pyrrole monomer takes place at a reaction time of 20 min and this step repeats after a time period of 4 h followed by filtering and washing of solution, a number of times in order to yield PPy with highest purity. The black colored powder (PPy powder) obtained is then oven dried at 60 °C for 8 h. in order to yield better results. Electrochemical synthesis of PPy is conventionally been carried out in a system consisting of molecular solvent/electrolyte say for example, acetonitrile/lithium perchlorate. Pringle and coworkers in [54] reported the dual applicability of ionic liquids for the electrochemical cycling of the PPy films serving the purpose of both growth medium and electrolyte. Significant modifications in surface morphologies of PPy film can be made possible by making use of ionic liquids for enhancing the electrochemistry. The widely used ionic liquids acting as growth medium are 1-butyl-3-methylimidazolium hexafluorophosphate, 1-ethyl-3-methylimidazolium bis(trifluoromethanesulfonyl) amide and N,N-butylmethylpyrrolidinium bis(trifluoromethanesulfonyl) amide.

2.2 *Polyalinine*

Polyaniline (PANI) as shown in Fig. 7, is a member of conducting polymer family can be distinguished from other members in having possession of characteristics like uncomplicated synthesis mechanism and high environmental constancy. Presence of conductivity endorsing emeraldine salt and different oxidation states fosters applicability of PANI in the disciplines of electronics and engineering [2] but its utilization is not widely recognized as it becomes unstable at neutral pH and higher temperatures. A pH of 7 and temperature exceeding 150 °C results in dedoping of polymer chain thereby limiting the conductivity of the polymer. PANI exhibit different featuring properties. The chemical structure, physical morphology and oxidation-reduction states are prepared in accordance with the synthesis procedure requirement, types of oxidant used, type and amount of dopant and solvent involved, temperature and time regimes [13] for their extensive use in different disciplines.

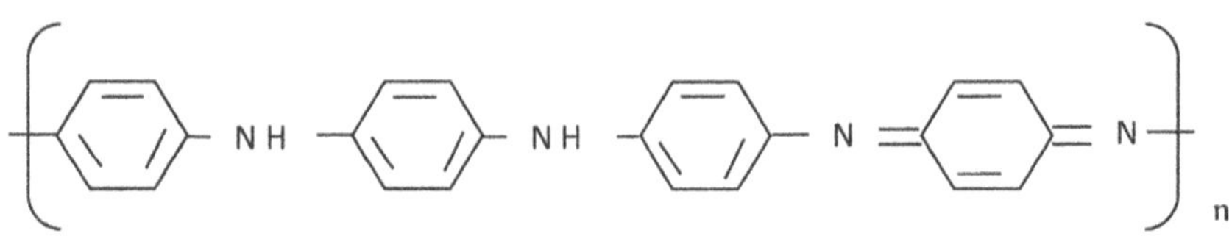

Fig. 7 Chemical structure of Polyaniline

Synthesis of PANI occurs by polymerization of aniline either chemically or electrochemically in the presence of specific dopants.

1. Electrochemical synthesis of PANI

Electrochemical polymerization is considered to be the best method for the synthesis of PANI as compared to the chemical method because of its ability to produce resultant products in the purest form and free from admixtures. There is no requirement of special procedures for the purification of PANI from the solvent or unreacted monomer and the polymerization simply takes place in an electrolyte solution containing an acid [13]. The electrochemical polymerization is favorable as it involves direct deposition of polymer on the electrode making the analysis easy. As stated by different researchers, electrochemical polymerization is a three steps process. In the first step, monomer placed at anode is oxidized resulting in the formation of soluble oligomers which are deposited through nucleation and growth process in the second step. The third and last step involves chain propagation by solid state polymerization [26]. Generally, synthesis of electro conducting polymers by electrochemical route strongly depends upon the framework of the doping anion, electrode material and the temperature at which polymerization occur [26, 55].

2. Synthesis by Chemical polymerization

Chemical synthesis of PANI takes place in acidic medium by making use of an oxidizing agent., Ammonium persulfate, cerium nitrate, hydrogen peroxide, potassium dichromate, etc. are the frequently used oxidants in the synthesis process [24, 76]. Among the wide variety of oxidants used, the most common is ammonium persulfate with the highest yield of about 90% [13]. Polyaniline preparation in emeraldine salt form is basically done by performing the process of polymerization in an acidic medium. In nutshell, chemical synthesis can be regarded as the simplest approach in the preparation of conducting polymers possessing variable physicochemical properties and supermolecular structures [13].

2.3 Polythiophenes Derivatives

Polythiophene shows high potential for environmental uses, as its conductivity goes beyond 100 S/cm which possesses chemical as well as thermal stability [39]. The Polythiophenes can be easily made-up by direct polymerization of monomers. After that, it could be exercised in the immobilization of enzymes as well as aptamers into the polythiophene layer [44]. The ploythiophenes can be synthesized by electrochemical synthesis as well as chemical methods. In recent times, Polythiophene polymer got more attention in synthesis search as well as industrial sectors due to its less band gap energy, soaring environmental immovability, superior thermal steadiness, and beyond that, the exhilarating properties such as: outstanding electronic, optical and semiconducting activities as well as improved mechanical strength, rigidity and

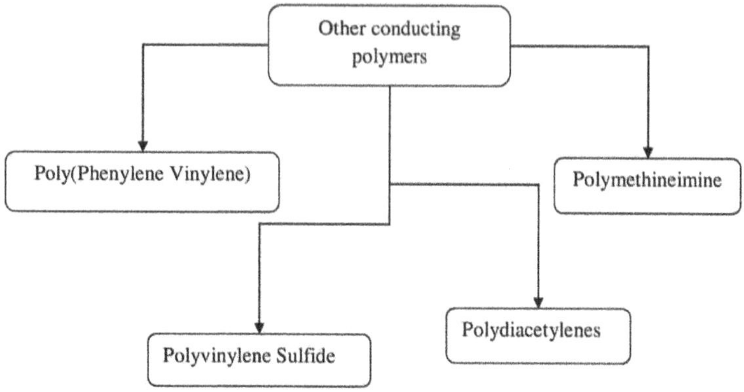

Fig. 8 Other conducting polymers [47]

simplicity in the process gathered considerable notice to the composites of polythiophene [56]. They are extensively used in solar cells as they are potentially efficient in terms of better contact formation with metal electrodes and are also stable at ambient conditions. It also shows outstanding optical properties due to its transparent properties moreover, it is considered as a tremendous essential conducting polymer as in the backbone it contains conjugated double bonds. Polythiophene conducting polymers, can be tainted to enhance their conducting capabilities by the creation of polarons and bipolarons in the backbone through their redox states [56].

2.4 Other Conducting Polymers

Beyond the above discuses conducting polymers, there are other conducting polymers, shows capabilities for various environmental applications (Fig. 8).

Poly (phenylene vinylene) shows higher conductivity level in comparison with polyphenylene or polyacetylene. The maximum occupied ¼ band of poly(phenylene vinylene) is approximately 2.8 eV [47]. On the other hand, Polydiacetylens are generally formed though solid-state reaction of single crystal. It also shows there is higher ionization potential value in polymethineimine than polyacetylene i.e. 8 eV [12].

3 Recommendations and Future Prospects

Conducting polymer based nanoadsorbents have been successfully used for the removal of large number of heavy metal and dyes from waste water. These nanoadsorbents are eco-friendly as their extensive use cause negligible environmental pollution. Still, processing and commercialization of these nanoadsorbents are not properly addressed. Also, detailed investigation behind the improved properties of conducting polymer based nanoadsorbents is lacking [6]. Therefore, for the commercial operation of these effective nanoadsorbents materials, the selection of the functional groups for activation of materials in the nanoadsorbent have to be properly decided to obtain enhanced kinetic sorption [10]. Adsorbent durability as well as the easiness in regeneration with required degree of recovery is an important parameter which need to be worked on. Detailed investigation of the behaviour of multiple component system is required as the interactions between the metal ions would play a major role in deciding the removal selectivity of the desired component. In particular, still, more intensive research work is required to make these nanoadsorbents work efficiently and for their commercialization.

4 Conclusion

Treatment of wastewater is the need of the hour because of decline in the availability of the potable water, stringent waste water disposal guidelines. Hence there is an urgent need for new and advanced technologies for water purification that ensures suitability of water for drinking purpose as well as eliminate the pollutants or contaminants present in water. Among the different water treatment techniques adopted, adsorption is considered most effective technique. Flexible and simple design, ease of operation, insensitivity to toxic pollutants, low cost in terms of installation, operation and maintenance, makes adsorption process as most effective water treatment technique. As mentioned above, adsorption process is affected by a number of different factors including adsorbent dosages, temperature, solution pH, contact time and chemical nature of the adsorbent. Further, the adsorption efficacy is enhanced by the use of nanotechnology as it has emerged as an effective tool for efficiently removing toxic materials from water or wastewater. Much attention is being paid to the use of nanomaterials as adsorbers. Nano-engineered materials known as nanoadsorbents such as polymer based, carbon based, inorganic and organic based nanoadsorbents offer the potential for water and wastewater purification over the conventional methods because of their efficiencies, cost effectiveness and environmental friendliness. To be very specific, conducting polymer based nanoadsorbents have high electrical conductance, are environmentally stable and are potential candidates for the adsorption and exclusion of metals and dyes present in water or wastewater which are toxic to the living creatures and harmful for the aquatic environment. Among the conducting polymers, PANI and its derivatives, because of the presence of huge amount of amine and imine functional groups in the polymeric chain, are used as effective nanoadsorbers of heavy metal ions/dyes.

References

1. Akpor, O.B., Otohinoyi, D.A., Olaolu, D.T., Aderiye, B.I.: Pollutants in wastewater effluents: impacts and remediation processes. Int. J. Environ. Res. Earth Sci. **3**(3), 050–059 (2014)
2. Aksu, Z., Kutsal, T.: A bioseparation process for removing lead (II) ions from waste water by using C. vulgaris. J. Chem. Technol. Biotechnol. **52**(1), 109–118 (1991)
3. Alemayehu, T., Himariam, B.: Synthesis and characterization of conducting polymers : a review paper. Int. J. Recent Res. Phys. Chem. Sci. **1**(1), 24–28 (2014)
4. Amin, M.T., Alazba, A.A., Manzoor, U.: A review of removal of pollutants from water/wastewater using different types of nanomaterials. Adv. Mater. Sci. Eng. (2014)
5. Anjum, M., Miandad, R., Waqas, M., Gehany, F., Barakat, M.A.: Remediation of wastewater using various nano-materials. Arab. J. Chem. (2016)
6. Arunachalam, P.: Polymer-based Nanocomposites for Energy and Environmental Applications, pp. 185–202. King Saud University, Riyadh, Saudi Arabia (2018)
7. Ashraf, S., Siddiqa, A., Shahida, S., Qaisar, S.: Titanium-based nanocomposite materials for arsenic removal from water: a review. Heliyon **5**(5), e01577 (2019)
8. Awasthi, A., Jadhao, P., Kumari, K.: Clay nano-adsorbent: structures, applications and mechanism for water treatment. SN Appl. Sci. **1**(9), 1076 (2019)
9. Bahraeian, S., Abron, K., Pourjafarian, F., Majid, R.A.: Study on synthesis of polypyrrole via chemical polymerization method. In: Advanced Materials Research, vol. 795, pp. 707–710. Trans Tech Publications Ltd. (2013)
10. Bhatia, M., Babu, R.S., Sonawane, S.H., Gogate, P.R., Girdhar, A., Reddy, E.R., Pola, M.: Application of nanoadsorbents for removal of lead from water. Int. J. Environ. Sci. Technol. **14**(5), 1135–1154 (2017)
11. Bhattacharya, S., Saha, I., Mukhopadhyay, A., Chattopadhyay, D., Chand, U.: Role of nanotechnology in water treatment and purification: potential applications and implications. Int J Chem Sci Technol **3**(3), 59–64 (2013)
12. Bloor, D., Chance, R.R. (eds.): Polydiacetylenes: synthesis, structure and electronic properties, vol. 102. Springer (2013)
13. Boeva, Z.A., Sergeyev, V.G.: Polyaniline: synthesis, properties, and application. Polymer Sci. Ser. C **56**(1), 144–153 (2014)
14. Bora, T., Dutta, J.: Applications of nanotechnology in wastewater treatment—a review. J. Nanosci. Nanotechnol. **14**(1), 613–626 (2014)
15. Cao, Y., Zhang, S., Wang, G., Li, T., Xu, X., Deng O., Zhang, Y., Pu, Y.: Enhancing the soil heavy metals removal efficiency by adding HPMA and PBTCA along with plant washing agents. J. Hazard. Mater. **339**, 33–42 (2017)
16. Chatterjee, A., Deopura, B.L.: Carbon nanotubes and nanofibre: an overview. Fibers Polym. **3**(4), 134–139 (2002)
17. Culebras, M., Gómez, C.M., Cantarero, A.: Review on polymers for thermoelectric applications. Materials **7**(9), 6701–6732 (2014)
18. Diallo, M.S., Christie, S., Swaminathan, P., Johnson, J.H., Goddard, W.A.: Dendrimer enhanced ultrafiltration. 1. Recovery of Cu(II) from aqueous solutions using PAMAM dendrimers with ethylene diamine core and terminal NH_2 groups. Environ. Sci. Technol. **39**(5), 1366–1377 (2005)
19. Dinesha, B.L., Sharanagouda, H., Udaykumar, N., Ramachandr, C.T., Dandekar, A.B.: Removal of pollutants from water/waste water using nano-adsorbents: a potential pollution mitigation. Int. J. Curr. Microbiol. Appl. Sci. **6**(10), 4868–4872 (2017)

20. Dixit, R., Wasiullah, M.D., Pandiyan, K., Singh, U.B., Sahu, A., Shukla, R., Singh, B.P., Rai, J.P., Sharma, P.K., Lade, H., Paul, D.: Bioremediation of heavy metals from soil and aquatic environment: an overview of principles and criteria of fundamental processes. Sustainability **7**, 2189–2212 (2015)
21. Elango, G., Rathika, G., Elango, S.: Physico-chemical parameters of textile dyeing effluent and its impacts with case study. Int. J. Res. Chem. Environ. **7**, 17–24 (2016)
22. Evelyn, M.I., Tyav, T.T.: Environmental pollution in Nigeria: the need for awareness creation for sustainable development. J. Res. Forestry Wildlife Environ. **4**(2), 92–105 (2012)
23. Firdaus, M.L., Krisnanto, N., Alwi, W., Muhammad, R., Serunting, M.A.: Adsorption of textile dye by activated carbon made from rice straw and palm oil midrib. Aceh Int. J. Sci. Technol. **7**(1), 1–7 (2017)
24. Gomes E.C., Oliveira, M.A.S.: Chemical polymerization of aniline in hydrochloric acid (HCl) and formic acid (HCOOH) media. Differences between the two synthesized polyanilines. Am. J. Polym. Sci. **2**(2), 5–13 (2012)
25. Gupta, V.K., Agarwal, S., Saleh, T.A.: Chromium removal by combining the magnetic properties of iron oxide with adsorption properties of carbon nanotubes. Water Res. **45**(6), 2207–2212 (2011)
26. Heinze, J., Frontana-Uribe, B.A., Ludwigs, S.: Electrochemistry of conducting polymers—Persistent models and new concepts. Chem. Rev. **110**(8), 4724–4771 (2010)
27. Holkar, C.R., Jadhav, A.J., Pinjari, D.V., Mahamuni, N.M., Pandit, A.B.: A critical review on textile wastewater treatments: possible approaches. J. Environ. Manage. **182**, 351–366 (2016)
28. Iftekhar, S., Ramasamy, D.L., Srivastava, V., Asif, M.B., Sillanpää, M.: Understanding the factors affecting the adsorption of Lanthanum using different adsorbents: a critical review. Chemosphere **204**, 413–430 (2018)
29. Ilin, I., Kalinina, O., Iliashenko, O., Levina, A.: Sustainable urban development as a driver of safety system development of the urban underground. Procedia Eng. **165**, 1673–1682 (2016)
30. Jia, S., Xing, G., Xu, Q., Shi, Z.: Influence of magnetic field on carbon nano-materials produced in liquid arc. In: 2006 1st IEEE International Conference on Nano/Micro Engineered and Molecular Systems, pp. 959–963. IEEE (2006)
31. Järup, L.: Hazards of heavy metal contamination. Br. Med. Bull. **68**(1), 167–182 (2003)
32. Kaushal, A., Singh, S.K.: Removal of heavy metals by nanoadsorbents: a review. J. Environ. Biotechnol. Res. **6**(1), 96–104 (2017)
33. Kavithayeni, V., Geetha, K., Akash Prabhu, S.: A review on dye reduction mechanism using nano adsorbents in waste water. Adsorption **43**, 44 (2019)
34. Khajeh, M., Laurent, S., Dastafkan, K.: Nanoadsorbents: classification, preparation, and applications (with emphasis on aqueous media). Chem. Rev. **113**(10), 7728–7768 (2013)
35. Khin, M.M., Nair, A.S., Babu, V.J., Murugan, R., Ramakrishna, S.: A review on nanomaterials for environmental remediation. Energy Environ. Sci. **5**(8), 8075–8109 (2012)
36. Kurniawan, T.A., Sillanpää, M.E., Sillanpää, M.: Nanoadsorbents for remediation of aquatic environment: local and practical solutions for global water pollution problems. Crit. Rev. Environ. Sci. Technol. **42**(12), 1233–1295 (2012)
37. Lajayer, B.A., Ghorbanpour, M., Nikabadi, S.: Heavy metals in contaminated environment: destiny of secondary metabolite biosynthesis, oxidative status and phytoextraction in medicinal plants. Ecotoxicol. Environ. Saf. **145**, 377–390 (2017)
38. Le, T.H., Kim, Y., Yoon, H.: Electrical and electrochemical properties of conducting polymers. Polymers **9**(4), 150 (2017)
39. Liu, H., Ge, J., Ma, E., Yang, L.: Advanced biomaterials for biosensor and theranostics. In: Biomaterials in Translational Medicine, pp. 213–255. Academic Press (2019)
40. Long, J., Li, H., Jiang, D., Luo, D., Chen, Y., Xia, J., Chen, D.: Biosorption of strontium (II) from aqueous solutions by Bacillus cereus isolated from strontium hyperaccumulator Andropogon gayanus. Process Saf. Environ. Prot. **111**, 23–30 (2017)
41. Lu, H., Wang, J., Stoller, M., Wang, T., Bao, Y., Hao, H.: An overview of nanomaterials for water and wastewater treatment. Adv. Mat. Sci. Eng. (2016)

42. Mahmud, H.N.M.E., Huq, A.O., Binti Yahya, R.: The removal of heavy metal ions from wastewater/aqueous solution using polypyrrole-based adsorbents: a review. RSC Adv. **6**(18), 14778–14791 (2016)
43. Malana, M.A., Qureshi, R.B., Ashiq, M.N.: Adsorption studies of arsenic on nano aluminium doped manganese copper ferrite polymer (MA, VA, AA) composite: kinetics and mechanism. Chem. Eng. J. **172**(2–3), 721–727 (2011)
44. Mehmood, U., Al-Ahmed, A., Hussein, I.A.: Review on recent advances in polythiophene based photovoltaic devices. Renew. Sustain. Energy Rev. **57**, 550–561 (2016)
45. Mehrizad, A., Zare, K., Dashti, K. H., Dastmalchi, S., Aghaie, H., Gharbani, P.: Kinetic and thermodynamic studies of adsorption of 4-chloro-2-nitrophenol on nano-TiO_2 (2011)
46. Mehta, K., Sata, P., Saraswat, A., Mehta, D.: Nanomaterials in water purification. Int. J. Adv. Eng. Res. Dev. In: Conference of Nanotechnology & Applications in Civil Engineering-2018, vol. 5, Special Issue 3 (2018)
47. Mishra, A.K.: Conducting polymers: concepts and applications. J. Atom. Mol. Condens. Nano Phys. **5**(2), 159–193 (2018)
48. Mohan, D., Pittman, C.U., Jr.: Arsenic removal from water/wastewater using adsorbents—A critical review. J. Hazard. Mater. **142**(1–2), 1–53 (2007)
49. Morosanu, I., Teodosiu, C., Paduraru, C., Ibanescu, D., Tofan, L.: Biosorption of lead ions from aqueous effluents by rapeseed biomass. New Biotechnol. **39**, 110–124 (2017)
50. Naarmann, H.: Polymers Ullmann's Encyclopedia of Industrial Chemistry Electrically Conducting (2000). https://doi.org/10.1002/14356007.a21_429
51. Ossman, M.E.: Similarity removal of heavy metals from aqueous solutions using advanced materials, with emphasis of synthetic and nanomaterials. Water Desal. Res. J. **1**(1), 1–44 (2017)
52. Pacheco, S., Medina, M., Valencia, F., Tapia, J.: Removal of inorganic mercury from polluted water using structured nanoparticles. J. Environ. Eng. **132**(3), 342–349 (2006)
53. Popa, C., Petrus, M.: Heavy metals impact at plants using photoacoustic spectroscopy technology with tunable CO_2 laser in the quantification of gaseous molecules. Microchem. J. **134**, 390–399 (2017)
54. Pringle, J.M., Efthimiadis, J., Howlett, P.C., Efthimiadis, J., MacFarlane, D.R., Chaplin, A.B., Forsyth, M.: Electrochemical synthesis of polypyrrole in ionic liquids. Polymer **45**(5), 1447–1453 (2004)
55. Pron, A., Rannou, P.: Processible conjugated polymers: from organic semiconductors to organic metals and superconductors. Prog. Polym. Sci. **27**(1), 135–190 (2002)
56. Ramesan, M.T., Suhailath, K.: Role of nanoparticles on polymer composites. In: Micro and Nano Fibrillar Composites (MFCs and NFCs) from Polymer Blends, pp. 301–326. Woodhead Publishing (2017)
57. Rana, R.S., Singh, P., Kandari, V., Singh, R., Dobhal, R., Gupta, S.: A review on characterization and bioremediation of pharmaceutical industries' wastewater: an Indian perspective. Appl. Water Sci. **7**(1), 1–12 (2017)
58. Salleh, M.A.M., Mahmoud, D.K., Karim, W.A.W.A., Idris, A.: Cationic and anionic dye adsorption by agricultural solid wastes: a comprehensive review. Desalination **280**(1–3), 1–13 (2011)
59. Science History Institute: vol. 13, p. 28. (2019). Accessed on 07 Nov 2019 (2019)
60. Sherene, T.: Mobility and transport of heavy metals in polluted soil environment. In: Biol. Forum Int. J. **2**(2), 112–121 (2010)
61. Siddique, K., Rizwan, M., Shahid, M.J., Ali, S., Ahmad, R., Rizvi, H.: Textile wastewater treatment options: a critical review. In: Enhancing Cleanup of Environmental Pollutants, pp. 183–207. Springer, Cham (2017)
62. Sostaric, T.D., Petrovic, M.S., Pastor, F.T., Loncarevic, D.R., Petrovic, J.T., Milojkovic, J.V., Stojanovic, M.D.: Study of heavy metals biosorption on native and alkali-treated apricot shells and its application in wastewater treatment. J. Mol. Liq. **259**, 340–349 (2018)
63. Sun, B., Zhang, X., Yin, Y., Sun, H., Ge, H., Li, W.: Effects of sulforaphane and vitamin E on cognitive disorder and oxidative damage in lead-exposed mice hippocampus at lactation. J. Trace Elem. Med Biol. **44**, 88–92 (2017)

64. Tan, I.A.W., Chan, J.C., Hameed, B.H., Lim, L.L.P.: Adsorption behavior of cadmium ions onto phosphoric acid-impregnated microwave-induced mesoporous activated carbon. J. Water Process Eng. **14**, 60–70 (2016)
65. Tebyetekerwa, M., Wang, X., Wu, Y., Yang, S., Zhu, M., Ramakrishna, S.: Controlled synergistic strategy to fabricate 3D-skeletal hetero-nanosponges with high performance for flexible energy storage applications. J. Mater. Chem. A **5**(40), 21114–21121 (2017)
66. Tyagi, I., Gupta, V.K., Sadegh, H., Ghoshekandi, R.S., Makhlouf, A.S.H.: Nanoparticles as adsorbent; a positive approach for removal of noxious metal ions: a review. Sci. Technol. Dev. **34**(3), 195–214 (2017)
67. Verma, M., Schneider, J.S.: Strain specific effects of low level lead exposure on associative learning and memory in rats. Neurotoxicology **62**, 186–191 (2017)
68. Vijayaraghavan, K., Palanivelu, K., Velan, M.: Biosorption of copper(II) and cobalt(II) from aqueous solutions by crab shell particles. Biores. Technol. **97**(12), 1411–1419 (2006)
69. Wang, S., Ang, H.M., Tadé, M.O.: Novel applications of red mud as coagulant, adsorbent and catalyst for environmentally benign processes. Chemosphere **72**(11), 1621–1635 (2008)
70. Wang, X., Han, J., Xu, L., Zhang, Q.: Spatial and seasonal variations of the contamination within water body of the Grand Canal, China. Environ. Pollut. **158**, 1513–1520 (2010)
71. Wilson, K., Yang, H., Seo, C.W., Marshall, W.E.: Select metal adsorption by activated carbon made from peanut shells. Biores. Technol. **97**(18), 2266–2270 (2006)
72. Yang, K., Xing, B.: Desorption of polycyclic aromatic hydrocarbons from carbon nanomaterials in water. Environ. Pollut. **145**(2), 529–537 (2007)
73. Zare, E.N., Motahari, A., Sillanpää, M.: Nanoadsorbents based on conducting polymer nanocomposites with main focus on polyaniline and its derivatives for removal of heavy metal ions/dyes: a review. Environ. Res. **162**, 173–195 (2018)
74. Zhang, Z., Hou, Z., Yang, C., Ma, C., Tao, F., Xu, P.: Degradation of n-alkanes and polycyclic aromatic hydrocarbons in petroleum by a newly isolated Pseudomonas aeruginosa DQ8. Biores. Technol. **102**(5), 4111–4116 (2011)
75. Zhang, C., Yu, K., Li, F., Xiang, J.: Acute toxic effects of zinc and mercury on survival, standard metabolism, and metal accumulation in juvenile ridgetail white prawn, *Exopalaemon carinicauda*. Ecotoxicol. Environ. Saf. **145**, 549–556 (2017)
76. Zilberman, M., Titelman, G.I., Siegmann, A., Haba, Y., Narkis, M., Alperstein, D.: Conductive blends of thermally dodecylbenzene sulfonic acid-doped polyaniline with thermoplastic polymers. J. Appl. Polym. Sci. **66**(2), 243–253 (1997)

Chemical, Gas and Optical Sensors Based on Conducting Polymers

Subramanian Nellaiappan, K. S. Shalini Devi, Stalin Selvaraj, Uma Maheswari Krishnan, and Jatinder Vir Yakhmi

Abstract Sensing technology has evolved from large systems with immovable components to flexible wearable devices capable of non-invasive and location-independent detection. With the demand for portable systems capable of real-time monitoring and accurate detection at ambient conditions increasing in every sector, there arises a need to transform or replace the conventional sensing elements with superior alternatives. Conducting polymers have unique features of processability and flexibility that make them prime candidates for developing new and advanced wearable devices. They represent a promising class of materials for sensing applications whose true potential is yet to be harnessed for device fabrication. Conducting polymers possess several advantages for use as sensing materials. These include their ability to respond to chemical and gaseous species through a change in their conductance at ambient conditions, large scale production and tunable electrical properties. This chapter focuses on the advances made in the development of conducting polymer-based sensors to detect chemical molecules, gaseous analytes, and optical detection of molecules of interest. The use of hybrid conducting polymer systems incorporating nanomaterials and metal oxides towards sensing pollutants, pharmaceuticals, microbes, volatile gases in the environment and breath, and clinically relevant biomarkers has been discussed in detail. The challenges involved in conducting polymer-based systems and the potential for the evolution of this class of sensors by integrating emerging technologies are also presented.

S. Nellaiappan · K. S. Shalini Devi · S. Selvaraj · U. M. Krishnan
Centre for Nanotechnology and Advanced Biomaterials, SASTRA Deemed University, Thanjavur 613401, India

S. Nellaiappan · S. Selvaraj · U. M. Krishnan
School of Chemical and Biotechnology, SASTRA Deemed University, Thanjavur 613401, India

U. M. Krishnan (✉)
School of Arts, Science and Humanities, SASTRA Deemed University, Thanjavur 613401, India
e-mail: umakrishnan@sastra.edu

J. V. Yakhmi
Physics Group, Bhabha Atomic Research Centre, Mumbai, India

© Springer Nature Switzerland AG 2021

S. Shahabuddin et al. (eds.), *Advances in Hybrid Conducting Polymer Technology*, Engineering Materials, https://doi.org/10.1007/978-3-030-62090-5_8

Keywords Conducting polymers · Chemical sensors · Gas sensing · Breath analysis · Hybrid systems · Optical sensors

1 Introduction

Sensors are devices that can qualitatively detect the presence of a species of interest commonly denoted as the analyte or can quantitatively measure the amount of the analyte present in the given sample. The sensor typically consists of a sensing element whose properties vary in the presence of a specific analyte leading to its detection or quantification and a transducing element that converts the change in property to a measurable output [1]. Sensors have become an integral part of everyday life. They have been extensively used in healthcare organizations for clinical diagnosis, in industries for detection of leaks, spills, and to monitor reactions, in the analysis of samples for purity and adulteration, in automobiles, aircraft and in space shuttles to assess various performance parameters and environmental parameters. Sensors have been classified based on the transducers employed as optical, electrical, electrochemical, calorimetric, and piezoelectric sensors [2]. The sensors can have either a chemical molecule or a biomolecule as the sensing element. Thus, biosensors can be considered a sub-type of chemical sensors with higher selectivity but poor stability. Among the different types of sensors, the most extensively employed transduction mechanisms are the electrical, electrochemical, and optical sensors [3]. Between the use of a chemical moiety or a biological molecule as the sensing element, chemical sensors have the advantage of superior stability, unlike the biomolecules that are susceptible to loss of function due to changes in pH, ionic environment, and temperature [4]. Sensors have been used to detect a diverse range of analytes. These include gaseous analytes, ions, small molecules, pathogens, and biomolecules.

Among these, sensors for the detection of gases and volatiles are more commonly encountered in industries as well as environment monitoring [5]. The use of interfacial layers on the sensing element confers improved surface area, analyte capture and in the case of electrical and electrochemical sensors, have also enhanced electron transfer at the surface of the sensing element [6]. Sensors with such interfacial layers exhibit superior sensitivity, detection range, and rapid response time [7]. Several materials have been explored as effective interfacial layers in sensors. Amongst these, metal oxide-based sensors have been most widely used. But these sensors are limited by their high operating temperatures [8]. Replacing metal oxide interfaces by conducting polymers have reduced the operating temperatures of the sensors without compromising on the sensing characteristics [9]. Further, conducting polymers have also served as immobilization matrices to retain sensing molecules on the sensor surface [10]. Owing to their electrical conductivity, conducting polymers also have been employed as transducing elements [11]. Thus, it is evident that conducting polymers can contribute to significant improvement in the sensing characteristics of different types of sensors. The present chapter discusses the salient advances made in the sensing technology involving chemical, gaseous, or optical sensors with conducting polymers (Fig. 1).

2 Overview of Hybrid Conducting Polymers-Based Chemical Sensors

Conducting polymers are generally organic polymers and after the revolutionary work of MacDiarmid in 1977, have elicited considerable interest due to their intrinsic electrical conductivity, which is quite similar to metal conductors [12, 13]. The unique delocalized pi electron system over the conjugated polymeric backbone confers conducting properties to this class of polymers. Conducting polymers are generally capable of responding through either a reversible or irreversible change in the physicochemical properties of polymers for even small variations in the external environment. The responses can be induced through various stimuli such as ionic strength, temperature, pH, mechanical forces, and light irradiation, electric, magnetic and acoustic fields, and specific analytes, which have been involved in changing polymer chain conformation and solubility [14]. A wide range of organic conducting polymers like polyaniline (PANI), polypyrrole (Ppy), polythiophene, poly(aminophenylboronic acid), polyfluorene, polyazulene, polyterthiophene and polyaminonaphthalenes, etc., have been explored for different applications (Fig. 2). However, under ambient conditions, these polymers exhibit very poor intrinsic conductivity and are made conducting by the addition of small amounts of dopants such as carbon nanoparticles (CNP's), metal nanoparticles (MNps), and dyes [15]. These dopants can be either oxidizing (remove electrons) or reducing (introduce electrons) that enable the mobility of the delocalized pi electrons throughout the conjugated polymer backbone [16]. The extent of doping determines the magnitude of conductivity of the polymer. Additional properties such as mechanical strength and selectivity have been also incorporated through doping with metal-organic frameworks (MOFs) and enzymes. Doping results in oxidation or reduction leading to the formation of polarons that confer electrical conductivity. The mechanical strength and the processability of conducting polymers (CPs) have also been improved by doping the conducting polymers with nanotubes or other nanomaterials [10]. These doped conducting polymers which are functionalized via chemical oxidation, electrochemical deposition, or through the entrapment of specific enzymes are collectively termed as "Hybrid Conducting Polymers (HCPs)" (Fig. 3). HCPs behave as the most promising candidates for detecting various analytes in the field of sensors due to their high chemical stability, excellent conductivity, mechanical strength, and large surface area [17]. Hence, these polymeric materials are currently employed in the manufacturing of sensor devices, mobile phone displays as well as in manufacturing new color screens for videos and television sets [18]. Conducting polymers exhibit good affinity for various chemical species and hence have been explored as a sensor material.

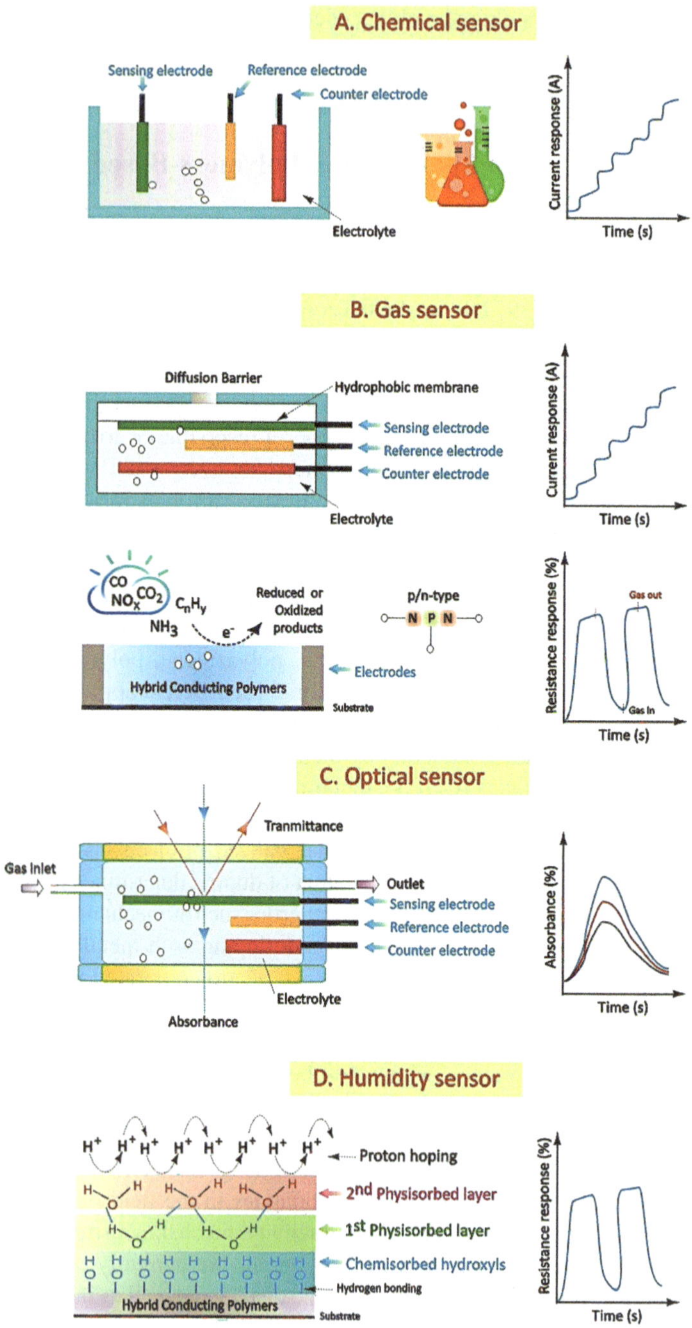

Fig. 1 Schematic representation of various sensing strategies involving chemical, gaseous or optical sensors based on conducting polymers

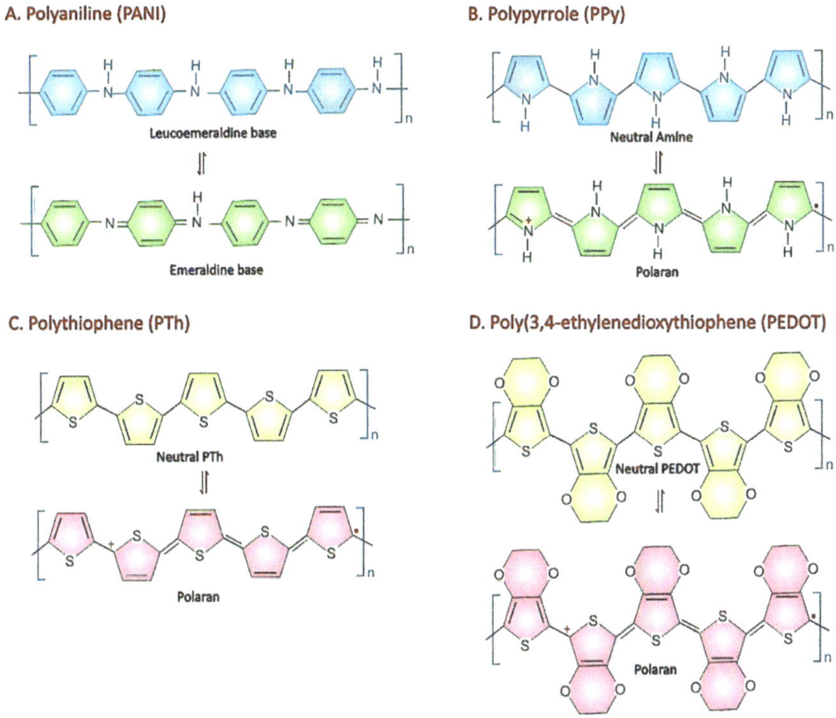

Fig. 2 Structures of precursors of commonly used conducting polymers (CPs)

In the context of sensing, the CPs have been used independently as well as in combination with biomolecules and redox mediators. The motive behind combining CPs and redox mediators, receptors, and enzymes is to avoid issues of sensitivity and selectivity, which are frequently encountered while performing analyte quantification at lower levels [19]. Chemical sensors based on HCPs are classified into four categories, namely, enzymatic, non- enzymatic, aptamer-based sensors, and immunosensors [20]. Chemically synthesized CPs have been widely employed for modifying sensor surfaces. Electropolymerization of few polymers of unusual materials has been reported in literature [21]. The important factors to be taken care of while modifying the electrodes with CPs are the thickness of the coating, control on the active area, and adhesion of the layers on the surface. However, these parameters are difficult to regulate in drop-casted CP films [22]. Electrochemical deposition of the CP films has been successfully achieved in conducting surfaces. This method enables better control over the film thickness. Other methods such as layer-by-layer deposition and Langmuir-Blodgett technique have been reported to obtain a controlled thickness of the CP-based sensing layers [23].

The combination of CPs and enzymes have been especially beneficial for sensing applications as they possess several advantages such as ease of deposition, increased

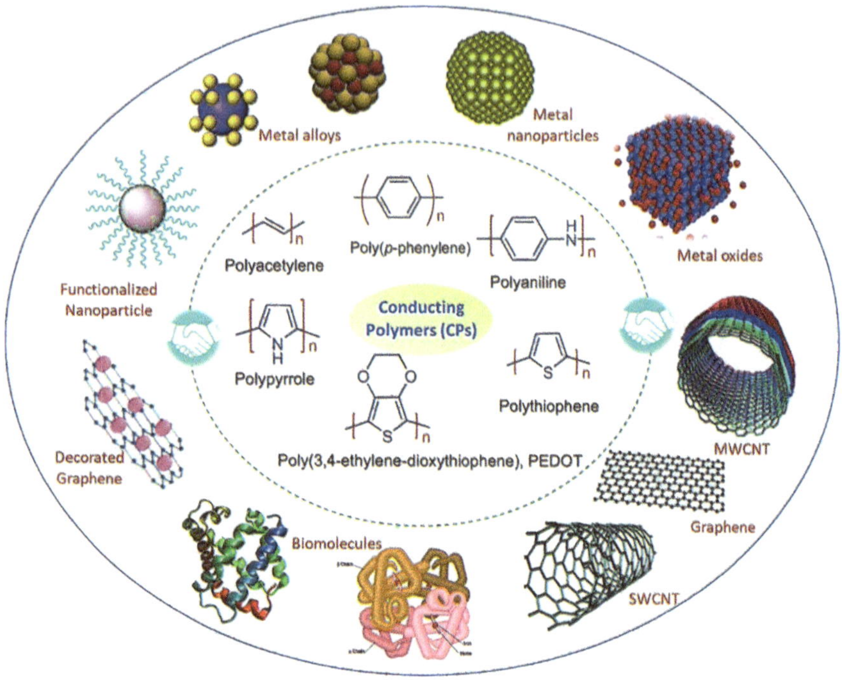

Fig. 3 Schematic representation of hybrid conducting polymeric nano-composites (HCPs)

conductivity of the polymer, which enables faster electron transfer due to participation of the redox-mediating centers of the enzyme molecules present on the electrode surface [24]. Electrochemical detection of several analytes such as glucose [25, 26], H_2O_2, phenol, amino acids, organophosporus pesticides, NADH, neurotransmitters, uric acid, ethanol, epinephrine, lactate and simultaneous detection of dual and multi-analytes have been extensively investigated using conducting polymers.

Different combinations of CPs and dopants have been explored for quantifying a wide range of analytes. The performances of these sensors are assessed by comparing the sensing parameters such as sensitivity, selectivity, limit of detection (LOD), and limit of quantification. In an interesting work, the pH responsiveness of polyaniline (PANI) was exploited for the development of a flexible micro-pH sensor. The PANI film was coated over interdigitated electrodes formed through etching a polyimide substrate. The electrodes supplied a constant potential to the PANI film that was made conducting through doping with dodecylbenzene sulfonic acid. The resistance of the PANI film changed with pH. The sensor detected pH between the ranges 5.45–8.62. The flexible platform for the sensor could be useful in monitoring wound healing that is accompanied by subtle changes in pH [27].

However, it has been found that several conducting polymers exhibit poor electron mobility due to the presence of a large number of defects and disordered chains.

This limitation can be addressed through the incorporation of appropriate nanostructured materials that intercalate between the polymer chains thereby improving the physicochemical properties of the conducting polymers.

3 Nanostructured Conducting Polymers for Sensor Applications

The reduction in the size of the conducting polymers to nano-dimensions can be beneficial for sensing as it results in an enormous increase in the surface area. However, in most cases, the preparation of conducting polymers at the nano-dimensional level is a greater challenge due to the unstable nature of the polymers. Still, several attempts have been made to prepare conducting polymers in the nano-dimensions through solid-phase template synthesis where mesoporous silica, anodically etched alumina, electrospun polymeric nanofibers, etc., were employed as the solid templates [28]. Molecular templated synthesis has been reported as a cost-efficient method for obtaining nanostructured conducting polymers through the use of liquid crystals, surfactants, and polyelectrolytes that form self-assembled nanostructures, which serve as a template for the formation of nano-dimensional conducting polymers. Template-free strategies have also been explored for the formation of nanoparticles of conducting polymers, which are based on the spontaneous self-organization of the precursors into nanostructures [29]. However, this method requires elaborate design strategies to obtain the appropriate precursors. Different types of nanostructures of the conducting polymers have been obtained using these methods. These include nanofibers, nanorods, and nanotubes that have a high aspect ratio, which augers well for sensing applications [30]. The nanostructures of the conducting polymers are generally deposited on the electrode surface by drop-casting. Alternately, the deposition of the conducting polymer layers has also been achieved through electropolymerization. The electropolymerized films exhibit nano-dimensional features that improve surface area apart from exhibiting excellent adhesion to the sensor surface. For example, electrodeposited polyaniline on glassy carbon electrodes forming nanojunctions has been reported for the detection of glucose with rapid response time below 200 ms [31]. Recently, a combination of PANI and reduced graphene oxide coated over screen-printed carbon electrode was employed for non-enzymatic quantification of glucose. The electrocatalytic property of the conducting polymer-based nanocomposite was instrumental in obtaining the response towards the analyte. The linear range of this sensor was between 1 and 1000 μM with a detection limit of 30 nM [32]. In another study, polypyrrole nanowires were employed for quantification of nitrate ions with a detection limit of 1.52 μM [33]. Conducting polymers decorated with nanoparticles have also been reported in literature for sensing applications. Polypyrrole films doped with multi-walled carbon nanotubes have been reported for sensitive detection of dopamine and DNA sequences [34]. Dopamine is an important neurotransmitter whose concentrations are dysregulated during neurological disorders [35]. In a recent study, a soft templating technique was employed successfully using sodium cholate micelles as the template for the synthesis of silver nanoparticle decorated polypyrrole nanorods. The conducting properties of polypyrrole were ensured by acid treatment post-synthesis. The nanocomposite was deposited over a glassy carbon electrode and employed for quantification of dopamine and an ultralow detection limit of 0.005 nM was achieved. The sensor exhibited linearity over two ranges—0.005–3 nM and 0.78–50 μM [36]. The system was demonstrated to be biocompatible and hence applications involving implantable sensors could be envisaged in the future. Polyvinylimidazole polymer, when deposited over carbon nanotubes along with glucose oxidase, was found to quantify glucose over a wide linear range of 0.5–6.0 mM [37]. A beta-cyclodextrin modified polyacetylaniline-carbon nanotube composite deposited on a glassy carbon electrode was used to quantify hydroquinone between 0.001 and 5 mM. The combination of the CP and carbon nanotubes enhanced the sensitivity and sensing range while the beta-cyclodextrin served as an efficient capture agent [38]. Polythiophene (PTh) incorporating α-Fe_2O_3 coated over a glassy carbon electrode was successfully employed to detect methanol with a wide range of detection between 5 and 1000 mM and superior reproducibility and stability [39]. In a novel attempt, PEDOT nanotubes were fabricated and used for quantification of the nerve gas agent sarin mimic dimethyl methylphosphonate with an ultralow detection limit of 10 ppt [40]. This strategy could be extended for real-time deployment in public transport places.

4 Enzymatic and Non-enzymatic Chemical Sensors Based on Conducting Polymers

A biorecognition component combined with HCP will capture the target analyte mainly in solution phase, and enhance the electron transfer and redox activity of the biomolecules. Electrochemical techniques like potentiometric, impedimetric, pulse voltammetric, and amperometric methods have been widely used for the detection of different analytes using CP-based sensors [41]. To enhance the electrocatalytic activity, CPs have been explored as an immobilization platform with various enzymes. For instance, a thin layer biosensor was constructed on a graphite electrode by electropolymerizing 2,5-di(furan-2-yl)thiazolo[5,4-d]thiazole monomer followed by immobilization of glucose oxidase (GOx) for the quantification of glucose. The detection limit and sensitivity were reported as 12.8 μM and 65.44 $\mu A\ mM^{-1} cm^{-2}$ respectively. This polymeric biosensor was employed for analyzing tea and beverage samples with good recovery values. [42]. Another portable biosensor using glucose oxidase immobilized on PANI containing multiwalled nanotubes and gold nanoparticles was fabricated in which a high charge transfer between the electrode and analyte was achieved for effective detection of glucose [43].

In another report, horseradish peroxidase immobilized on terthiophene-3′-carboxylic acid polymer prepared by electropolymerization was used for quantification of H_2O_2. The enzyme-CP combination was responsible for the H_2O_2 reduction reaction in the absence of any additional redox mediators [44]. Similarly, dopamine detection was achieved using layer-by-layer assembly of Prussian blue-modified monoamine oxidase B incorporated on polyethyleneimine prepared via chemical deposition on indium tin oxide electrode. The polymer prepared was encapsulated on liposomes and acted as a polycation which helped in the oxidation of amines, which in turn oxidized the dopamine. The amperometric response of the prepared sensor showed high sensitivity of 0.33 ± 0.02 μA cm^{-2}/mmol L^{-1} with a detection limit of 0.86 mmol L^{-1} [45]. In another interesting application, the detection of the organophosporus pesticide malathion was achieved through covalent linking using poly-3,4-ethylenedioxythiophene (PEDOT) in combination with carboxylated multi-walled carbon nanotubes by electropolymerization method on fluoride incorporated tin oxide. Chronoamperometry technique was adopted for quantification and the results were compared with HPLC. It was found that the sensor exhibited an ultra-low detection limit of 0.5 fM [46]. Another sensor employing the enzyme acetylcholinesterase was fabricated by immobilizing the enzyme on polyanisidine for the detection of carbaryl pesticide [47]. Other pesticides like paraoxon and trichlorfon were quantified using a sensor that used poly(4,7-di(furan-2-yl)benzo[c] [1, 2, 5] thiadiazole polymer prepared via co-precipitation method [48]. Molecularly imprinted polypyrrole films blended with zinc oxide nanorods were successfully employed for quantification of bisphenol S between 2.5 and 12.5 μM [49]. Table 1 tabulates some enzymatic and non-enzymatic sensors that use CPs, along with their preparation method, analyte detected, sensitivity, and detection limit.

Apart from detecting conventional molecules through the CP-based sensors, few novel applications have been envisaged. In an interesting application of CP-based sensors, a blend of PEDOT and Ppy was electrodeposited over a glass film coated with indium tin oxide that was used to culture *Escherichia coli* bacteria using appropriate growth conditions. The bacterial proliferation under aerobic and anaerobic conditions was monitored using cyclic voltammetry and the current recorded was used as a measure of bacterial proliferation [79]. This sensor may have interesting applications in monitoring wound healing and biofilm formation in the future. In a similar strategy, films of PEDOT:PSS were employed for the detection of the uropathogenic bacteria *Salmonella spp* [80]. The extracellular secretions from the bacterial culture altered the redox potential between the electrodes coated with the CP films. The magnitude and potential shift were characteristic for *Salmonella* and other uropathogenic bacterial species thereby making this a versatile and specific diagnostic system (Fig. 4). Films of polypyrrole and PANI incorporating a ruthenium complex was deposited on a glass substrate containing gold interdigitated electrodes. Additional layer-by-layer assembly of azobenzene-containing polymer and polyallylyl hydrochloride was also carried out on the electrode. The system was employed as an artificial tongue to detect four different tastes, namely, sweet, salt, sour and bitter represented by sucrose, sodium chloride, HCl, and quinine respectively, by measuring the impedance changes [81]. This sensor could be a potential replacement for human tasting in the food and

Table 1 List of enzyme-free and enzymatic hybrid conducting polymer-based chemical sensors reported for the eletrochemical detection of analytes

HCPs used	Bio-recognition element	Synthetic procedure	Analytes	Technique	Real sample	References
Thiophene derivatives copolymer	Glucose oxidase	Electrochemical Polymerization	Glucose	Amperometry	Pear, Peach, Apricot	[25]
Polypyrrole	Glucose oxidase/gluteralde-hyde	Electrochemical	Glucose	Chronopotentiometry	–	[26]
3-methyl thiophene/thiophene-3-acetic acid	Glucose oxidase	Electrochemical copolymerization	Glucose	Amperometry	–	[50]
5,2′:5′, 2*f*″ terthiophene-3′-carboxylic acid polymer	Horseradish peroxidase	Electrochemical polymerization	H_2O_2	Amperometry	–	[44]
Poly o-anisidine	Acetyl cholinesterase	Electrochemical adsorption	Carbaryl	Amperometry	–	[51]
Poly-3,4 ethylenedi-oxy-thiophene (PEDOT)/carboxyl-ated multi-walled carbon nanotubes (c-MWCNTs)	Aptamers	Electrochemical adsorption	Malathion	Chronoamperometry	Lettuce sample	[46]
PEDOT/ZrO_2 nanoparticles	Non-enzymatic	Electrochemical Stripping	Methyl parathion	Square wave voltammetry	Tap water sample	[52]
Poly([2,2″;5″ 2″]-terthiophene-3′-carb-aldehyde) (PTT)	Acetyl cholinesterase	Chemical synthesis	Malathion	Cyclic voltammetry, Amperometry, electro-chemical impedance spectroscopy	Parsley leaves samples	[53]
Poly(4,7-di(furan-2-yl)benzo[c] [1, 2, 5] -Thiadiazole)/magnetic nanoparticles	Acetyl cholinesterase	Co-precipitation method	Paraoxon and trichlorfon	CV, Amperometry	Tap water samples	[54]
Polypyrrole hydrogel	Non-enzymatic	Chemical polymerization	Ascorbic Acid, Dopamine, Uric Acid	Cyclic voltammetry, differential pulse voltammetry, electro-chemical impedance spectroscopy	Human Urine, Yinqiao tablet	[55]

(continued)

Table 1 (continued)

HCPs used	Bio-recognition element	Synthetic procedure	Analytes	Technique	Real sample	References
Graphene (GR) and poly 4-amino-3-hydroxy-1-naphthalene sulfonic acid (p-AHNSA)/Screen- printed electrodes	Non-enzymatic	Electrochemical polymerization	Dopamine, 5-hydroxy tryptamine	Cyclic voltammetry, Square wave voltammetry	Human Urine and blood samples, Pharma-cological formula-tions	[56]
Zinc oxide/copper oxide poly pyrrole nano-composite	Non-enzymatic	Electrochemical cveroxidation	Ascorbic Acid, Dopamine, Uric Acid	Cyclic voltammetry, differential pulse voltammetry	Injectable medicine and Human Urine	[57]
ZnO decorated polyaniline rGO nanoflowers	Non-enzymatic	Electrochemical Polymerization	Dopamine, Uric Acid	Cyclic voltammetry, differential pulse voltammetry	Human Serum, Human Urine, dopamine injection solution	[58]
Polythiopene derivatives	Non-enzymatic	Ultrasonication	Dopamine, Uric, Trypto-phan	Cyclic voltammetry, differential pulse voltammetry	Blood Serum, Urine	[59]
Poly(3,4-ethylenedi-oxythio-phene)/MoS_2 nanosheets	Non-enzymatic	Electrodeposition	Ascorbic Acid, Dopamine, Uric Acid	Cyclic voltammetry, differential pulse voltammetry	Urine	[45]
Polyethylene imine	Monoamine oxidase B	Chemical deposition	Dopamine	Amperometry	–	[45]
α-Fe_2O_3/polyaniline nano-composite	Non-enzymatic	Chemical synthesis	Hydrazine	Cyclic voltammetry, amperometry	–	[60]
Fe_3O_4/poly-pyrrole/GO/GCE	Non-enzymatic	Ultrasonication	Hydrazine	Cyclic voltammetry, differential pulse voltammetry	Tap water	[61]
Polythiopene-ZnO nano-composite	Non-enzymatic	Chemical oxidative polymerization	Hydrazine	Cyclic voltammetry, amperometry	–	[62]

(continued)

Table 1 (continued)

HCPs used	Bio-recognition element	Synthetic procedure	Analytes	Technique	Real sample	References
NiHCF/Poly-pyrrole-poly styrene sulfonic acid	Glucose oxidase/Lactate oxidase	Electro-polymerization	Glucose, lactate	Cyclic voltammetry, electro-chemical impedance spectroscopy	–	[63]
Polyproline-amino functional-ized magnetic mesoporous silica-beta cyclodextrin nanohybrid on MCM-41	Non-enzymatic	Electro-polymerization	L-Trypto-phan, L-tyrosine, L-phenyl-alanine, L-methionineL-Proline, L-cysteine, ascorbic acid, uric acid, Dopamine	Cyclic voltammetry, differential pulse voltammetry	–	[64]
Polybrilliant cresyl blue/Graphene composite	Non-enzymatic	Two-step Electro-polymerization	Epi-nephrine	Cyclic voltammetry, electro-chemical impedance spectroscopy	Pharma-ceutical samples	[65]
Graphene nanoplatelets functional-ized with silver nanorods and polymer-ization of 4-amino-1-1′-azobenzene-3,4-disulfonic acid dye	Non-enzymatic	Electro-polymerization	Dopamine and Ascorbic Acid	Cyclic voltammetry, amperometry	Human serum, urine sample, Celin-500 vitamin-C tablets	[66]
Polythio-phene g-poly(ethyl-ene glycol) with lateral amino groups	Laccase	Suzuki condensation polymerization	Phenolic compo-unds	Cyclic voltammetry, amperometry	Artificial waste water	[67]
Poly(o-anisidine)/graphene nano-composites	Non-enzymatic	Chemical oxidative polymerization	Nicotin-amide adenine di nucleotide (NADH)	Cyclic voltammetry, electro-chemical impedance spectroscopy	–	[68]
Polythionine/Screen printed electrodes	Non-enzymatic	Electro-polymerization	NADH and H_2O_2	Cyclic voltammetry, flow injection analysis	–	[69]
poly (3,4-ethylendioxy thiophene)/Thionine	Lactate dehydroge-nase	Ultrasonication	Lactate, NADH	Scanning electro-chemical microscopy	–	[70]
Carbon nanotubes/Gold nanoparticles/Poly neutral red	Non-enzymatic	Electro-polymerization	NADH	Cyclic voltammetry, amperometry	–	[71]

(continued)

Table 1 (continued)

HCPs used	Bio-recognition element	Synthetic procedure	Analytes	Technique	Real sample	References
Poly(3,4-ethylenedi-oxy thiophene) (PEDOT)/Poly (brilliant green)/Carbon nanotubes	Alcohol oxidase/de-hydrogenase	Electro-polymerization	Ethanol	Cyclic voltammetry, differential pulse voltammetry	Sangria, vermouth, Vodka, whisky, Red wine, white wine	[72]
Gold-Silver bimetallic nanoparticles/Poly-L-Cysteine/reduced graphene oxide/Nafion	Alcohol dehydrogen-ase/Gluter-aldehyde	Electro-polymerization	NADH and Ethanol	Cyclic voltammetry, amperometry	Commer-cial beverages	[73]
Reduced graphene oxide/poly pyrrole/cadmium oxide nano-composite	Non-Enzymatic	Electrochemical over oxidation	Xanthine	Cyclic voltammetry, differential pulse voltammetry	Trout Fish meat samples	[74]
Nano-Silver/Zinc oxide/Poly pyrrole/Pencil graphite electrode	Xanthine Oxidase	Electro-polymerization	Xanthine	High performance liquid chromato-graphy, Amperometry, electrochemical impedance spectroscopy	Sea bass	[75]
Thiophene-3-boronic acid and 3-aminophenyl boronic acid/Pencil graphite electrode	Apo-cholesterol oxidase	Electro-polymerization	Chole-sterol	Amperometry	Human Urine	[76]
Poly(GMA-co-VFc)/MWCNT nano-composite	Xanthine Oxidase	Co-precipitation	Xanthine	Cyclic voltammetry, amperometry	Fish samples	[77]
Chitosan-polypyrrole-gold bionano-composite	Xanthine Oxidase	Hydrolysis	Xanthine	Cyclic voltammetry, Amperometry, electro-chemical impedance spectroscopy	Fish, Chicken, Beef samples	[78]

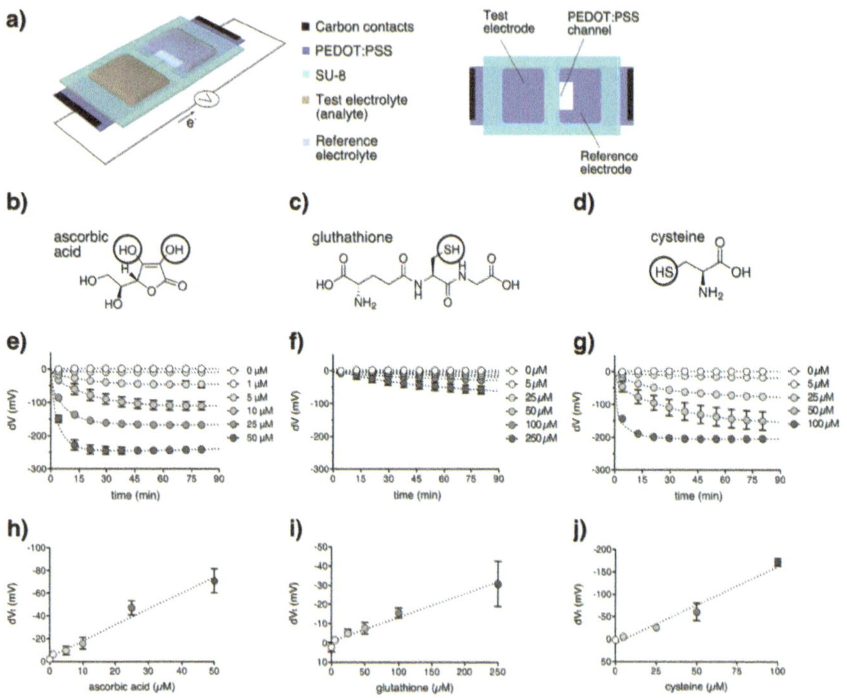

Fig. 4 Conducting polymer-based electrochemical sensor developed for sensing redox-active species. **a** Schematic representation of PEDOT:PSS-based sensor comprising two electrodes made from PEDOT:PSS and a channel. The hydrophobic photoresist SU-8 was used to pattern electrolyte reservoirs on the top. The analyte was introduced on the working electrode and the reference electrolyte was placed on the reference electrode. **b–d** Chemical structures of the different electroactive species with their redox-active groups highlighted by a circle: **b** ascorbic acid, **c** glutathione, and **d** cysteine. **e–g** The plot of potential difference versus time for: **e** ascorbic acid, **f** glutathione, and **g** cysteine as recorded using electrochemical sensors. Dotted lines denote mean values (MV) of potential difference recorded for each concentration over time ($n \geq 3$). Error bars represent MV ± SD from data with $k = 100$. **h–j** The plot of potential difference at the detection time (dVt) with the analyte concentration for **h** ascorbic acid, **i** glutathione, and **j** cysteine. Error bars represent MV ± SD. The dotted line represents a linear fit ($R^2 = 0.975$) (Adapted from Ref. [80])

pharmaceutical industries. DNA sensors based on CPs have emerged as a new strategy towards the diagnosis of genetic mutations. The probe DNA strand is immobilized on the conducting polymer coated on the transducer surface. On the addition of the sample, hybridization of the sample strand with the probe strand occurs, which alters the conductivity of the underlying CP. This could be used for the detection of the oligonucleotide sequence specifically [82]. Many CPs such as polypyrrole, PANI, and polythiophene have been used for the detection of specific DNA sequences through cyclic voltammetry or impedance analysis [83].

5 Gas Sensors Based on Conducting Polymers

The porous nature of the conducting polymers favour permeation of gaseous analytes making them ideal substrates for quantification of volatile molecules. The interaction with the gases leads to changes in the electrical conductivity of the polymers that can be correlated with the concentration of the gaseous analyte [84]. Several organic conducting polymers such as polypyrrole (PPy), polyaniline (PANI), polythiophene (PTh), poly-(3,4-ethylenedioxythiophene) (PEDOT), etc., have been employed for various applications including gas sensors, volatile organic compounds, and moisture or humidity sensors [85–89]. As in the case of chemical sensors, many researchers have developed new hybrid or functional nanocomposite materials by incorporating carbon nanomaterials and metal or metal oxide nanoparticles with polymeric matrices to enhance and improve the fundamental and technological perspectives of these materials [90–95]. For instance, carbon nanomaterials including single-walled (SWCNT) and multiwalled carbon nanotubes (MWCNT), graphene and graphene oxide (GO), carbon nanofibers (CNFs), metal nanoparticles such as gold, silver, palladium, and platinum, with different compositions and dimensions, and metal oxides such as zinc oxide (ZnO), tungsten oxide (WO_3), tin oxide (SnO_2), iron oxide (Fe_2O_3/Fe_3O_4), titanium oxide (TiO_2), indium oxide (In_2O_3), etc., have been extensively employed for the development of various gas sensors or transducers for environmental monitoring [90–95]. These additives would increase the mechanical and electrical properties of the composites and also improve the physical properties. These composite nanomaterials have opened up a new arena in the realm of gas sensing.

The rapid modernization and industrialization coupled with a huge increase in the population have lead to a drastic increase in environmental pollution with the discharge of harmful gases such as nitrogen oxides (NO_x), ammonia (NH_3), carbon monoxide (CO), carbon dioxide (CO_2), sulfur dioxide (SO_2), hydrocarbons, volatile organic compounds (VOCs), etc., [88, 92, 96]. Acute exposure to these harmful gases poses serious risks to human health [92, 96]. There are several types of gas sensors that are intended for various applications. These include the electrochemical sensors, biosensors, immunosensors, and chemiresistive sensors [96]. Among these, the electrochemical gas sensors have been classified based on their mode of transduction, i.e., amperometric, potentiometric, and conductometric (chemiresistive) sensors [88, 96, 97]. In amperometric sensors, the transducer current signal is proportional to the concentration of the targeted analytes, which are electroactive species that readily undergo oxidation or reduction at specific applied potentials [88]. In the case of potentiometric sensors, the selective sensor membrane is in contact with the analyte solution where the cell potential is measured at zero current input [94]. Conductometric gas sensors (simply called chemiresistors) seem to be alternative gas sensors where the chemical interaction of gases with active sensing materials, i.e., adsorption and desorption induces a change in the electrical conductivity or resistivity of the sensor upon gas exposure [96]. Amongst the different types, the chemiresistive gas sensor is considered as one of the best low-cost transducers due to its simplicity, rapid response times, and easy fabrication. In general, the sensor performance is completely evaluated by several parameters that include sensitivity, selectivity, response and recovery times, and detection limit. Some of the most popular CP containing hybrid nanocomposite based gas sensors are tabulated in Table 2.

As mentioned earlier, the selection of fabrication or synthesis method and transduction techniques is a critical factor to achieve the desired gas sensing abilities of the HCP. The hybrid conducting polymer composites have been synthesized by various approaches. In chemical methods, controlled reactivity of molecular precursors, hydro- or solvothermal conditions, assembly or intercalation mode of formation and/or dispersion, have been found to influence the sensing characteristics of the resultant HCPs. Electrochemical methods of synthesis involve incorporation of metal nanoparticles during electro-polymerization, electro-deposition of metal nanoparticles, or reduction of metal salts on the polymer matrix. Apart from the classical electrodeposited polymers, the functionalization of polymeric side chains with sulfonated anionic group, tetrafluoroborate, perchlorate have been carried out to facilitate the interaction of gas analytes with the functional polymers, which enhance the gas sensing abilities of the parent conducting polymers [104]. The classification of hybrid conducting polymer-based gas sensors can be summarized as follows: (i) electrodeposited conducting polymers, (ii) electrodeposited film as a mixture of a polymer and an organic material, (iii) electrodeposited films as a mixture of a polymer and metal nanoparticles or metal oxides. Table 3 lists the different classes of HCP-based gas sensors reported in recent literature for the detection of various harmful gases such as nitrogen oxides (NO_x), ammonia (NH_3), carbon dioxide (CO_2), carbon mono-oxide (CO), and hydrocarbons (HCHO, CH_3CHO, C_nH_x, $CHCl_3$).

6 Hybrid Conducting Polymers for Ammonia Detection

Ammonia (NH_3) is a gaseous pollutant commonly encountered in industry and needs to be constantly monitored for environmental and air pollution control. High dosages of NH_3 (>300 ppm) can be detected easily due to its dense pungent smell, however, the detection of its low concentrations (about 1 ppm) needs highly sensitive and selective gas sensors [98]. In this context, several combinations of CPs have been investigated owing to their ability to interact with gaseous analytes. For instance, in situ dilute polymerization of a thin film nanocomposite consisting of polyaniline (PANI) on flexible poly(styrene-butadiene-styrene) (SBS) was investigated for NH_3 sensing and was found to respond to concentrations as low as 0.1 ppm at room temperature [105]. This hybrid polymer composite film showed excellent mechanical stability and could be utilized as a highly sensitive and flexible ammonia sensor platform. The vapor deposition polymerization technique was employed to prepare porous PPy on non-conductive polymethylmethacrylate (PMMA). The composite film was used for the sensitive and rapid detection of NH_3 [106]. Combinations of CP with carbon nanotubes have been extensively employed for quantification of ammonia gas. For instance, highly conducting hybrid polymer nanocomposites PANI-MWCNT,

Table 2 Various strategies adopted for ideal gas sensor design based on hybrid conducting polymeric nanocomposites

Hybrid nanocomposite	Gas sensing technique	Role of the gas sensing matrix	References
Insulating polymer with carbon nanomaterials like polymethyl methacrylate (PMMA), polypyrrole (PPy), ethylcellulose with carbon nanotubes (CNTs), carbon nanofibers (CNF), etc.	Chemiresistive	The insulating polymer is a gas penetrable matrix; Carbon nanomaterials will increase the percolation conductivity	[98]
Insulating polymer with conducting polymer like polyvinyl alcohol and polypyrrole (PVA-PPy, etc.)	Chemiresistive (humidity sensors)	The insulating polymer is a gas penetrable matrix; Conductive polymer improves the conductivity	[99]
Conducting polymer with carbon nanomaterials (CNT, CNF, GO, etc.)	Electrochemical	The polymer is a gas permeable matrix; Carbon nanomaterials act as conducting component	[100]
Conducting polymer with nanometals (Au, Ag, Pt, Pd, etc.)	Electrochemical/chemiresistive	The polymer is a gas penetrable matrix; Nano-dimensional noble metal particles act as catalysts	[101]
Conducting polymer with nano metal oxides (WO_3, ZnO, SnO_2, TiO_2, Fe_2O_3/Fe_3O_4, SiO_2, etc.)	Electrochemical/chemiresistive	Polymer is a gas penetrable matrix; Nanostructured metal oxides act as catalyst	[102]
Conducting polymer with carbon nanomaterials and metal or metal oxides as surface functionalized systems	Electrochemical/chemiresistive	Polymer is a gas sensing matrix; Carbon nanomaterials act as conducting component; Noble metals/metal oxides act as catalysts/promoters (surface reactivity enhancers)	[103]

Table 3 Gas sensing characteristics of various hybrid conducting polymer-based nanocomposites reported in the literature

Gas analyte	Hybrid polymers	Fabrication methods	Technique	Concentration range	Detect-ion limit	References
NH_3	PANI/poly(styrene-butadiene-styrene)	In situ dilute polymerization	Electrical resistance	0.1–100 ppm	0.1 ppm	[105]
	PPy/polymethyl methacrylate (PMMA)	Vapor deposition polymerization	Electrical resistance	30–850 ppm	30 ppm	[106]
	PANI-MWCNT/SWCNT thin film	Oxidative polymerization	Electrical resistance	0.5–5000 ppm	0.5 ppm	[107]
	PANI-Graphene composite	Chemical oxidative polymerization	Electrical resistance	1–6400 ppm	1 ppm	[108]
	PPy-SWCNT	Chemical oxidative polymerization	Electrical resistance	10–800 ppm	10 ppm	[109]
	PPy-reduced graphene oxide	In situ chemical oxidative polymerization	Electrical resistance	1–10 ppm	1 ppm	[110]
	PEDOT/poly(styr-ene sulfonic acid)-MWCNT	Reinforced polymerization	Electrical resistance	1–50 ppm	1 ppm	[111]
	PANI/TiO_2	Electrostatic self-assembly and in situ chemical oxidative polymerization	Electrical resistance	23–141 ppm	23 ppm	[112]
	PANI/SnO_2 and In_2O_3	Electrostatic self-assembly and in situ chemical oxidative polymerization	Electrical resistance	23–141 ppm	23 ppm	[113]
	PANI/WO_3	In situ chemical oxidative polymerization	Electrical resistance	5–100 ppm	5 ppm	[114]

(continued)

Table 3 (continued)

Gas analyte	Hybrid polymers	Fabrication methods	Technique	Concentration range	Detect-ion limit	References
	PPy/ZnO or SnO_2	Template-based method	Electrical resistance	10–80 ppm	10 ppm	[115]
	PPy/ZnO-TiO_2	Electrospinning followed by vapor-phase polymerization	Electrical resistance	0.5–450 ppm	0.5 ppm	[116]
	PANI/Pd	Electrochemical copolymerization	Electrical resistance	10–500 ppm	10 ppm	[117]
	PPy/Pd	Thermal dynamic refluxing and vapor-phase polymerization	Electrical resistance	50–5000 ppm	50 ppm	[118]
NO_X	PANI/WO_3	Vacuum thermal evaporation technique	Electrical resistance	n.a.	n.a.	[119]
	PPy/ZnO	In situ chemical oxidative polymerization	Electrical resistance	1000, 1500, 2000 ppm	n.a.	[120]
	PTh/SnO_2	In situ chemical oxidative polymerization	Electrical resistance	10–200 ppm	10 ppm	[121]
	Polyhexylthio-phene (PHTh), PHTh−PEDOT copolymer, sulfonated PANI, PANI−SnO_2, PPy−SnO_2, PEDOT-SnO_2, PHTh−SnO_2	In situ self-assembled layer-by-layer approach	Electrical resistance	10–800 ppb	10 ppb	[122]
	PANI-Au-Nafion	Electro-chemical polymerization method	Ampero-metric	20–100 ppm	20 ppm	[123]

(continued)

Table 3 (continued)

Gas analyte	Hybrid polymers	Fabrication methods	Technique	Concentration range	Detect-ion limit	References
CO	PANI-MWCNT	In situ chemical oxidative polymerization	Electrical resistance	500 to 1000 ppm	500 ppm	[124]
	PANI-TiO_2, PANI−SnO_2	In situ self-assembled layer-by-layer approach	Electrical resistance	1–1000 ppm	1 ppm	[125]
CO_2	PPy	Chemical oxidative polymerization	Electrical resistance	100, 400 and 700 ppm	100 ppm	[126]
	PEDOT/polyethy-lenimine (BPEI)	Chemical oxidative polymerization	Ampero-metric	1000–3000 ppm	1000 ppm	[127]
	PANI/graphene on electrospun polystyrene (PS)	In situ electro-polymerization method followed by using plasma treatment	Electrical resistance	20–100 ppm	20 ppm	[128]
	PANI/NaO_2	Ex-situ chemical route	Electrical resistance	100–4000 ppm	100 ppm	[129]
Aroma-tic hydro-carbons	PANI/MWCNT	In situ electro-polymerization method	Electrical resistance	200−1000 ppm	200 ppm	[130]

n.a not available

PANI-SWCNT, and PANI nanogranules-polyvinyl pyrrolidone (~ 100 nm thick) on gold microelectrodes were successfully employed for sensitive detection of NH_3 (< 0.5 ppm) and other organic vapors such as methanol, ethanol, and acetone [107]. Yet another PANI-based nanocomposite with reduced graphene oxide thin film showed higher sensitivity with wide linearity from 1 to 6400 ppm of NH_3 [108].

In another case, a hybrid composite of PPy with SWCNT thin film was used for NH_3 sensing in the range 10–800 ppm [109]. The hybrid PPy with reduced graphene oxide thin film also detected NH_3 up to 10 ppm (Fig. 5), which is higher than that of pure PPy [110]. The incorporation of graphene nanosheets over PANI [108], SWCNT [109], and reduced graphene oxide over PPy [110] might increase the surface-to-volume ratio than that of their pure polymers. A mixture of the electrically conducting polymer poly(3,4-ethylenedioxythiophene)–polystyrene sulfonic acid (PEDOT:PSS) reinforced with MWCNT was developed for sensing of trace levels (1–50 ppm) of NH_3 gas [111].

Apart from carbon nanomaterials, nano metals and metal oxide-based hybrid conducting polymer composites have also been employed for NH_3 sensing. Nanocomposites comprising a thin film of PANI with TiO_2, SnO_2, or In_2O_3 nanoparticles were fabricated on a silicon substrate by in situ self-assembly method and have been investigated as NH_3 sensors (23−141 ppm) at an elevated temperature of 60°C [112, 113]. As a result of the compatibility of the energy bandgap, the hybrid

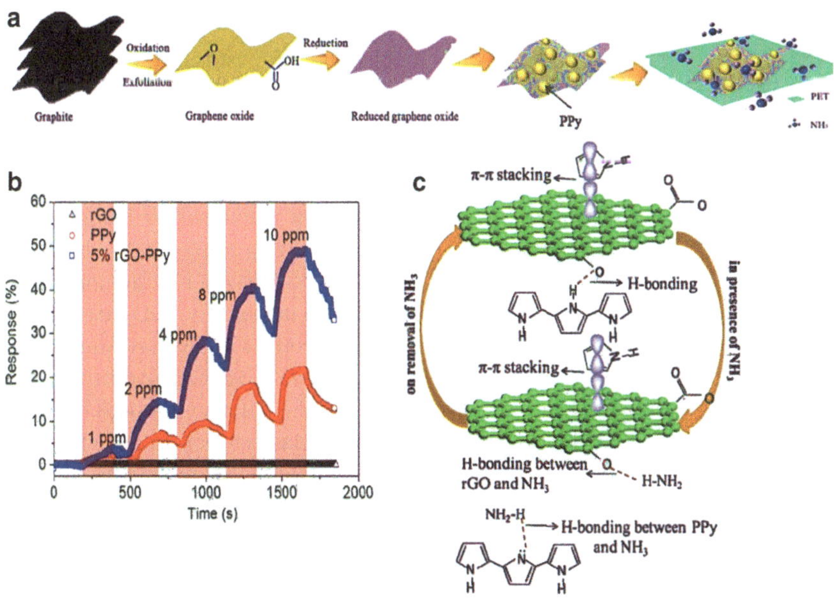

Fig. 5 **a** In situ chemical oxidation polymerization mechanism reported for PPy on rGO, **b** the real-time dynamic response to 1–10 ppm NH_3 recorded for rGO, Ppy, and hybrid polymer-reduced graphene oxide PG5 hybrids, **c** Schematic depiction of the interaction of NH_3 with PPy/rGO hybrid (Adapted from Ref. [110])

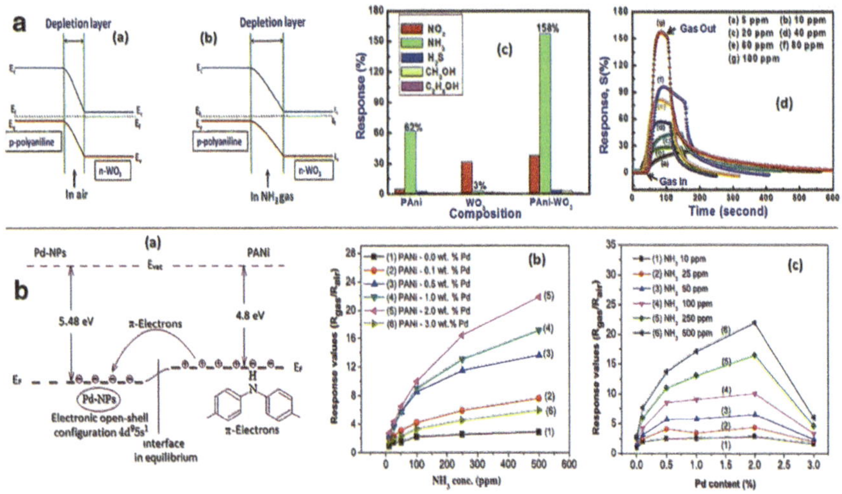

Fig. 6 **a** Schematic representation of energy band diagram proposed for PANI-WO_3 hybrid nanocomposites in the presence of **a** air and **b** NH_3 gas. Response to 5–100 ppm of NH_3 gas of (b) PANI-WO_3 (10–50%) hybrid nanocomposites and **c** PANI-WO_3 (50%) hybrid nanocomposites (Adapted from Ref. [114]). **b** Schematic depiction of probable NH_3 gas sensing mechanism of PANI/Pd (**a**). Effect of Pd content on the response of sensor for various NH_3 concentrations (**b**, **c**) (Adapted from Ref. [117])

PANI-TiO_2 sensors showed the most favourable sensing towards NH_3 [112]. Hybrid PANI-WO_3 nanocomposite sensor in which WO_3 nanoparticles were entrapped on the PANI matrix showed reasonable response for 5 ppm NH_3 and attained saturation after 100 ppm concentration at room temperature (Fig. 6a) [114]. Besides, few more ammonia gas sensors based on hybrids of PANI-ZnO [131] and PANI-Fe_2O_3 [132] have been reviewed recently. One-dimensional template-based synthesized PPy-SnO_2 and PPy-ZnO hybrid nanocomposites have been employed for the detection of NH_3 [115]. It was found that the response of PPy-ZnO to ammonia (10–80 ppm) was much pronounced and reproducible than that of PPy-SnO_2 and pristine PPy. A highly porous PPy-coated ZnO/TiO_2 nanohybrid was also investigated for enhanced NH_3 sensing performance [116]. To improve the conductive behavior of polymers, decorating or doping of nanometals over the polymer matrices has been effectively employed for highly efficient ammonia gas sensors. In an interesting study, osmosis-based synthesis of gold (Au) nanoparticles was reported for doping with PANI. The hybrid nanocomposite when employed for quantification of NH_3 in the presence of organic vapors at room temperature showed a detection limit of 10.8 ppm for NH_3 [133]. Electrochemical copolymerization was used for decorating palladium (Pd) nanoparticles on the PANI surface. This sensor exhibited superior NH_3 gas-sensing performance (Fig. 6B) [117]. A one-pot approach based fabrication of PPy-Au nanocomposites [134], and thermal dynamic refluxing followed by vapor-phase polymerization techniques have been employed to obtain Pd-encapsulated-PPy [118] that were demonstrated to display rapid response to NH_3.

Exhaled breath analysis has garnered interest in recent years for the diagnosis of diseased conditions. Ammonia gas has been cited as a potential gaseous marker for liver disorders. Hence, rapid and sensitive detection is desirable for quantifying NH_3 from exhaled breath. With this objective, a porous nanofibrous network of Eu^{3+} doped PANI was developed. The sensor exhibited a rapid response and recovery time of 5 s when exposed to NH_3 vapors [135].

7 Hybrid Conducting Polymers for Nitrogen Oxide (NO_x) Detection

Oxides of nitrogen (NO_x: NO and NO_2) are responsible for several adverse effects like global warming and ozone depletion to our atmosphere. Inhalation of these gaseous oxides causes serious respiratory problems to humans. As the demand has increased for monitoring these gases, electrochemical sensors have emerged at the forefront for the detection of NO_x gases. Several of these sensors have explored the use of conducting polymers and their composites with metal oxides or carbon nanomaterials for NOx detection. Thin films of hybrid PANI-WO_3 [119] and PPy-ZnO [120] nanocomposites have been reported for the detection of NO_x gases. The PPy-ZnO hybrid sensor showed superior selectivity and reversibility to NO_x at higher concentrations of 1000, 1500, and 2000 ppm at various operating temperatures (30–90 °C) [120]. In situ polymerization of PTh through chemical oxidation was accomplished to prepare PTh-SnO_2 hybrid sensor that was used for the detection of NO_2 (10–200 ppm). The synergetic effect of the combination of PTh and SnO_2 resulted in enhanced sensitivity and thermal stability [121]. In situ layer-by-layer self-assembled series of ultrathin films made up of pristine polyhexylthiophene (PHTh), PHTh−PEDOT copolymer, sulfonated PANI, PANI−SnO_2, PEDOT-SnO_2, PPy−SnO_2, PHTh−SnO_2, were utilized by different groups for NO_2 gas sensing [122]. The obtained results from these hybrid nanocomposites provide a good platform for selective recognition of NO_2 (10–800 ppb) with excellent changes in conductivity at room temperature. Nanometal-doped hybrid conducting polymer film of PANI-Au-Nafion has also been investigated as an amperometric NO_2 gas sensor [123], where diffusion of NO_2 on porous polymer matrix resulted in an increase in cathodic current in the concentration range from 0 to 100 ppm (Fig. 7). Copper phthalocyanine incorporated conducting polymer films when investigated for their sensitivity and selectivity towards NO_2 quantification revealed that the polythiophene-copper phthalocyanine composites exhibited the best response when compared to the PANI and polypyrrole due to formation of a charge-transfer complex [136].

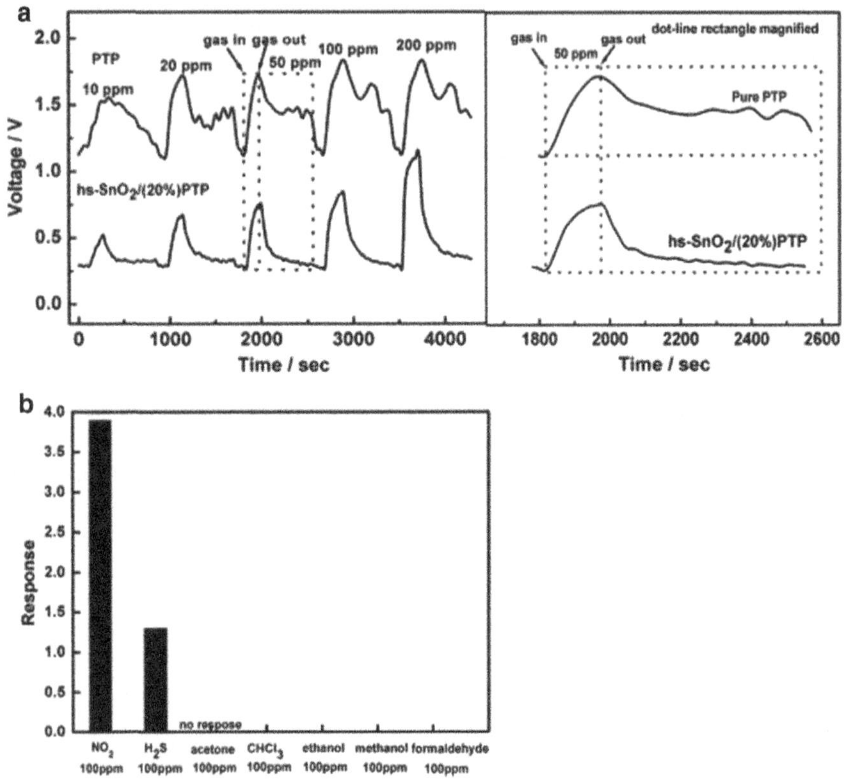

Fig. 7 **a** Recovery-response of the NO_2 sensor with different concentrations using PTP and hs-SnO_2/(20%)PTP hybrids. **b** Various gases with constant 100 μM concentration were recorded for the hs-SnO_2/(20%)PTP hybrids (Adapted from Ref. [137])

8 Hybrid Conducting Polymers for Carbon Monoxide (CO) Detection

CO is well-known for its deadly effect on human life by inhibiting oxygen transport in human blood. Owing to its high toxicity and odourless nature, there exists a need to develop a robust and rapid detection system for CO [138]. Several CO gas sensors have been fabricated using conducting polymers. In one such study, PANI-MWCNT hybrid composite, which quantifies CO as a function of change in the resistivity values at room temperature, was reported [124]. To improve the sensor performance, electrodeposited Pt-Ni alloy on PANI-MWCNT-based hybrid composites have been employed for electrochemical quantification of CO [125]. This sensor facilitates bifunctional catalysis by showing enhanced catalytic oxidation and reduction of adsorbed CO gas. Further, the combination of conducting polymer with metal oxides have been reported to exhibit low detection limits up to 1 ppm CO gas [139]. Ultrathin films of conducting PANI-SnO_2 and PANI-TiO_2 hybrid nanocomposites synthesized by self-assembly method have been utilized for the quantifying CO gas over a wide concentration range (1–1000 ppm) (Fig. 8) [140]. The resistivity of the PANI–TiO_2 system exhibited saturation after exposure to 800 ppm CO, which is superior to the concentrations reported for PANI–SnO_2 films. It is evident from the literature reports that there exists much scope for further studies on HCP-based sensors for CO detection and continuous monitoring.

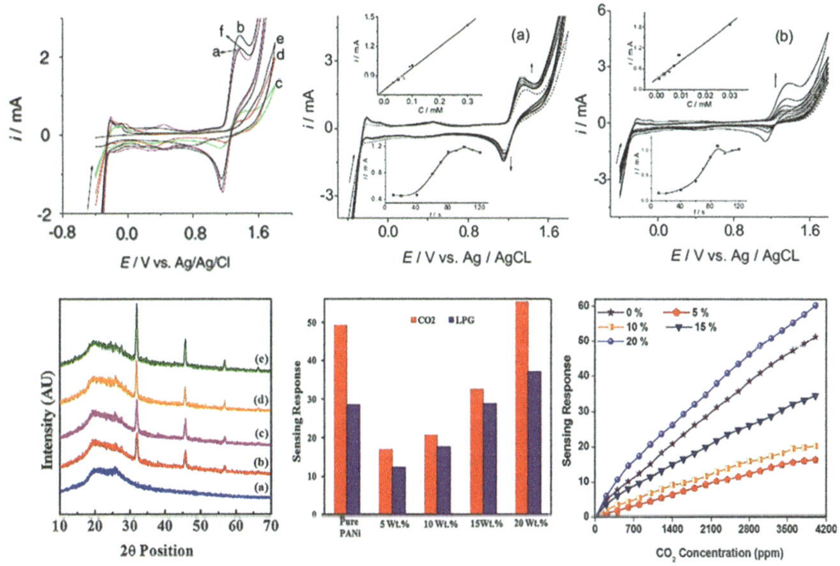

Fig. 8 **a** CVs of 0.1 mM CO at Ni/PANI/MWCNTs/WGE (e), Pt–Ni (1:1)/MWCNTs/WGE (f), Pt/PANI/MWCNTs/WGE (a), Pt–Ni/PANI/MWCNTs/WGE made in solution concentration ratio of Pt:Ni of 1:1 (b), 1:3 (c), and 1:6 (d) using constant potential and CV step. Buffer: 0.5 M $HClO_4$. Scan: 50 mV s^{-1}. **B** CVs of CO (0.01, 0.03, 0.05, 0.07, 0.09, 0.3, 0.5, and 0.7 mM) at Pt/PANI/MWCNTs/WGE; the upper inset depicts the dependence of reduction peak currents of CO on the concentration. The lower inset shows the accumulation effect of 0.1 mM CO at Pt/PANI/MWCNTs/WGE. (C) CVs of CO (1.0, 3.0, 5.0, 7.0, 9.0, 30.0, and 50.0 μm) at Pt–Ni/PANI/MWCNTs/WGE made in the solution concentration ratio of Pt–Ni of 1:1 using constant potential and CV step; the upper inset shows the dependence of oxidation peak currents of CO on the concentration. The lower inset shows the accumulation effect of 0.1 mM CO at Pt–Ni/PANI/MWCNTs/WGE. Buffer: 0.5 M $HClO_4$. Scan: 50 mV s^{-1} (Adapted from Ref. [140]). **d** XRD of (a) Pure PANI, (b) 5 wt% PANI/NaO_2, (c) 10 wt% PANI/NaO_2, (d) 15 wt% PANI/NaO_2 and (e) 20 wt% PANI/NaO_2. **e** The response of the chemiresistors towards CO_2 and LPG at 303 K. **f** Chemiresistor response as a function of the concentration of CO_2 at 303 K

9 Hybrid Conducting Polymers for Carbon Dioxide (CO_2) Detection

Carbon dioxide (CO_2) is an important product from cellular respiration involving the metabolism of carbohydrates, fats, and amino acids. CO_2 plays various roles in the human body and regulates the pH of blood as well as influences the oxygen affinity of hemoglobin [141]. However, if its concentrations in blood exceed a threshold limit, it could turn fatal. Conducting polymers have been demonstrated to possess the unique ability to detect CO_2 at higher concentrations. Ppy polymerized through chemical oxidation in the presence of an oxidant ($FeCl_3$) at two different weight ratios, has been employed for detection of 100, 400 and 700 ppm of CO_2 at room temperature [126]. In another study, a resistive-type CO_2 gas sensor has been fabricated using PEDOT copolymerized with branched polyethyleneimine (BPEI) layers for sensing CO_2 in the range of 1000–3000 ppm [127]. Incorporation of graphene over electropolymerized PANI nanocomposite hybrid deposited on polystyrene by using plasma treatment resulted in a flexible and highly sensitive CO_2 (20–100 ppm) gas sensor [128]. The nanocomposite additive significantly increases the surface area, electrical conductivity, and sensitivity towards CO_2 gas. An ex situ chemical route approach for the fabrication of PANI doped with sodium superoxide (NaO_2) has been adopted for CO_2 gas sensing at room temperature [129]. The sensor showed selective and significant sensitive fast response towards CO_2 gas on 20 wt% PANI/NaO_2 hybrid composite film (Fig. 9). Thus, it is evident that unlike metal oxide-based

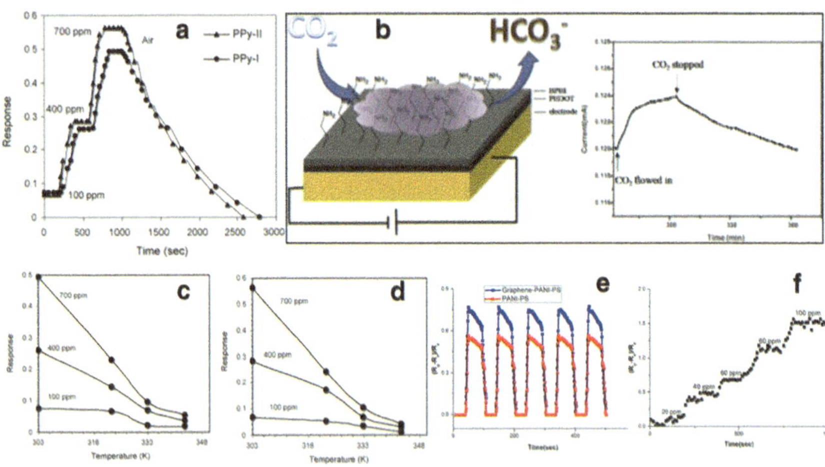

Fig. 9 **a** Dynamic response of PPy-I and PPy-II sensors to 100, 400 and 700 ppm CO_2 at room temperature (Adapted from Ref. [142]). **b** Schematic illustration of the PEDOT–BPEI film sensor device and its corresponding response to CO_2 gas; **c**, **d** Resistance of (a) PPy-I and (b) PPy-II films in air, 100, 400 and 700 ppm CO_2 at room temperature (303 K), 323, 333 and 343; **e** Repeatability of PANI-coated and graphene–PANI-coated fiber CO_2 gas sensors. **f** The sensitivity of graphene–PANI-coated sensor to different CO_2 concentrations (Adapted from Ref. [127])

gas sensors that have high operating temperatures, CP-based CO_2 sensors can be deployed for room temperature monitoring of the gas, especially in public places.

10 Hybrid Conducting Polymers for Hydrocarbon Vapor Detection

Gaseous hydrocarbons and volatile organic compounds pose a grave health risk to humans. For the detection of hydrocarbon vapors, a conducting polymer composite of PANI-MWCNT prepared by an in situ electrochemical polymerization technique in mass ratio 4:1 [130] was reported, which showed response for aromatic hydrocarbon vapors in the range of 200–1000 ppm. After gas exposure, the conductivity of the PANI-MWCNT composite had increased due to dipole–dipole interactions. In an interesting study, a layer-by-layer nanocomposite film of PANI-MWCNT was employed as switchable chemiresistive sensors for methanol (CH_3OH), ethanol (C_2H_5OH) and toluene ($C_6H_5CH_3$) vapors with good reproducibility upon chemical cycling [143]. Another hybrid system used as a selective gas sensor for formaldehyde (HCHO) and acetaldehyde (CH_3CHO) vapors was fabricated through intercalation and ion-exchange reaction between PANI and MoO_3 resulting in a hybrid PANI/MoO_3 nanocomposite [144]. The fabricated film showed appreciable electrical resistivity for 50 ppm of both HCHO and CH_3CHO vapors at room temperature. It has been reported in an independent study that the nanocomposite of PANI/Pd can detect CH_3OH vapors in the linear range of 1–2000 ppm [145], (Fig. 10) selectively. Volatile halogenated organic compounds (VHOCs) produced as an industrial emission pose a great threat to the depletion of the ozone layer and human health due to their ability to cause genetic aberrations and cancer. Computational studies have revealed that polythiophene interacts with a higher affinity to halogenated volatile compounds through dipole-dipole and halogen bonding interactions. The affinities to various halogenated gases follow the order: $Cl_2 > HCl > CCl_4 > CH_2Cl_2 > CHCl_3$. The interactions alter the conducting properties of polythiophene which can be explored for the quantification of these hazardous gases [146]. A PANI-copper nanoparticle composite has been reported for the quantification of chloroform vapors well below 100 μM. The pristine PANI did not exhibit such high response to chloroform and it has been postulated that the interactions between the analyte gas and metal clusters in the polymeric matrix could be a major reason for the observed differences in the sensing characteristics [147].

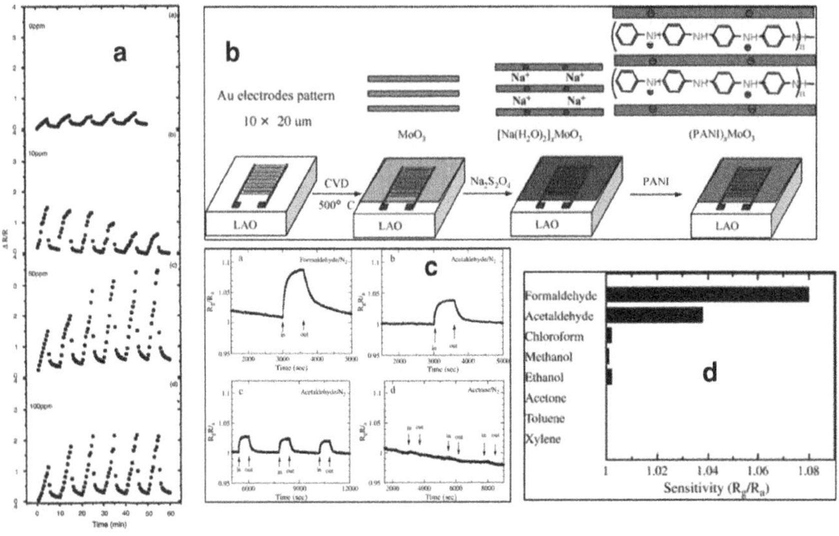

Fig. 10 **a** The responses of the nanocomposite on exposure to various chloroform concentrations: (a) 0 ppm (neat hexane), (b) 10 ppm; (c) 50 ppm, (d) 100 ppm. The sample was exposed to chloroform vapors for 5 min and the to dry air for 5 min alternately (Adapted from Ref [148]); **b** Schematic illustration of the preparation process of the (PANI)xMoO_3 thin film (LAO: $LaAlO_3$ (100) substrate, PANI: polyaniline, CVD: chemical vapor deposition; **c** Response signal Rg/Ra of the (PANI)xMoO_3 thin films to 50 ppm of various vapors at 30 °C: (a) formaldehyde, (b) acetaldehyde, (c) acetaldehyde, showing the reversible and reproducible response, (d) acetone. **d** The magnitude of the response signal Rg/Ra of the (PANI)xMoO_3 thin film at 30 °C upon exposure to various VOCs with a concentration of 50 ppm (Adapted from Ref [149])

11 Conducting Polymers for Quantification of Other Gaseous Analytes

Conducting polymer-based nanocomposites have also been used to detect other organic vapours such as acetone, chloroform as well as humidity. Sodium cholate-mediated template synthesis of polypyrrole yielded nanorods that were decorated with silver nanoparticles. This nanocomposite was used for the detection of different organic vapours such as methanol, ethanol, propanol, and acetone. While pristine polypyrrole films were found to exhibit superior sensitivity to methanol owing to hydrogen-bonding interactions with the polymer surface, the silver decorated polypyrrole nanorods displayed greater sensitivity towards acetone. The interaction of acetone over the polypyrrole surface was found to disrupt the silver-polypyrrole interactions resulting in an increase in the resistance, which was taken as a measure of the concentration of acetone. This acetone sensor exhibited a recovery time of 160–300 s and has been envisaged as a potential breath analyzer for the diagnosis of diabetes from the breath of individuals [150]. However, other organic vapours like methanol and ethanol also elicited a considerable response from the sensor that limits the clinical applicability of this version of the sensor. Further refinements to this strategy could probably lead to the development of clinically relevant technology. A novel attempt to extrapolate the measurement of relative humidity as a diagnostic tool for respiratory disorders such as asthma, chronic obstructive respiratory disorders, apnoea, etc., was reported recently. The study involved the fabrication of a multi-walled carbon nanotube-PANI composite on a flexible polyethylene terephthalate substrate. The system exhibited good response depending on the type of dopant employed in addition to excellent reproducibility and less hysteresis loss [151]. PANI nanofibers were explored for detection of relative humidity and it was found that the nanofibers exhibited a biphasic response to the presence of moisture. This 'reverse' behaviour was found to be absent in conventional PANI films suggesting that the high surface area-to-volume ratio of the nanofibers caused an abnormal increase in the film resistance due to swelling in water. In contrast, nano-aggregates of PANI deposited over screen-printed carbon electrodes exhibited a stable response to relative humidity suggesting that the stability of the nanostructure has a profound influence on the sensing performance of CP [152]. In an interesting variant, pre-determined cracks were introduced on thin films of the conducting polymer PEDOT:PSS deposited over flexible polydimethylsiloxane substrates. These engineered cracks interfered with the conduction pathways thereby enhancing the response to vapors of methanol, ethanol, or moisture [153]. The system exhibited rapid response time below 5 s and continued to perform for longer durations when compared to its crack-free counterpart. This study introduces the concept of conducting polymers with engineered cracks that could be explored for breath analysis. Thin films of PANI blended with functionalized single-walled carbon nanotubes have been reported for quantification of hydrogen sulfide (H_2S gas) by monitoring the changes in electrical resistance of the thin nanocomposite film on exposure to the analyte gas. The sensor exhibited the highest resistance at 50 °C. The carbon nanotubes distributed on the polymer matrix served to retain the H_2S gas molecules on the PANI surface. The H_2S gas removes the electrons from PANI, which increases the conductivity of the p-type conducting polymer that is quantified as a function of the analyte concentration [154]. Another study had explored the influence of film thickness, surface morphology, and ions incorporated on the sensing characteristics of polypyrrole films towards ethanol vapours in detail (Fig. 11) [155]. The systematic studies reported could be extrapolated for other types of conducting polymers-gaseous analyte combinations to achieve an optimum response from the sensor. The aging of olive oil was detected by monitoring the aromatic vapors collected in the headspace of the sample compartment using an array of four sensors each containing a specific type of conducting polymer. The different polymers employed were PANI, polypyrrole doped with Tiron or p-toluene sulfonate and PDOTT (poly(3,3″-dioctyl-[2,2′;5′,2″]terthiophene)). Among the four, polypyrrole doped with Tiron and PDOTT films outperformed the others during the initial screening while PANI/HCl systems were not efficient. The sensor successfully distinguished between three different olive oil samples [156].

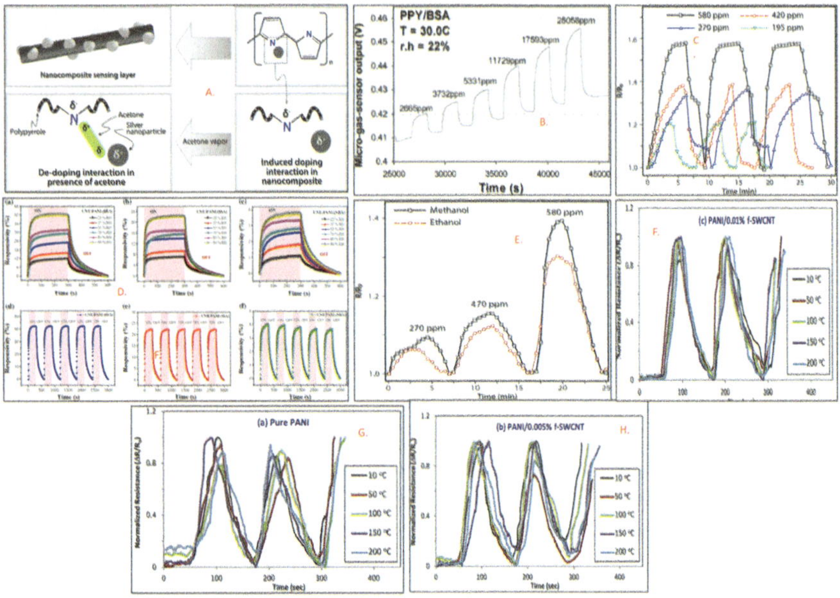

Fig. 11 **a** Proposed mechanism of acetone sensing by PPY-Ag nanocomposite layer; **b** Acetone sensing properties of AAPY-S-3 nanocomposite at different concentrations (Adapted from Ref. [157]); **c** Methanol and ethanol sensing properties of AAPY-S-3 nanocomposite at different concentrations; **d** Single-cycle responsivity plot with RH and d–f repeatability plot for HCl-, SSA- and NBA-doped CNT–PANI (Adapted from Ref. [158]); **e** Typical response of a PPY micro-gas-sensor to pulses of ethanol vapor at a fixed temperature and relative humidity. **f** The variation of normalized resistance with time for: (a) pure PANI, (b) PANI/0.005% f-SWNT, and (c) PANI/0.01% f-SWNT

12 Optical Sensing Applications of Conducting Polymers

An optical sensor is a device that transfers the light rays into electrical and/or electronic signals. The basic feature of an optical sensor is the ability to measure the changes in the intensity of one or more beam of light rays during sensing different analytes [100]. Optical sensors are convenient for the detection in real samples and are extensively employed in clinical and analytical laboratories. Conducting polymers have been used in many ways to augment the performance of optical sensors. They display electrochromic and gasochromic effects owing to the extensively conjugated pi bonds and low band gaps [159]. CPs have been employed as direct optical sensors that generate a response on the introduction of the analyte. They have also been employed as indirect optical sensors that use the conducting polymers as a matrix to immobilize fluorescent labels for quantification of the analyte. The latter strategy has been more extensively employed with a wide range of luminescent molecules used in combination with CPs for the detection of a specific analyte. Among the different types of CPs, polydiacetylenes (PDAs) exhibit inherent properties of absorption and fluorescence emission due to the pi-pi* transitions within the polymer network [16].

Several studies have employed PDAs for the detection of bacteria from samples by monitoring the change in color of the PDA membrane from blue to red upon binding of the bacteria and subsequent interactions with the molecules released by the captured bacteria [160]. Modification of the PDA with glycolipids has also been reported to improve the capture efficiency. Several bacterial species such as *Salmonella enterica*, *Bacillus megaterium*, *Bacillus cereus*, and *Escherichia coli* have been detected by monitoring the change in optical properties. However, the specificity of the sensor towards specific bacterial strains is poor [16]. Changes in the intensity of the absorption or emission has been used for quantifying the analyte of interest. Another report had employed poly(3-alkoxy-4-methyl-thiophene), a chromic conducting polymer, to complex with a single-stranded aptamer sequence that has a specific binding affinity to human α-thrombin, a key protein involved in the coagulation pathway. The complexation between the polymer and the aptamer sequence results in a specific emission in the visible region that is dependent upon the nature of the alkali metal ion present in the medium. Upon binding to the thrombin, a conformation change is induced in the aptamer that forms a G-quartet stabilized by hydrogen bonding between the four guanosine residues involved in the tetraplex formation. This conformation change results in a shift in the fluorescence emission wavelength that is used for quantifying the thrombin. The sensor exhibited high sensitivity and could detect thrombin at concentrations as low as 2 fM [161]. Polyaniline has been employed for gas sensing using changes in the absorption brought about by the interaction of the NH_3 gas on the PANI surface. The principle of quantification is based on the different oxidation states in which PANI exists, namely the emeraldine base, leucoemeraldine base (reduced form), and the pernigraniline base (oxidized form), which are altered due to interaction with the analyte gas resulting in the shift in absorbance. The sensor exhibited 90% of the response within 15 s after exposure to the sample gas and was reversible [162]. Another study compared the ammonia sensing efficiency of three conducting polymers, namely, PEDOT:PSS, PANI, and polypyrrole by monitoring the change in the absorbance produced on exposure to NH_3. Under identical conditions, the sensing performance of the polymers was ranked as follows: PEDOT:PSS > PANI > polypyrrole [159]. Films formed using PANI and its derivatives have been also employed to detect pH changes between 5.4 to 10.5 based on the changes in absorbance brought about by the changes in the PANI oxidation states at various pH [163].

A composite employing graphene quantum dots incorporated on the surface of the polypyrrole matrix was recently investigated for quantification of the hormone epinephrine. The fluorescence property of the graphene quantum dots was enhanced by surface passivation mediated by the amine functionalities on the polymer. The addition of the analyte resulted in quenching of the fluorescence in a concentration-dependent manner, which was used for quantification of epinephrine in the sample over a wide concentration range of 0.7–400 μM. The sensor exhibited the ability to detect epinephrine in human serum and hence holds promise for clinical diagnostic applications [164]. Another approach employed a modified optic fiber sensor where the outer cladding was replaced by the composite of PANI and poly(vinyl alcohol). The changes in pH alter the absorption and refractive index of the cladding that enabled the detection of pH over a wide range from 2 to 9 units. The response time of the sensor was below 22 s while the recovery time was about 35 min. The reproducibility of the results suggests that this system could find use in analytical laboratories as well as in water quality monitoring units [165]. Along similar lines, a composite of PANI and poly(ethylene glycol) was used for monitoring the growth kinetics of *Escherichia coli* bacteria. The pH responsiveness of PANI results in changes in its bandgap that is manifested by a change in the absorbance and the color of the film [166]. This strategy holds promise to be extended for the detection of multiple bacterial species or biomolecules. Conducting polymers have also been used in optodes that are used as sensing devices and contain a combination of a sensing element, polymer, and transducing elements. Polyoctylthiophene nanospheres modified with the chromo-ionophore N-octadecanoyl-Nile blue were fabricated as optode material. The optode displayed sensitive detection of potassium and calcium ions at pH 9 and 9.2 respectively. The chromo-ionophore underwent pH-dependent protonation at acidic pH that enabled detection in acidic media also. In the absence of the CP, the chromoionophore lost its sensitivity beyond pH 8. Thus, this combination was effective in sensing ions over a wide pH range, which could be invaluable for industrial and analytical applications [167].

Polypyrrole and Polyaniline polymers have also been used for the detection of heavy metals from contaminated water. PANI synthesized by the chemical method was used to quantify arsenic, lead, mercury, and cadmium in the samples using the evanescent wave absorbance [169]. Similarly, a composite comprising polypyrrole, chitosan, silver nanoparticles on indium tin oxide has been reported for the quantification of cadmium, lead, and mercury from water samples using surface plasmon resonance spectroscopy [170].

In recent years, the conducting nature of polyaniline and polypyrroles has been used for developing electrochemical-based surface plasmon resonance (EC-SPR) systems for the detection of various biomolecules such as immunoglobulin G. SPR technique uses the refractive index change at a solid-liquid interface that can be effectively used for studying polymer films, interfaces and several kinetic processes. Also, the combination of electrochemical and SPR technique is widely used for the in situ studies on the optical and electrical properties of conducting polymeric films as well as for immunosensing. The EC-SPR property of poly(3,4-ethylenedioxythiophene) (PEDOT) ultrathin films has been studied using enhanced photoluminescence spectroscopy. The photoluminescence property of PEDOT is improved when the polymer is de-doped under an applied potential. The photoluminescence intensity of PEDOT is controlled by the potential as well as the angular position in SPR refractive index experiments and plays a key role in optical sensing applications [171]. The electroactivity of PANI at neutral pH has been effectively investigated for NADH sensing through the SPR technique [172]. PANI films deposited over indium tin oxide (ITO) substrate have been used for the detection of NH_3 gas using SPR. The interaction of the gas with PANI results in a lowering of the refractive index of the polymer and a shift in the resonance frequency, which was used for quantification of the analyte gas. The thickness of the ITO film that served as the plasmonic layer was found to influence the sensing performance and a 60 nm thick layer was found to be optimum

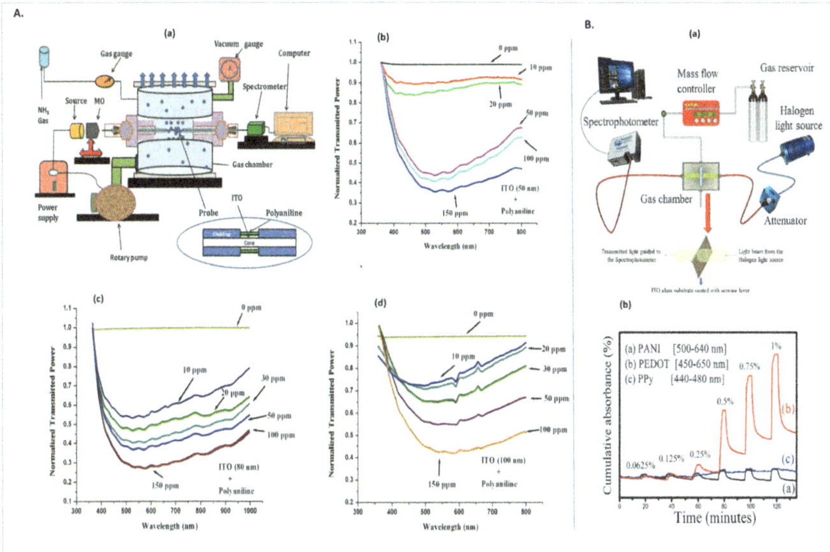

Fig. 12 **a** Schematic representation of the experimental set-up of the fiber optic sensor for ammonia gas (a). SPR spectra of the fiber optic probe for (b) ITO (50 nm), (c) ITO (80 nm), (d) ITO (100 nm), and polyaniline coatings recorded for different concentrations of ammonia gas (Adapted from Ref. [168]). **b** General set-up for the optical NH_3 gas sensor (a) and Dynamic response of (a) PANI, (b) PEDOT, and (c) PPy towards different concentrations of NH_3 gas, integrated over their respective wavelength range (b) (Adapted from Ref. [159])

(Fig. 12) [168]. Electrochromic sensors that exhibit a change in the optical properties after undergoing reduction or oxidation have been fabricated using organic conducting polymers such as PANI, PEDOT, and polypyrrole. These sensors have been used for quantification of different gaseous analytes such as ammonia, H_2S, methanol, humidity, hydrogen, ethanol, and some biologically relevant molecules such as glucose and ascorbic acid [173]. PANI-ZnO nanocomposites were employed for the detection of acetic acid vapors based on the reduction in the photoluminescent properties of the composite due to the interaction with the analyte. The sensor quantified acetic acid between 1 and 10 ppm with a response time of 30–50 s [174].

13 Challenges in HCP-Based Sensors

Hybrid conducting organic/inorganic nanocomposites has opened up new avenues for applications in environmental air quality monitoring and pollution control at ambient temperatures, but selectivity remains a major concern. Achievement of good selectivity and better sensitivity at the ppt level requires excellent adsorption of the given target analyte on the surface of the nanocomposites. Perhaps, this can be resolved effectively by detailed studies and a better understanding of a suitable dopant (mainly carbon and metal/metal oxide nanostructures, e.g. graphene, Pt, Ni, Au, Ag, WO_3, ZnO, SnO_2, TiO_2, etc.) that may improve the charge migration toward electrode surfaces. With rapid advances made in engineering and technology in the context of measurement devices, it is now possible to achieve sensitivities in the ppt level. Another challenge in the use of CP-based sensors is the stability of the conducting polymers that tend to deteriorate with time thereby limiting the lifetime and performance of the sensors. The nature and amount of dopant introduced in the CPs are additional factors that influence sensing performance. The design of appropriate synthetic strategies and doping protocols is essential to achieve a stable sensor response over an extended period. Concerted efforts to develop specific-target-analyte hybrid nanocomposites of CPs and understanding the sensing mechanism on the interface between the conducting polymeric host materials and dopant organic/inorganic nanostructures are required for realizing the true potential of CPs towards sensing applications.

14 Conclusion

Sensors have found applications in all walks of life and attempts to improve the sensing parameters have fuelled the search for efficient materials that can serve as sensing or transducing elements. Conducting polymers have found favor for sensor fabrication owing to their unique conducting properties and ability to serve as immobilization matrices. The interaction of analyte molecules with the conducting polymers results in changes in their conductivity and optical properties that have been well utilized for quantification of the analyte of interest. Further, these polymeric networks could be doped or blended with other molecules, ions, or particles that improve their sensing property. The flexible nature of these conducting polymers has now opened up new avenues for their use as wearable sensors. However, there are few limitations of conducting polymer-based sensors that need to be addressed for their extensive commercialization. The adhesion of the conducting polymer film differs on various substrates and this reduces the structural stability of the device. Further, the lack of specificity towards the analytes limits their use in clinical diagnosis. The thickness of the polymer film is another parameter that could influence the sensor output owing to the difference in the diffusion profiles. Nevertheless, the merits of conducting polymers far outweigh their demerits and with emerging fabrication techniques such as 3D screen printing, additive manufacturing, and nanostructuring, the conducting polymer-based sensors have not only exhibited improved sensing performances but also have become part of novel applications such as touch sensor, pressure sensing, wearable sensors, and breath analysis.

Acknowledgements The authors would like to acknowledge the Research Council of United Kingdom (funding from RCUK grant MR/P027881/1) and SASTRA Deemed University for infrastructural support.

References

1. Windmiller, J.R., Wang, J.: Wearable electrochemical sensors and biosensors: a review **25**(1), 29–46 (2013)
2. Stetter, J.R., Penrose, W.R., Yao, S.: Sensors, chemical sensors, electrochemical sensors, and ECS. J. Electrochem. Soc. **150**(2), S11 (2003)
3. Hall, E.A.H.: Overview of biosensors. In: Biosensors and Chemical Sensors, pp. 1–14. American Chemical Society (1992)
4. Justino, C.I.L., et al.: Recent developments in recognition elements for chemical sensors and biosensors. TrAC Trends Anal. Chem. **68**, 2–17 (2015)
5. Bourgeois, W., et al.: The use of sensor arrays for environmental monitoring: interests and limitations. J. Environ. Monit. **5**(6), 852–860 (2003)
6. Shrivastava, S., Jadon, N., Jain, R.: Next-generation polymer nanocomposite-based electrochemical sensors and biosensors: a review. TrAC Trends Anal. Chem. **82**, 55–67 (2016)
7. Chen, H.-I., Chou, Y.-I., Chu, C.-Y.: A novel high-sensitive Pd/InP hydrogen sensor fabricated by electroless plating. Sens. Actuators B: Chem. **85**(1), 10–18 (2002)
8. Kannan, P.K., et al.: Recent developments in 2D layered inorganic nanomaterials for sensing. Nanoscale **7**(32), 13293–13312 (2015)
9. Partridge, A.C., Jansen, M.L., Arnold, W.M.: Conducting polymer-based sensors. Mater. Sci. Eng., C **12**(1), 37–42 (2000)
10. Naveen, M.H., Gurudatt, N.G., Shim, Y. B.: Applications of conducting polymer composites to electrochemical sensors: a review. Appl. Mater. Today **9**, 419–433 (2017)
11. Yoon, H., Jang, J.: Conducting-polymer nanomaterials for high-performance sensor applications: issues and challenges. Adv. Funct. Mater. **19**(10), 1567–1576 (2009)
12. MacDiarmid, A.G.: Synthetic metals: a novel role for organic polymers (nobel lecture). Angew. Chem. Int. Ed. **40**, 2581–2590 (2001)
13. Chandrasekhar, P.: Conducting Polymers and Other New Electronically Conductive Materials Including Carbon Nanotubes and Graphene: Fundamentals and Applications. Springer, Berlin (2018)
14. Hu, J., Liu, S.: Responsive polymers for detection and sensing applications: current status and future developments. Macromolecules **43**(20), 8315–8330 (2010)
15. Tourillon, G., Garnier, F.: Effect of dopant on the physicochemical and electrical properties of organic conducting polymers. J. Phys. Chem. **87**(13), 2289–2292 (1983)
16. Arshak, K., et al.: Conducting polymers and their applications to biosensors: emphasizing on foodborne pathogen detection. IEEE Sens. J. **9**, 1942–1951 (2009)
17. Iqbal, S., Ahmad, S.: Recent development in hybrid conducting polymers: synthesis, applications and future prospects. J. Ind. Eng. Chem. **60**, 53–84 (2018)
18. Persaud, K.C.: Polymers for chemical sensing. Mater. Today **8**(4), 38–44 (2005)
19. Ivanova, E.V., et al.: Evaluation of redox mediators for amperometric biosensors: Ru-complex modified carbon-paste/enzyme electrodes. Bioelectrochemistry **60**(1), 65–71 (2003)
20. Wenjuan, Y., et al.: Electrogenerated trisbipyridyl Ru(II)-/nitrilotriacetic-polypyrene copolymer for the easy fabrication of label-free photoelectrochemical immunosensor and aptasensor: application to the determination of thrombin and anti-cholera toxinantibody. Biosens. Bioelectron. **42**, 556–562 (2013)
21. Lange, U., Roznyatovskaya, N.V., Mirsky, V.M.: Conducting polymers in chemical sensors and arrays. Anal. Chim. Acta **614**(1), 1–26 (2008)
22. Teles, F.R.R., Fonseca, L.P.: Applications of polymers for biomolecule immobilization in electrochemical biosensors. Mater. Sci. Eng., C **28**(8), 1530–1543 (2008)
23. Miasik, J.J., Hooper, A., Tofield, B.C.: Conducting polymer gas sensors. J. Chem. Soc. Faraday Trans. Phys. Chem. Condens. Phases **82**(4), 1117–1126 (1986)
24. Ates, M.: A review study of (bio)sensor systems based on conducting polymers (1873–0191 (Electronic))

25. Pilo, M., et al.: Design of amperometric biosensors for the detection of glucose prepared by immobilization of glucose oxidase on conducting (Poly)thiophene films. J. Anal. Methods Chem. **2018**, 1849439 (2018)
26. Jugović, B., et al.: Electrochemical science: polypyrrole-based enzyme electrode with immobilized glucose oxidase for electrochemical determination of glucose. Int. J. Electrochem. Sci. **11**, 1152–1161 (2016)
27. Li, Y., et al.: Flexible pH sensor based on a conductive PANI membrane for pH monitoring. RSC Adv. **10**, 21–28 (2020)
28. Bhadra, S., et al.: Preparation of nanosize polyaniline by solid-state polymerization and determination of crystal structure. Polim. Int. **58**(10), 1173–1180 (2009)
29. Li, C., Bai, H., Shi, G.: Conducting polymer nanomaterials: electrosynthesis and applications. Chem. Soc. Rev. **38**(8), 2397–2409 (2009)
30. Viter, R., et al.: Enhancement of electronic and optical properties of ZnO/Al_2O_3 nanolaminate coated electrospun nanofibers. J. Phys. Chem. C **120**(9), 5124–5132 (2016)
31. Forzani, E.S., et al.: A conducting polymer nanojunction sensor for glucose detection. Nano Lett. **4**, 1785–1788 (2004)
32. Kailasa, S., et al.: High sensitive polyaniline nanosheets (PANINS) @rGO as non-enzymatic glucose sensor. J. Mater. Sci.: Mater. Electron. **31**, 2926–2937 (2020)
33. Zhang, X.-L., et al.: Improvement of amperometric sensor used for determination of nitrate with polypyrrole nanowires modified electrode. Sensors **5**, 580–593 (2005)
34. Li, J., Wei, W., Luo, S.: A novel one-step electrochemical codeposition of carbon nanotubes-DNA hybrids and tiron doped polypyrrole for selective and sensitive determination of dopamine. Microchim. Acta **171**, 109–116 (2010)
35. Berk, M., et al.: Dopamine dysregulation syndrome: implications for a dopamine hypothesis of bipolar disorder. Acta Psychiatr. Scand. **116**(s434), 41–49 (2007)
36. Adhikari, A., et al.: Selective sensing of dopamine by sodium cholate tailored polypyrrole-silvernanocomposite. Synth. Met. **260**, 116296–116309 (2020)
37. Gao, Q., et al.: A biosensor prepared by co-entrapment of a glucose oxidase and a carbon nanotube within an electrochemically deposited redox polymer multilayer. Bioelectrochemistry **81**, 109–113 (2011)
38. Kong, B., et al.: Voltammetric determination of hydroquinone using β-Cyclodextrin/Poly(N-Acetylaniline)/carbon nanotube composite modified glassy carbon electrode. Anal. Lett. **40**, 2141–2150 (2007)
39. Harrazab, F., et al.: Conducting polythiophene/α-Fe_2O_3 nanocomposite for efficient methanol electrochemical sensor. Appl. Surf. Sci. **508**, 145226 (2020)
40. Kwon, O.S., et al.: Multidimensional conducting polymer nanotubes for ultrasensitive chemical nerve agent sensing. Nano Lett. **6**, 2797–2802 (2012)
41. Bakker, E., Qin, Y.: Electrochemical sensors. Anal. Chem. **78**(12), 3965–3984 (2006)
42. Soylemez, S., et al.: A multipurpose conjugated polymer: electrochromic device and biosensor construction for glucose detection. Org. Electron. **65**, 327–333 (2019)
43. Maity, D., Minitha, C.R, Rajendra Kumar, R.T.: Glucose oxidase immobilized amine terminated multiwall carbon nanotubes/reduced graphene oxide/polyaniline/gold nanoparticles modified screen-printed carbon electrode for highly sensitive amperometric glucose detection. Mater. Sci. Eng. C **105**, 110075 (2019)
44. Kong, Y.-T., Boopathi, M., Shim, Y.-B.: Direct electrochemistry of horseradish peroxidase bonded on a conducting polymer modified glassy carbon electrode. Biosens. Bioelectron. **19**(3), 227–232 (2003)
45. Miyazaki, C.M., et al.: Monoamine oxidase B layer-by-layer film fabrication and characterization toward dopamine detection. Mater. Sci. Eng., C **58**, 310–315 (2016)
46. Kaur, N., Thakur, H., Prabhakar, N.: Multi walled carbon nanotubes embedded conducting polymer based electrochemical aptasensor for estimation of malathion. Microchem. J. **147**, 393–402 (2019)
47. Shelke, U.N., et al.: Synthesis of poly o-anisidine based sensor for amperometric detection of pesticide carbaryl. Current Pharma Res. **9**, 3208–3215 (2019)

48. Cancar, H.D., et al.: A novel acetylcholinesterase biosensor: core-shell magnetic nanoparticles incorporating a conjugated polymer for the detection of organophosphorus pesticides. ACS Appl. Mater. Interfaces. **12**, 8058–8067 (2016)
49. Viter, R., et al.: Photoelectrochemical bisphenol S sensor based on ZnO-nanoroads modified by molecularly imprinted polypyrrole. Macromol. Chem. Phys. **221**, 1900232 (2019)
50. Kuwahara, T., et al.: Immobilization of glucose oxidase and electron-mediating groups on the film of 3-methylthiophene/thiophene-3-acetic acid copolymer and its application to reagentless sensing of glucose. Polymer **46**(19), 8091–8097 (2005)
51. Shelke, U.N., K.S.P., Deshmukh, V.B., Folane, S.A., Gade, V.K.: Synthesis of poly o-anisidine based sensor for amperometric detection of pesticide. *carbaryl*. Current Pharma Res. **9**(4), 3208–3215 (2019)
52. Chen, X., et al.: Development and validation of HPLC-MS/MS method for the simultaneous determination of 8-Hydroxy-2′-deoxyguanosine and twelve cosmetic phenols in human urine. Chromatographia **82**(9), 1415–1421 (2019)
53. Guler, M., Turkoglu, V., Kivrak, A.: Electrochemical detection of malathion pesticide using acetylcholinesterase biosensor based on glassy carbon electrode modified with conducting polymer film. Environ. Sci. Pollut. Res. **23**(12), 12343–12351 (2016)
54. Dzudzevic Cancar, H., et al.: A novel acetylcholinesterase biosensor: core-shell magnetic nanoparticles incorporating a conjugated polymer for the detection of organophosphorus pesticides. ACS Appl. Mater. Interfaces. **8**(12), 8058–8067 (2016)
55. Wang, M., et al.: Highly dispersed conductive polypyrrole hydrogels as sensitive sensor for simultaneous determination of ascorbic acid, dopamine and uric acid. J. Electroanal. Chem. **832**, 174–181 (2019)
56. Raj, M., et al.: Graphene/conducting polymer nano-composite loaded screen printed carbon sensor for simultaneous determination of dopamine and 5-hydroxytryptamine. Sens. Actuat. B: Chem. **239**, 993–1002 (2017)
57. Ghanbari, K., Hajheidari, N.: ZnO–CuxO/polypyrrole nanocomposite modified electrode for simultaneous determination of ascorbic acid, dopamine, and uric acid. Anal. Biochem. **473**, 53 62 (2015)
58. Ghanbari, K., Moloudi, M.: Flower-like ZnO decorated polyaniline/reduced graphene oxide nanocomposites for simultaneous determination of dopamine and uric acid. Anal. Biochem. **512**, 91–102 (2016)
59. Narouei, F.: An electrochemical sensor based on conductive polymers/graphite paste electrode for simultaneous determination of dopamine, uric acid and tryptophan in biological samples. Int. J. Electrochem. Sci. **12**, 7739–7753 (2017)
60. Harraz, F.A., et al.: Highly sensitive amperometric hydrazine sensor based on novel α-Fe_2O_3/crosslinked polyaniline nanocomposite modified glassy carbon electrode. Sens. Actuat. B: Chem. **234**, 573–582 (2016)
61. Yang, Z., et al.: One-pot synthesis of Fe_3O_4/polypyrrole/graphene oxide nanocomposites for electrochemical sensing of hydrazine. Microchim. Acta **184**(7), 2219–2226 (2017)
62. Faisal, M., et al.: Polythiophene/ZnO nanocomposite-modified glassy carbon electrode as efficient electrochemical hydrazine sensor. Mater. Chem. Phys. **214**, 126–134 (2018)
63. Kotanen, C.N., et al.: Fabrication and in vitro performance of a dual responsive lactate and glucose biosensor. Electrochim. Acta **267**, 71–79 (2018)
64. Hasanzadeh, M., et al.: Electropolymerization of proline supported beta-cyclodextrin inside amino functionalized magnetic mesoporous silica nanomaterial: one step preparation, characterization and electrochemical application. Anal. Bioanal. Electrochem **10**, 77–97 (2018)
65. Ding, M., et al.: An electrochemical sensor based on graphene/poly(brilliant cresyl blue) nanocomposite for determination of epinephrine. J. Electroanal. Chem. **763**, 25–31 (2016)
66. Sharma, V., et al.: Graphene nanoplatelets-silver nanorods-polymer based in-situ hybrid electrode for electroanalysis of dopamine and ascorbic acid in biological samples. Appl. Surf. Sci. **449**, 558–566 (2018)

67. Akbulut, H., et al.: Polythiophene-g-poly(ethylene glycol) with lateral amino groups as a novel matrix for biosensor construction. ACS Appl. Mater. Interfaces. **7**(37), 20612–20622 (2015)
68. Sangamithirai, D., et al.: Investigations on the performance of poly(o-anisidine)/graphene nanocomposites for the electrochemical detection of NADH. Mater. Sci. Eng., C **55**, 579–591 (2015)
69. Baskar, S., Chang Jl Fau-Zen, J.-M., Zen, J.M.: Simultaneous detection of NADH and H(2)O(2) using flow injection analysis based on a bifunctional poly(thionine)-modified electrode (1873–4235 (Electronic))
70. Warren, S., et al.: Scanning electrochemical microscopy imaging of poly (3,4-ethylendioxythiophene)/thionine electrodes for lactate detection via NADH electrocatalysis. Biosens. Bioelectron. **137**, 15–24 (2019)
71. Sahin, M., Ayranci, E.: Electrooxidation of NADH on modified screen-printed electrodes: effects of conducting polymer and nanomaterials. Electrochim. Acta **166**, 261–270 (2015)
72. Barsan, M.M., et al.: New CNT/poly(brilliant green) and CNT/poly(3,4-ethylenedioxythiophene) based electrochemical enzyme biosensors. Anal. Chim. Acta **927**, 35–45 (2016)
73. Aydoğdu Tığ, G.: Highly sensitive amperometric biosensor for determination of NADH and ethanol based on Au-Ag nanoparticles/poly(L-Cysteine)/reduced graphene oxide nanocomposite. Talanta **175**, 382–389 (2017)
74. Ghanbari, K., Nejabati, F.: Construction of novel nonenzymatic Xanthine biosensor based on reduced graphene oxide/polypyrrole/CdO nanocomposite for fish meat freshness detection. J. Food Measur. Character. **13**(2), 1411–1422 (2019)
75. Sahyar, B.Y., et al.: Electrochemical xanthine detection by enzymatic method based on Ag doped ZnO nanoparticles by using polypyrrole. Bioelectrochemistry **130**, 107327 (2019)
76. Dervisevic, M., et al.: Amperometric cholesterol biosensor based on reconstituted cholesterol oxidase on boronic acid functional conducting polymers. J. Electroanal. Chem. **776**, 18–24 (2016)
77. Dervisevic, M., et al.: Construction of novel xanthine biosensor by using polymeric mediator/MWCNT nanocomposite layer for fish freshness detection. Food Chem. **181**, 277–283 (2015)
78. Dervisevic, M., et al.: Novel electrochemical xanthine biosensor based on chitosan–polypyrrole–gold nanoparticles hybrid bio-nanocomposite platform. J. Food Drug. Anal. **25**(3), 510–519 (2017)
79. Saito, M., et al.: A microbial platform based on conducting polymers for evaluating metabolic activity. Anal. Chem. **20**, 12793–12798 (2019)
80. Butina, K., et al.: Electrochemical sensing of bacteria via secreted redox active compounds using conducting polymers. Sens. Actuat. B: Chem. **297**, 126703–126709 (2019)
81. Riul, A., et al.: Artificial taste sensor: efficient combination of sensors made from langmuir−blodgett films of conducting polymers and a ruthenium complex and self-assembled films of an azobenzene-containing polymer. Langmuir **1**, 239–245 (2002)
82. Peng, H., et al.: Conducting polymers for electrochemical DNA sensing. Biomaterials **30**(11), 2132–2148 (2009)
83. Peng, H., et al.: Conducting polymers for electrochemical DNA sensing. Biomaterials **30**, 2132–2148 (2009)
84. Janata, J., Josowicz, M.: Conducting polymers in electronic chemical sensors. Nat. Mater. **2**, 19–24 (2003)
85. Chang, J.B., et al.: Printable polythiophene gas sensor array for low-cost electronic noses. J. Appl. Phys. **100**, 014506 (2006)
86. Jun, H.K., et al.: Electrical properties of polypyrrole gas sensors fabricated under various pretreatment conditions. Sens. Actuators, B **96**, 576–581 (2003)
87. Saxena, V., Malhotra, B.D.: Prospects of conducting polymers in molecular electronics. Curr. Appl. Phys. **3**, 293–305 (2003)
88. Stetter, J.R., Li, J.: Amperometric gas sensors—a review. Chem. Rev. **108**, 352–366 (2008)

89. Xie, D., et al.: Fabrication and characterization of polyaniline based gas sensor by ultra-thin film technology. Sens. Actuators, B **81**, 158–164 (2002)
90. Tang, C., Chen, N., Hu, X.: Conducting polymer nanocomposites: recent developments and future prospects. In: K. V, K. S, and S. H (eds.) Conducting Polymer Hybrids. Springer Series on Polymer and Composite Materials. Springer, Cham (2017)
91. Hatchett, D.W., Josowicz, M.: Composites of intrinsically conducting polymers as sensing nanomaterials. Chem. Rev. **108**, 746–769 (2008)
92. Kaushik, A., et al.: Organic−inorganic hybrid nanocomposite-based gas sensors for environmental monitoring. Chem. Rev. **115**, 4571–4606 (2015)
93. Liu, Y., Kumar, S.: Polymer/carbon nanotube nanocomposite fibers—a review. ACS Appl. Mater. Interfaces **6**, 6069–6087 (2014)
94. Rhazi, M.E., et al.: Recent progress in nanocomposites based on conducting polymer: application as electrochemical sensors. Int. Nano Lett. **8**, 79–99 (2018)
95. Yu, X., et al.: Fabrication technologies and sensing applications of graphene-based composite flms: advances and challenges. Biosens. Bioelectron. **89**, 72–84 (2017)
96. Wong, Y.C., et al.: Conducting polymers as chemiresistive gas sensing materials: a review. J. Electrochem. Soc. **167**, 037503 (2020)
97. Korotcenkov, G., Cho, B.K.: Metal oxide composites in conductometric gas sensors: achievements and challenges. Sens. Actuat. B **244**, 182–210 (2017)
98. Pandey, S.: Highly sensitive and selective chemiresistor gas/vapor sensors based on polyaniline nanocomposite: a comprehensive review. J. Sci. Adv. Mater. Dev. **1**(4), 431–453 (2016)
99. Fratoddi, I., et al.: Chemiresistive polyaniline-based gas sensors: a mini review. Sens. Actuat. B: Chem. **220**, 534–548 (2015)
100. Hatchett, D.W., Josowicz, M.: Composites of intrinsically conducting polymers as sensing nanomaterials. Chem. Rev. **108**(2), 746–769 (2008)
101. El Rhazi, M., et al.: Recent progress in nanocomposites based on conducting polymer: application as electrochemical sensors. Int. Nano Lett. **8**(2), 79–99 (2018)
102. Ray, C., Pal, T.: Recent advances of metal–metal oxide nanocomposites and their tailored nanostructures in numerous catalytic applications. J. Mater. Chem. A **5**(20), 9465–9487 (2017)
103. Mallakpour, S., Khadem, E.: Carbon nanotube–metal oxide nanocomposites: fabrication, properties and applications. Chem. Eng. J. **302** (2016)
104. Lakard, B., et al.: Gas sensors based on electrodeposited polymers. Metals **5**, 1371–1386 (2015)
105. Wang, X., et al.: Nanostructured polyaniline/poly(styrene-butadiene-styrene) composite fiber for use as highly sensitive and flexible ammonia sensor. Synth. Met. **233**, 86–93 (2017)
106. Han, G., Shi, G.: Porous polypyrrole/polymethylmethacrylate composite film prepared by vapor deposition polymerization of pyrrole and its application for ammonia detection. Thin Solid Films **515**, 6986–6991 (2007)
107. Lobotka, P., et al.: Thin polyaniline and polyaniline/carbon nanocomposite films for gas sensing. Thin Solid Films **519**, 4123–4127 (2011)
108. Wu, Z., et al.: Enhanced sensitivity of ammonia sensor using graphene/polyaniline nanocomposite. Sens. Actuator. B **178**, 485–493 (2013)
109. Van Hieu, N., et al.: Thin film polypyrrole/SWCNTs nanocomposites based NH_3 sensor operated at room temperature. Sens. Actuat. B **140**, 500−507 (2009)
110. Sun, J., et al.: Facile preparation of polypyrrole-reduced graphene oxide hybrid forenhancing NH_3 sensing at room temperature. Sens. Actuat. B **241**, 658–664 (2017)
111. Sharma, S., et al.: MWCNT-conducting polymer composite based ammonia gas sensors: a new approach for complete recovery process. Sens. Actuat. B **194**, 213–219 (2014)
112. Tai, H., et al.: Fabrication and gas sensitivity of polyaniline−titanium dioxide nanocomposite thin film. Sens. Actuat. B, **125**, 644−650 (2007)
113. Tai, H., et al.: Preparation, characterization and comparative NH_3-sensing characteristic studies of PANI/inorganic oxides nanocomposite thin films. J. Mater. Sci. Technol. **26**, 605–613 (2010)

114. Kulkarni, S.B., et al.: Enhanced ammonia sensing characteristics of tungsten oxide decorated polyaniline hybrid nanocomposites. Org. Electron. **45**, 65–73 (2017)
115. Joulazadeh, M., Navarchian, A.H.: Ammonia detection of one-dimensional nano-structured polypyrrole/metal oxide nanocomposites sensors. Synth. Met. **210**, 404–411 (2015)
116. Wang, Y., et al.: Ammonia gas sensor using polypyrrole-coated TiO_2/ZnO nanofibers. Electroanalysis **21**, 1432–1438 (2009)
117. Hien, H.T., et al.: Elaboration of Pd-nanoparticle decorated polyaniline films for roomtemperature NH_3 gas sensors. Sens. Actuat. B **249**, 348–356 (2017)
118. Hong, L., Li, Y., Yang, M.: Fabrication and ammonia gas sensing of palladium/polypyrrole nanocomposite. Sens. Actuat. B **145**, 25–31 (2010)
119. Kaushik, A., et al.: Hybrid cross-linked polyaniline-WO_3 nanocomposite thin film for NOx gas sensing. J. Nanosci. Nanotechnol. **9**, 1792–1796 (2009)
120. Geng, L., et al.: The preparation and gas sensitivity study of polypyrrole/zinc oxide. Synth. Met. **156**, 1078–1082 (2006)
121. Xu, M., et al.: Gas sensing properties of SnO_2 hollow spheres/polythiophene inorganic−organic hybrids. Sens. Actuat. B **146**, pp. 8–13 (2010)
122. Ram, M.K., Yavuz, O., Aldissi, M.: NO_2 gas sensing based on ordered ultrathin films of conducting polymer and its nanocomposite. Synth. Met. **151**, 77–84 (2005)
123. Do, J.-S., Chang, W.-B.: Amperometric nitrogen dioxide gas sensor based on Pan/Au/Nafion prepared by constant current and cyclic voltammetry methods. Sens. Actuat. B, **101**, 97−106 (2004)
124. Roy, A., et al.: Polyaniline-multiwalled carbon nanotube (PANI-MWCNT): room temperature resistive carbon monoxide (CO) sensor. Synth. Met. **245**, 182–189 (2018)
125. Jin, G.P., et al.: Electrodeposition of platinum−nickel alloy nanocomposites on polyaniline-multiwalled carbon nanotubes for carbon monoxide redox. J. Solid State Electrochem. **13**, 967–973 (2009)
126. Waghuley, S.A., et al.: Application of chemically synthesized conducting polymer-polypyrrole as a carbon dioxide gas sensor. Sens. Actuat. B **128**, 366–373 (2008)
127. Chiang, C.-J., et al.: In situ fabrication of conducting polymer composite film as a chemical resistive CO_2 gas sensor. Microelectron. Eng. **111**, 409–415 (2013)
128. Bhadra, J., et al.: Fabrication of polyaniline–graphene/polystyrene nanocomposites for flexible gas sensors. RSC Adv. **9**, 12496–12506 (2019)
129. Barde, R.V.: Preparation, characterization and CO_2 gas sensitivity of Polyaniline doped with Sodium Superoxide (NaO_2). Mater. Res. Bull. **73**, 70–76 (2016)
130. Li, W., Kim, D.: Polyaniline/multiwall carbon nanotube nanocomposite for detecting aromatic hydrocarbon vapors. J. Mater. Sci. Technol. **46**, 1857–1861 (2011)
131. Patil, S.L., et al.: Measurements on room temperature gas sensing properties of CSA doped polyaniline ZnO nanocomposites. Measurement **45**, 243–249 (2012)
132. Bandgar, D.K., et al.: Ultrasensitive polyaniline-iron oxide nanocomposite room temperature flexible ammonia sensor. RSC Adv. **5**, 68964–68971 (2015)
133. Venditti, I., et al.: Nanostructured composite based on polyaniline and gold nanoparticles: synthesis and gas sensing properties. Nanotechnology **24**, 155503–155510 (2013)
134. Zhang, J., et al.: One-pot fabrication of uniformpolypyrrole/Au nanocomposites and investigation for gas sensing. Sen. Actuators B **186**, 695–700 (2013)
135. Zhang, W., et al.: High performance tube sensor based on PANI/Eu^{3+} nanofiber for low-volume NH_3 detection. Anal. Chim. Acta **1093**, 115–122 (2020)
136. Radhakrishnan, S., Deshpande, S.D.: Conducting polymers functionalized with phthalocyanine as nitrogen dioxide sensors. Sensors **2**, 185–194 (2002)
137. Xu, M., et al.: Gas sensing properties of SnO_2 hollow spheres/polythiophene inorganic–organic hybrids. Sens. Actuators B: Chem. **146**(1), 8–13 (2010)
138. Barn, P., et al.: A review of the experimental evidence on the toxicokinetics of carbon monoxide: the potential role of pathophysiology among susceptible groups. Environ. Health **17**(1), 13 (2018)

139. Wong, Y.C., et al.: Review—conducting polymers as chemiresistive gas sensing materials: a review. J. Electrochem. Soc. **167**(3), 037503 (2019)
140. Ram, M.K., et al.: CO gas sensing from ultrathin nano-composite conducting polymer film. Sens. Actuat. B, **106**, 750–757 (2005)
141. Wells, R.M.G.: Chapter 6—Blood–gas transport and hemoglobin function: adaptations for functional and environmental hypoxia. In: Richards, J.G., Farrell, A.P., Brauner, C.J. (eds.) Fish Physiology, pp. 255–299. Academic Press (2009)
142. Waghuley, S.A., et al.: Application of chemically synthesized conducting polymer-polypyrrole as a carbon dioxide gas sensor. Sens. Actuators B: Chem. **128**(2), 366–373 (2008)
143. Lu, J., et al.: Polyaniline nanoparticle–carbon nanotube hybrid network vapour sensors with switchable chemo-electrical polarity. Nanotechnology **21**, 255501 (2010)
144. Wang, J., et al.: The preparation of polyaniline intercalated MoO_3 thin film and its sensitivity to volatile organic compounds. Thin Solid Films **514**, 329–333 (2006)
145. Athawale, A.A., Bhagwat, S.V., Katre, P.P.: Nanocomposite of Pd–polyaniline as a selective methanol sensor. Sens. Actuators, B **114**, 263–267 (2006)
146. Ashraf, A., et al.: Electronic structure of polythiophene gas sensors for chlorinated analytes. J. Mol. Model. **26**, 44 (2020)
147. Sharma, S., et al.: Chloroform vapour sensor based on copper/polyaniline nanocomposite. Sens. Actuat. B **85**, 131–136 (2002)
148. Sharma, S., et al.: Chloroform vapour sensor based on copper/polyaniline nanocomposite. Sens. Actuators B: Chem. **85**(1), 131–136 (2002)
149. Wang, J., et al.: The preparation of polyaniline intercalated MoO3 thin film and its sensitivity to volatile organic compounds. Thin Solid Films **514**(1), 329–333 (2006)
150. Adhikari, A., et al.: Synthesis of sodium cholate mediated rod-like polypyrrole-silver nanocomposite for selective sensing of acetone vapor. Nano-Struct. Nano-Obj. **21**, 10041 (2020)
151. Kundu, S., et al.: Relative humidity sensing properties of doped polyaniline-encased multiwall carbon nanotubes: wearable and flexible human respiration monitoring application. J. Mater. Sci. **55**, 3884–3901 (2020)
152. Zeng, F.-W., et al.: Humidity sensors based on polyaniline nanofibres. Sens. Actuators B: Chem. **143**, 530–534 (2010)
153. Sarkar, B., Satapathy, D.K., Jaiswal, M.: Nanostructuring mechanical cracks in a flexible conducting polymer thin film for ultra-sensitive vapor sensing. Nanoscale **11**, 200–210 (2019)
154. Suhaila, M.H., Abdullah, O.G., Kadhim, G.A.: Hydrogen sulfide sensors based on PANI/f-SWCNT polymer nanocomposite thin films prepared by electrochemical polymerization. J. Sci. Adv. Mater. Dev. **4**, 143–149 (2019)
155. Fang, Q., et al.: Micro-gas-sensor with conducting polymers. Sens. Actuators B: Chem. **84**, 66–71 (2002)
156. Stella, R., et al.: Characterisation of olive oil by an electronic nose based on conducting polymer sensors. Sens. Actuators B: Chem. **63**, 1–9 (2000)
157. Adhikari, A., et al.: Synthesis of sodium cholate mediated rod-like polypyrrole-silver nanocomposite for selective sensing of acetone vapor. Nano-Struct. Nano-Obj. **21**, 100419 (2020)
158. Kundu, S., et al.: Relative humidity sensing properties of doped polyaniline-encased multiwall carbon nanotubes: wearable and flexible human respiration monitoring application. J. Mater. Sci. **55**(9), 3884–3901 (2020)
159. HadiIsmaila, A., et al.: Optical ammonia gas sensor of poly(3,4-polyethylenedioxythiophene), polyaniline and polypyrrole: A comparative study. Synth. Met. **260**, 116294 (2020)
160. Xu, X., et al.: Controlled-temperature photothermal and oxidative bacteria killing and acceleration of wound healing by polydopamine-assisted Au-hydroxyapatite nanorods. Acta Biomater. **77**, 352–364 (2018)
161. Ho, H.-A., Leclerc, M.: Optical sensors based on hybrid aptamer/conjugated polymer complexes. J. Am. Chem. Soc. **126**, 1384–1387 (2004)

162. Jin, Z., Su, Y., Duan, Y.: Development of a polyaniline-based optical ammonia sensor. Sens. Actuators B: Chem. **72**, 75–79 (2001)
163. Pringsheim, E., Terpetschnig, E., Wolfbeis, O.S.: Optical sensing of pH using thin films of substituted polyanilines. Anal. Chim. Acta **357**, 247–252 (1997)
164. Le, T.H., et al.: Polypyrrole/graphene quantum dot composites as a sensor media for epinephrine. J. Nanosci. Nanotechnol. **20**, 4005–4010 (2020)
165. Khanikar, T., Singh, V.K.: PANI-PVA composite film coated optical fiber probe as a stable and highly sensitive pH sensor. Opt. Mater. **88**, 244–251 (2019)
166. Ko, Y., et al.: pH-responsive polyaniline/polyethylene glycol composite arrays for colorimetric sensor application. Sens. Actuators B: Chem. **305**, 127447 (2020)
167. Stelmach, E., et al.: Tailoring polythiophene cation-selective optodes for wide pH range sensing. Talanta **211**, 120663 (2020)
168. Mishra, K.S., Kumari, D., Gupta, B.D.: Surface plasmon resonance based fiber optic ammonia gas sensor using ITO and polyaniline. Sens. Actuators B: Chem. **171–172**, 976–983 (2012)
169. Chandra, S., Mukherji, S.: Conducting polymer-based optical sensor for heavy metal detection in drinking water. Proceedings of SPIE 10680, optical sensing and detection V, 106801A (9 May 2018)
170. Verma, R., Gupta, B.D.: Detection of heavy metal ions in contaminated water by surface plasmon resonance based optical fibre sensor using conducting polymer and chitosan. Food Chem. **166**, 568–575 (2015)
171. Baba, A., et al.: Optical properties of ultrathin poly(3,4-ethylenedioxythiophene) films at several doping levels studied by in situ electrochemical surface plasmon resonance spectroscopy. Langmuir **21**, 9058–9064 (2003)
172. Tian, S., et al.: Electroactivity of polyaniline multilayer films in neutral solution and their electrocatalyzed oxidation of β-nicotinamide adenine dinucleotide. Adv. Func. Mater. **13**, 473–479 (2003)
173. Celiesiute, R., et al.: Electrochromic sensors based on conducting polymers, metal oxides, and coordination complexes. Crit. Rev. Anal. Chem. **49**, 195–208 (2018)
174. Turemis, M., et al.: ZnO/polyaniline composite based photoluminescence sensor for the determination of acetic acid vapor. Talanta **211**, 120658 (2020)

Advances in Hybrid Conducting Polymer Technology for EMI Shielding Materials

Vineeta Shukla

Abstract With the increasing the consumption of gigahertz communication devices in modern technology, interference of electromagnetic (EM) wave become the major problem. Thus, the protection of communication and electronic devices against the EMI pollution become the worldwide concern. In this context, present chapter demonstrate the current status of conducting polymer based EMI shielding materials. The discovery of two dimensional material such as graphene, MoS_2, MXene put a forward step in direction of advancement of EMI shielding material in compare with traditional metallic filler based polymer composites. Combination of 2D materials with polymers not only facilitate the EMI shielding, but also beneficial for microwave absorption. The basic EMI shielding mechanism and the factor that influence the EMI performance have been discussed. After this, we have explored the major challenges in existing technology along with future opportunities to develop the advanced composites for EMI shielding and microwave absorption application.

Keywords Polymers · Two dimensional material · Absorption · Reflection

1 Introduction

1.1 Electromagnetic Interference (EMI) Pollution

In recent years, with the increasing the demand of automotive industries (i.e., integrated electrical circuits), electronic equipment and communication devices (e.g., cell phone, computer, Bluetooth, laptop), commercial appliances (i.e., microwave ovens, microwave circuits designing), defense safety techniques, high-quality information technology including mobile and satellite communication, electromagnetic (EM) wave radiation in gigahertz (GHz) range has conceived as an alarming danger. Because the interference of EM waves with the input signal of the electronic devices

V. Shukla (✉)
Department of Physics, Indian Institute of Technology, Kharagpur 721302, India
e-mail: vineeta@phy.iitkgp.ernet.in

© Springer Nature Switzerland AG 2021

S. Shahabuddin et al. (eds.), *Advances in Hybrid Conducting Polymer Technology*, Engineering Materials, https://doi.org/10.1007/978-3-030-62090-5_9

creates a noise, known an electromagnetic interference (EMI) pollution or a modern type of environmental pollution. EMI pollution not only weakens the performance of electronic gadgets, deceases the lifetime of electronic equipment but also hazardous for human health. It creates headaches, sleeping disorders, and trepidation. Nowadays, EMI become the global problem. Thus, government regulations have increased effort to overcome the EM radiation problem in the environment. In this direction, many efforts have been made and continued to prevent EMI pollution. For this purpose, EMI shielding materials are widely used, which protect the gigahertz devices from the EM inference from the surrounding. Traditionally metallic materials would be used for EMI prevention, but the main drawbacks of these materials are high density, lack of mechanical flexibility, costly processing, and prone to corrosion, etc. The low density, high electrical conductivity, good flexibility, and thermal stability are some desirable parameters for the good EMI shielding material. Therefore, polymer (conducting and nonconducting), carbon material like 2D graphene, 1D carbon nanotube, expanded graphite, and ceramic such as $BaTiO_3$, TiO_3 are in fashion for making the advance EMI shielding material. The excellent conductivity, high thermal, and mechanical stability are some benefits of using light weighted carbon materials. Moreover, the large surface area of 2D graphene and high dielectric constant favor the 2D carbon for making the EMI shielding material [1–4]. However, the high cost and multi-step synthesis process, poor magnetic properties, and high skin depth lead to the poor absorption of these materials. Even though electron polarization dominates over the Debye dipole relaxation and natural resonance phenomenon within carbon materials. On the other hand, transition metal oxides material such as ZnO, $BaTiO_3$, TiO_2 and SiO_2 suffer from lack of permittivity at GHz limits. Also, poor dispersion, processing related problems, and undesirable agglomeration that occurs during film processing hinder the use of these materials. Therefore, carbon materials or transition metal oxides are used with the addition of polymers matrix [5, 6]. Polymer composites come under two categories: (1) Conductive polymer composite (CPCs) and (2) intrinsically conductive polymers (ICPs) Generally, ICPs are organic polymers that are associated with metals. These polymers exhibit tunable electrical, magnetic, and optical properties. Further, good mechanical properties and ease of processing can be seen in these polymers. Some of the most studied ICPs are polyaniline (PANI) and polypyrrole (PPY), polyacetylene (PA), poly(p-phenylenevinylene) (PPV), poly(3,4-ethylene dioxythiophene) (PEDOT), polyfuran (PF) and other polythiophene (PTh), etc. In general, polymers can be considered as insulating materials whose resistance typically varies from 10^9 Ω.m and 10^{14} Ω.m). The insulating polymer matrices may be thermoplastics, thermoset, or elastomers. Inclusion of conductive fillers into these insulating polymer matrix, good electrical conductivity can be attained. For example, Chen et al. achieved satisfactory EMI performance (more than 30 dB) in Fe_3O_4@reduced graphene oxide (RGO) filler with Polystyrene (PS) thermoplastics in 9.8–12 GHz [7]. Similarly, Ghamdi et al. studied the Nitrile Butadiene rubber/magnetite nanocomposites for EMI shielding, obtained the effectiveness of 28–91 dB in the 1–12 GHz [8]. Darwish and coworkers obtained EMI shielding of ∽22 dB for Fe_3O_4/polyvinylpyrrolidone composite nanofibers at 8.2–12.4 GHz [9]. In this direction, a lot of research papers, book chapters, and review papers have been published that are based on polymer-based EMI shielding materials. But, the invention of new thin materials like MoS_2, MXene with extraordinary properties, has given a chance for exploring the more new efficient EMI shielding materials. Thus, the present chapter is based on the advancement of polymers composite for EMI and microwave absorption materials.

2 Chapter Details

In the present chapter, intrinsic and extrinsic polymer-based composites have been investigated for EMI shielding and microwave absorber performance. We have given the introduction of EMI shielding, and further, we have demonstrated the synthesis method of conducting and insulating polymer composites. After this, the role of metallic filler, carbonaceous materials, and other two dimensional materials with polymers matrices have been discussed. The last section of this chapter reveals the challenges and future opportunities for the advancement of EMI shielding that may open the door for the development of high-efficiency radar absorbing materials and wireless devices.

3 EMI Shielding Effectiveness

In EMI terminology, electromagnetic shielding effectiveness (SE or SE_T) is represented in decibels (dB) units that measure the material's ability to block the EM waves. The SE is given by the logarithm of the ratio of incident power (P_{In}) to transmitted power (P_{Tr}) of a plane wave (P), electric field (E), and magnetic field (H) strengths which are

$$SE_P = 10 * \log \frac{P_{In}}{P_{Tr}} \tag{1}$$

$$SE_E = 20 * \log \frac{E_{In}}{E_{Tr}} \tag{2}$$

$$SE_H = 20 * \log \frac{H_{In}}{H_{Tr}} \tag{3}$$

where subscripts In and Tr indicates the magnitude of the P, E, or H strength. Further, the ratio of the electric field strength (E) and the magnetic field strength (H) is known as wave impedance Z. Total shielding effectiveness is

$$SE_T = 10 * \log \frac{P_{In}}{P_{Tr}} = 20 * \log \frac{E_{In}}{E_{Tr}} = 20 * \log \frac{H_{In}}{H_{Tr}} \tag{4}$$

3.1 Mechanisms of EMI Shielding

When an EM wave falls on the EMI shielding material, three phenomena can be seen, namely absorption, reflection, and multiple-internal reflections, as shown in Fig. 1a. Due to the intrinsic impedance mismatch between EM wave and shielding materials, the EM wave gets reflected from the surface; also, some part gets transmitted through the material. It is attributed that the strength of the reflected and transmitted waves influenced by the impedance of the medium and material. The wave travel through the material strength of the transmitted waves decreases exponentially as it travels inside the material. When the strength remains 1/e value of initial value, then that traveled distance is known as skin depth (δ), as depicted in Fig. 1b. The skin depth (δ) is given by the following equation

$$\delta = \sqrt{\frac{1}{\pi \sigma \mu f}} \tag{5}$$

Inside the material, multiple-internal reflection occurs. After the reaching of EM transmitted wave from another surface, some parts of this wave get secondary reflections transmission. In term of reflection, absorption and multiple reflections, total shielding is given by the sum of all components i.e.

$$SE_T = SE_R + SE_A + SE_M \tag{6}$$

where SE_R, SE_A represents reflection and absorption from the shielding material. In term of conductivity (σ), frequency (f) and relative permeability (μ) and thickness (d), SE_R, SE_A and SE_M are

$$SE_R = 20 * \log \frac{Z_0}{4Z_{in}} = 39.5 + 10 * \log \frac{\sigma}{2f\pi\mu} \tag{7}$$

$$SE_A = 20 * \log e^{\frac{d}{\sigma}} = 8.7d\sqrt{f\pi\sigma\mu} \tag{8}$$

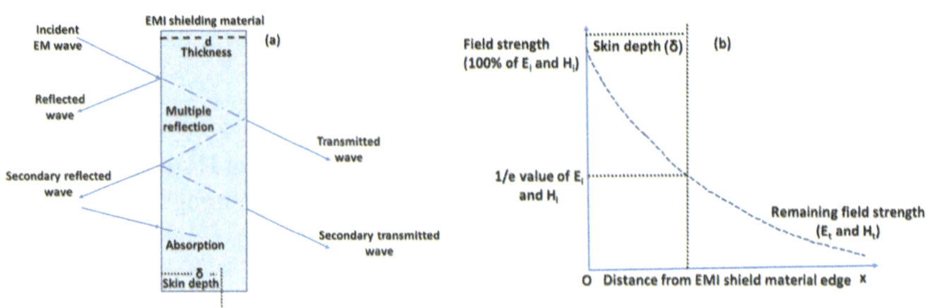

Fig. 1 A schematic representation of **a** EM wave interaction with the material, **b** skin depth plot

And

$$SE_M = 20 * \log(1 - e^{\frac{-2d}{\delta}}) \tag{9}$$

It is clear from equations that reflection and multiple-internal reflection losses depend on the impedance, thus have different values of plane wave, electric and magnetic field. In contrast, absorption loss does not contain impedance terms so that plane wave, electric and magnetic field will be the same for absorption loss. multiple-internal reflection losses can be neglected for $SE_A > 10$ dB or when the thickness of the material is larger than its skin depth (δ) Thus

$$SE_T = SE_R + SE_A \tag{10}$$

3.1.1 Requirement of EM Absorption

High conducting materials may encounter secondary EMI pollution due to skin depth. Thus, minimization of the reflection loss becomes important for ideal EMI shielding materials. This can be achieved only by increasing the absorption process. According to the transmission line theory, reflection loss (RL) can be given by

$$RL(dB) = 20 * \log \mid \frac{Z_{in} - Z_0}{Z_{in} + Z_0} \mid \tag{11}$$

where Z_0 is the intrinsic impedance of free space (377 Ω), while Z_{in} is the input impedance of the EMI shielding material. In term of permittivity and permeability, Z_{in} can be written as

$$Z_{in} = Z_0 \sqrt{\frac{\mu}{\epsilon}} \tan h[\frac{J(2\pi f d)}{c\sqrt{\mu\epsilon}}] \tag{12}$$

where Z_0 , Z_{in} are intrinsic and impedance input characteristics of free space and in the material. The term μ, ϵ refers relative the complex permittivity, permeability and terms c, d, and f denote velocity of light, the thickness of the absorber, and frequency in free space. The ideal condition for attaining the maximum absorption is Z_{in}=Z_0=377 Ω. Furthermore, the above ideal condition can be obtained at a particular matching thickness (t_m) along with matching frequency (f_m). It is noteworthy that ideal EM absorption desire broad effective width which is controlled by 1/4 wavelength equation [10]

$$t_m = nc/4f_m\sqrt{\epsilon\mu} \tag{13}$$

Term n and c refers to the refractive index and velocity of light. The RL~ -20 dB is believed to adequate for EM absorption, which shows the 99% microwave absorption from the materials. In the composites, increasing the effective permeability or decreasing the effective permittivity may reduce the impedance mismatching, and hence the choice of filler-matrix with a balanced combination of dielectric losses and/or magnetic loss is very important for integration the EMI performance. In addition to proper tuning between dielectric/magnetic losses, morphology, lightweight, low thickness, chemical and corrosion resistance, high flexibility, processing easiness, and cheapness are other essential factors that plays a crucial role in the fabrication of effective EMI material for technological application [11].

3.2 EMI Related Influencing Parameters for IPCs and CPCs

3.2.1 Complex Permittivity and Permeability

The permittivity and permeability are two very important parameters that influence the EMI shielding mechanism. The relative term complex permittivity $\epsilon = \epsilon'$+iϵ'' and complex permeability $\mu^* = \mu'$+iμ'' represent the dielectric and magnetic properties of the material. The ϵ' and μ' indicate the electric and magnetic energy storage, but the second term ϵ'' and μ'' shows the dielectric loss and magnetic loss. On other hands, tangent of dielectric/magnetic loss tells about the amount of losses i.e. tanδ_ϵ = $\frac{\epsilon''}{\epsilon'}$ and tanδ_μ = $\frac{\mu''}{\mu'}$, respectively.

In an EMI shielding material, dielectric loss is led by conductivity along with many polarization phenomena such as ionic, orientational, electronic, and interfacial polarization. Among all polarization, ionic and orientational polarization requires bound charges in the material, but interfacial polarization results from the accumulation of space charges from the interfaces which occur due to dissimilarity in electrical conductivity or dielectric constant at interfaces of two or more materials. Ionic polarization and electronic polarization works only at a very high-frequency region (above 1000 GHz); hence their effect can be neglected in the low microwave frequency region. The real and complex permittivity ϵ' and ϵ" are given by Cole-Cole equation

$$(\epsilon' - \frac{\epsilon_{st} + \epsilon_\infty}{2})^2 + (\epsilon")^2 = (\frac{\epsilon_{st} - \epsilon_\infty}{2})^2 \tag{14}$$

where ϵ_{st} and ϵ_∞ are the static and relative dielectric permittivity at the higher frequencies. The ϵ'' versus ϵ' Cole-cole plot will be a single semicircle for Debye relaxation. The Debye relaxation polarization mostly appears in hierarchical and multi interfaces composites. The occurrence of more than one semicircle or distorted semi-circle represents the more relaxation mechanism associated with interfacial polarization. However, semicircle may not appear for highly conducting material owing to conduction loss, which is given by $\epsilon'' = \sigma/2\pi\omega\epsilon_0$. The ϵ' and ϵ'' are related with frequency and conductivity by relation: $\epsilon'' = \sigma_{Ac}$-$\sigma_{Dc}(\epsilon' \times \omega)$ hence the ϵ' and ϵ'' decrease with increasing frequencies.

The magnetic loss arises from eddy current loss, domain wall loss, hysteresis loss, and residual loss. The eddy current loss (E_d) can be represented by following equation $\mu'' = (2\pi/3)\mu_0\mu'^2\sigma d^2 f$, where μ_o and d are the vacuum permeability and layer thickness of the absorber [12]. The eddy current losses decreases below a critical small size because of decreasing induced eddy voltage. Furthermore, anisotropy energy become dominant factor for small size particles due to the breaking of some exchange bonds. The eddy current loss (E_d) does not change with frequency and also can be neglected for highly conducting materials. The loss of EM energy in magnetic material due to mechanical rotation and movement of magnetic dipoles and domains represent the hysteresis loss in magnetic materials [12]. The magnetic field energy in lower frequencies is dissipated as heat attributable to re-alignment of magnetic dipoles. Residual losses results from the several relaxation processes. The contributing processes to residual losses are the resonance losses: domain wall resonance and the spin rotational resonance. Usually spin rotational resonance dominates at high frequencies while domain wall resonance works at lower frequencies.

3.2.2 Geometrical Factors

Geometrical factors such as shape, size, chirality, morphology, and orientation of filler in polymer matrices significantly influence the EMI performance [13]. Kim et al. saw the effect of changing the orientation angle of the carbon fiber from 0, 15, 30, 45, and 90° and showed that how the orientation angle effect to EMI shield performance [14]. Another researcher group, Li and coworkers, observed EMI shielding effectiveness of 27 dB for epoxy composite with 0.8 wt% thermally annealed GAs (TGAs) that enhances (~32 dB) along the radial direction and decreases (25 dB) along the axial direction [15]. Tian [16] et al. used the concept of chilarity and combined the molecular chirality using PANI and nanoscale chirality of helical carbon nanotube (HCNT). They modified PANI by chiral D-CSA and achiral HCl acid and prepared hybrid with HCNTs by in-situ polymerization, which can be denoted by PANI-CSA@HCNTs and PANI-HCl@HCNTs, respectively. In addition, many molar ratios of aniline to D-CSA i.e., 2.1, 1.1, and 1:2 for different chiralities, were used to prepare the PANI-CSA@HCNTs. Figure 2a–c shows the FE-SEM image of PANI@HCNT HCl doping, PANI@HCNT with 10-camphor sulfonic acid (d-CSA) doping and for aniline: dopant molar ratios of 1:1), and Fig. 2d–f depicts the TEM images of PANI@HCNT with HCl doping, PANI@HCNT with d-CSA doping and PANI@HCNT at aniline: dopant molar ratios of 1:1. The dashed lines (as shown in Fig. 2e, f) indicate the interface between the HCNTs and PANI nanorods. Using chiral dopant D-CSA and achiral HCl, the array of PANI nanorods successfully formed on the surface of HCNTs (Fig. 2a, b). TEM image for both PANI-CSA@HCNTs and PANI-HCl@HCNTs was similar except that PANI-CSA@HCNTs were longer than that PANI-HCl@HCNTs (Fig. 2d, f). Nevertheless, the morphology of PANI gets change from nanoparticle to regular nanorod by changing the ratio of aniline to D-CSA from 2:1 to 1:2 (Fig. 2c, f). Figure 3a, b, d, f shows the ϵ', ϵ'', μ', μ'' of complex permittivity and permeability, while Fig. 3c, f depicts the dielectric loss and magnetic loss tangent of HCNTs and D-CSA-doped PANI@HCNT for aniline to D-CSA ratios i.e., 2:1, 1:1, and 1:2. Pristine HCNT has low $\tan\delta_\mu$ and $\tan\delta_\epsilon$, hence leading impedance matching. In contrast, CSA-modified PANI molecular chain introduced a new chiral asymmetric center, which gives rise to the new dielectric and magnetic resonance peaks. Figure 4a–c depicts the 3D RL plot for D-CSA-doped PANI@HCNT

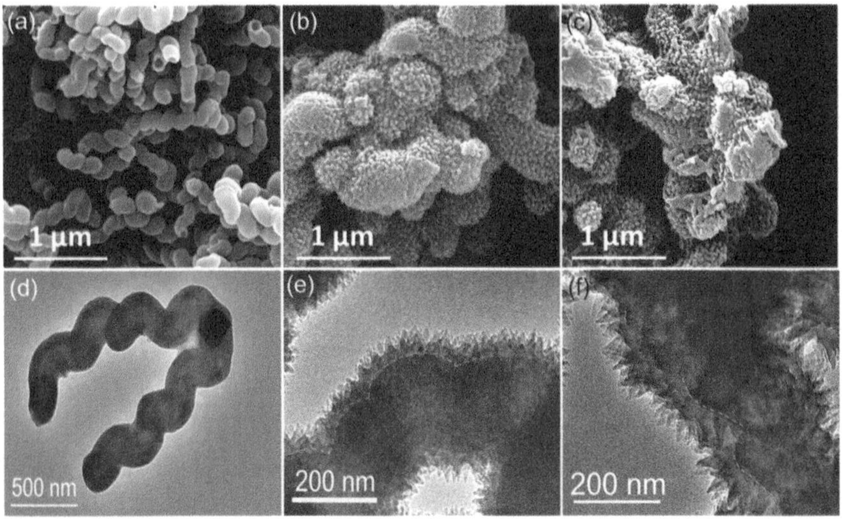

Fig. 2 FE-SEM image of **a** PANI@HCNT HCl doping, **b** PANI@HCNT with 10-camphor sulfonic acid (d-CSA) doping, **c** aniline: dopant molar ratios of 1:1) and TEM images of **d** PANI@HCNT with HCl doping, **e** PANI@HCNT with d-CSA doping, **f** PANI@HCNT at aniline: dopant molar ratios of 1:1. The dashed lines (Fig. **e** and **f**) indicate the interface between the HCNTs and PANI nanorods [16], reproduced by permission of the American Chemical Society

hybrids with aniline to D-CSA molar ratios of (a) 2:1, (b) 1:1, and (c) 1:2. The order of obtained RL value for different aniline to D-CSA ratio 1:1> 1:2 > 2:1 are −32.5 > −17 > −12.3, respectively. Moreover, the thickness variation (2–5 mm) followed the different RL values, as shown in Fig. 4d. Authors revealed that helical shaped PANI molecular chains with helical nanotube produce cross-polarization. The advantage of cross-polarization is the additional interactions with induced EM waves along with conventional multi-scaled relaxations such as interface polarization that improves the absorbing properties of materials.

3.2.3 Effect of Moisture Absorption

It is attributed that hydrogen-rich water enhances the electrical conductivity. Some researchers have reported that electrical conductivity enhances with increasing the weight gain (M%) resulting from moisture absorption. Thus dielectric constant increases with increasing the percentage of moisture absorption. In fact, with increasing the ionic element in water, such as hydrogen, the electrical properties of water improve significantly. Thus increase in water absorption gives rise to the improved electrical conductivity that affects the total EMI shielding effectiveness [13, 17].

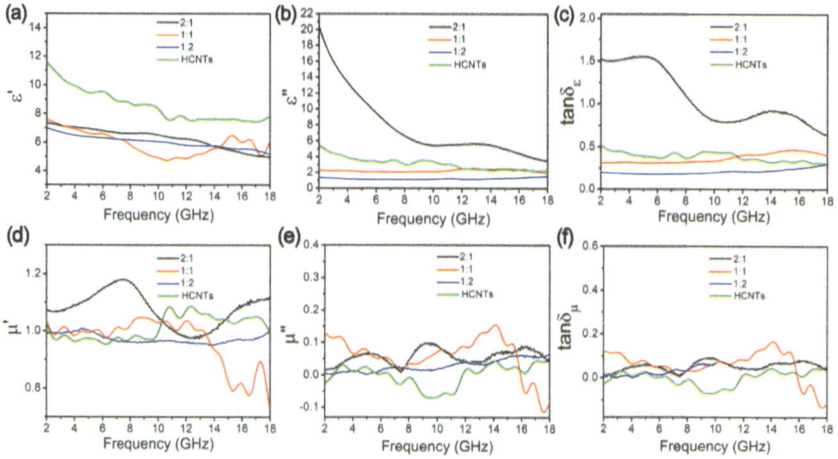

Fig. 3 **a–c**) indicates the real (ϵ') and imaginary parts ϵ'') of complex permittivity and dielectric loss tangent (tan δ_ϵ) while **d–f** shows the real (μ') and imaginary parts μ") of complex permeability and magnetic loss tangent (tanδ_μ) of HCNTs and D-CSA-doped PANi@HCNT hybrids with different molar ratios of aniline to D-CSA: 2:1, 1:1, and 1:2) [16], reproduced by permission of the American Chemical Society

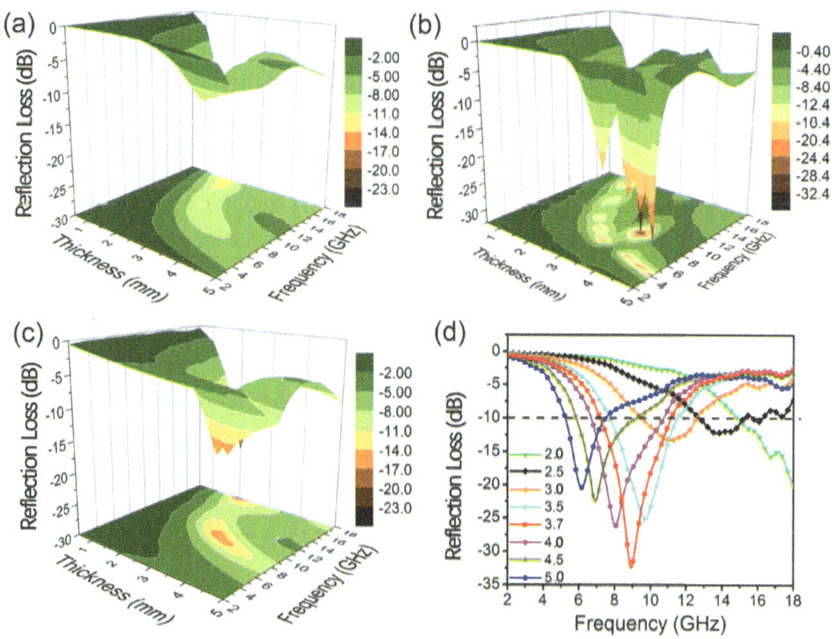

Fig. 4 RL 3D plot of D-CSA-doped PANI@HCNT hybrids with aniline to D-CSA molar ratios of **a** 2:1, **b** 1:1, and **c** 1:2 and **d** Calculated RL values at different thicknesses for PANI-CSA@HCNTs measured in the 2–18 GHz frequency range [16], reproduced by permission of the American Chemical Society

3.2.4 External Factors Like Temperature and Reaction Time and Irradiation

The external factors such as preparation time of material, annealing, irradiation of γ- radiation also affect the EMI properties. For instance, annealing increases the defects like vacancies, stone Wale and dangling bonds in the materials [18]. These defects create extra energy levels around the Fermi level and hence enhance the attenuation rather than reflection. Furthermore, reaction time and temperature also influence to EMI shielding and absorption properties. Lv and coworkers prepared FeCo and FeCo/ZnO composites at different temperature (130°C/140 °C)/12 h or time (15 h, 20 h)/150 °C. The reflection loss reached up to −31 dB with an effective frequency bandwidth 5.5 GHz for FeCo/ZnO that demonstrates the importance of the time and temperature on EMI performance [19]. Hota et al. [20] observed that with an increase in radiation dose and SWCNT loading in ethylene acrylic elastomer/millable polyurethane matrices, the absorption of electromagnetic radiation increases. It is anticipated that electrical conductivity contributes more towards the absorption even though the SWCNT filler is non-magnetic. Basically, absorption results from the movement of charge carrier, which leads the polarization of the dipoles at the interfaces, ohmic loss, along with multiple scattering at the interface of the shielding composites. With increasing the radiation dose SE_R decreases. The author predicted that irradiation increases the formation of conductive networks, which facilitates the hooping/tunneling of the charge carriers that causes the easy movement of the charge carrier and responsible for decrements in SE_R at different radiation dosages.

3.2.5 Percolation Conductivity

The addition of conducting filler in the insulating polymer increases the complex permittivity of the material. The conducting filler in insulating matrices works as a capacitor due to bound charges and enhance the electrical energy. When conducting fillers are distributed on insulating matrices, then the formation of a conducting network takes place that can be followed by the percolation threshold theory. Generally, electrical properties of any of material depending on the percolation threshold value of conductivity which is given by the formula

$$\sigma = \sigma_f (V_c - V)^n \quad for V < V_c \tag{15}$$

$$\sigma = \sigma_i (V - V_c)^{-m} \quad for V > V_c \tag{16}$$

where σ is the electrical conductivity of the materials, σ_f, σ_i is filler and insulating polymer conductivities, V is the volume fraction of filler, V_c is volume fraction at the percolation threshold, and n and m are the parameters that quantified the connections between the conducting fillers at the percolation threshold. In addition n, m describe the arrangement of the conductive filler in 2 dimensional (D) or 3D systems e.g., n>2 indicates 3D networks but, n<2 shows the 2D networks of conducting fillers in case of CPCs. A percolation threshold, interconnected conductive networks form within matrices that increases the charge hooping within materials. As a result, the insulating polymer started to work as a conducting polymer. This phase transition is called the percolation transition, and the critical conductivity is known as threshold conductivity. In general, percolation threshold can be believed to influence by shape, morphology, aspect ratio, and conductivity of filler along with the distribution of filler, the concentration of filler, and compatibility of filler with host matrix also affect to the percolation threshold.

3.2.6 Thickness

The absorption phenomenon in EMI shielding depends on the thickness of the materials. With increasing the thickness of the shield materials, absorption increases. Thus, the thickness can be considered as an effective parameter that may change the shielding effectiveness. In the microwave absorption process, minimal reflection loss RL_{min} occurs at the matching thickness (t_m)

$$t_m = n\frac{\lambda_m}{4} \tag{17}$$

where n = 1, 3, 5, 7, 9...), so that n = 1 corresponds to the first dip at low frequency. Corresponding to matching thickness, matching wavelength (λ_m) is given by is given by

$$\lambda_m = \frac{\lambda_0}{\sqrt{\mu\epsilon}} \tag{18}$$

The matching condition may be obtained due to cancellation of the incident wave and reflected waves from the the surface of the material. According to Eq. (17), reflection peaks shift in the lower frequency side with increasing the sample thickness. In a similar way, layer or coating thickness on the surface of nanoparticles (e.g., in core-shell structure) can alter the microwave absorption peak. It is attributed to EM wave dimensional resonance that enhances with the increasing the layers/coating thickness. Tian and coworkers prepared uniform polypyrrole@polyaniline core–shell composites. The thickness (30–120 nm) of PANI shells was controlled by modulating the weight ratio of aniline and PPy microspheres [21]. As prepared PPy@$PANI_x$, where x represented the mass ratio of aniline monomers to PPy microspheres were designated as PPy@PANI-0.8, and PPy@PANI-2.0 meant that 0.32 0.80 g of aniline monomers, respectively. Observation showed that PPy@PANI-0.8, PPy@PANI-1.2, and PPy@PANI-1.6 have the RL value –23.3 dB (15.8 GHz), –34.8 dB (13.9 GHz) and –31.5 dB (13.6 GHz) at same absorber thickness 2 mm, respectively. Similarly, different thickness of absorber produces the negative shifts to lower frequency with increased thickness from 2.0, 2.5, 3.0, 4.0, and 5.0 mm. Thus, maximum RL for –43.1 dB 7.1 GHz and –51.3 dB at 8.8 GHz was found with a thickness of 4.0 3.0 mm in case of PPy@PANI-0.8 and PPy@PANI-1.2, respectively. That shows the thickness as well as coating thickness, and both affect the absorbance performance.

4 Measurement Technique

The EMI Shielding effectiveness of shielding material is calculated from complex scattering parameters (S). The S parameters are S_{11} (or S_{22}), and S_{12} (or S_{21}) that can be obtained by the vector network analyzer. It is the two-port system, as depicted in Fig. 5. In this two port VNA, incident and transmitted waves are given by S_{12} (or S_{21}) and S_{11} (or S_{22}) respectively. These S parameters can be associated with coefficients of shielding mechanisms by transmittance T = $|S_{12}|^2$ = $|S_{21}|^2$ and reflectance R = $|S_{11}|^2$ = $|S_{22}|^2$. From these S parameters, one can calculate the EMI SE by following equation

$$SE_T(dB) = 10 * \log\left(\frac{1}{|S_{12}|^2}\right) = 10 * \log\left(\frac{1}{|S_{21}|^2}\right) = 10 * \log\left(\frac{1}{T}\right) \quad (19)$$

where T is the transmittance

$$SE_R = 10 * \log\left(\frac{1}{1-|S_{11}|^2}\right) = 10 * \log\left(\frac{1}{1-|S_{22}|^2}\right) = 10 * \log\left(\frac{1}{1-R}\right) \quad (20)$$

where R the reflectance

$$SE_A = 10 * \log\left(\frac{1-|S_{11}|^2}{|S_{12}|^2}\right) = 10 * \log\left(\frac{1-|S_{22}|^2}{|S_{21}|^2}\right) = 10 * \log\left(\frac{1-R}{T}\right) \quad (21)$$

Summation of the reflectance (R), transmittance (T) and absorbance (A) is always equal to 1;

$$R + T + A = 1 \quad (22)$$

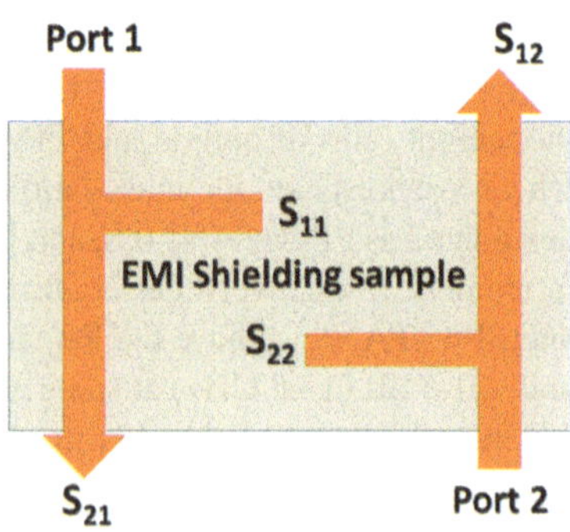

Fig. 5 A pictorial image of two port system in a filled transmission line

5 Polymer Composites for EMI Application

Polymer based composites have many advantage over the conventional metals e.g. polymers have flexibility, light-weighted, less toxic since less damaging to environment. In general, polymers can be classified into three categories: (1) Intrinsically conducting polymers (2) Coordination/inorganic conducting polymers (3) Extrinsically conducting polymers as explained in later subsections. The conducting polymers are widely used in rechargeable batteries, sensors for sensing such SO_2, NH_3, glucose, photo voltaic devices, telecommunication systems, micro-electronic devices, bio-medical applications etc. The conducting polymers are recognized as the conductive materials made by organic polymers of conjugated chain structure. The unique properties of conducting polymers offer an opportunity to produce a new kind of promising materials with tunable optical, electrical and magnetic property in wide range of interest for potential application in fields of scientific and technical benefits. For instance, Conducting polymers possess considerable chemical memory due to high reversible redox activities thus can be used to synthesize the industrial sensors like devices. The properties of these polymers highly influence by doping level, protonation level, ion size of dopant, and water content. However, they suffer from the flexibility and processing related problem. Alternatively, conducting filler-insulating polymer matrix based composites are used for protecting the high frequency devices from the EMI pollution. Figure 6 depicts the schematic representation the different types of conducting polymers which are explained in later subsections.

5.1 *Intrinsically Conductive Polymers*

Intrinsically Conductive polymers (ICPs) are organic polymers with conjugated chain structure. ICPs have tunable conductivity that varies between metal conductivity to

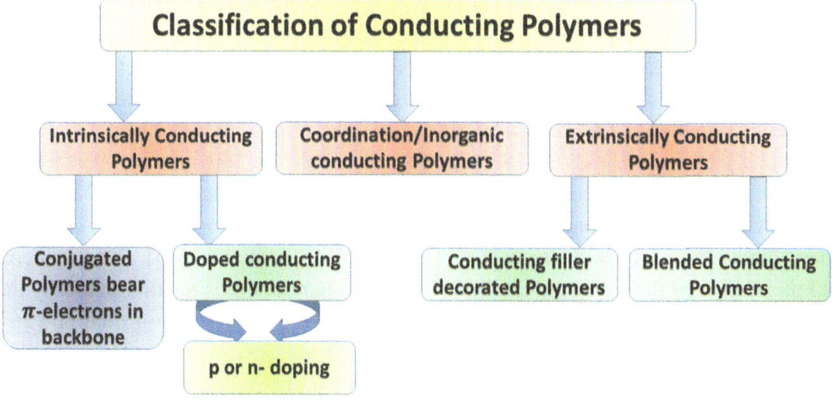

Fig. 6 A schematic representation of the type of conducting polymers

semiconductor conductivity. Thus, ICPs are further divided into two subcategories. The first one is the conducting polymers that have π-electrons in the polymer backbone. In this case, good conductivity in ICPs is led by the presence of conjugated π-electron along the polymer backbone that takes part in the charge transfer via the solid polymer in the presence of the external electrical field. Moreover, valence bands and conduction bands form as the result of the overlapping of orbitals of conjugated π electrons over the entire backbone that distends over the complete polymer molecule. It is attributed that conjugated π-electrons in polymers is responsible for the superior conductivity of polymers similar to metals. The second one is doped conducting polymers. Doping within polymers generates the positive/negative charge on the polymer backbone either by oxidation or reduction process that improves the conductivity of the ICPs polymers. The p-doping can be achieved by the oxidation process while n-doping can be obtained through the reduction process. When the conducting polymers undergo a gas phase or in solution through a charged transfer agent, called doped conducting polymers. Some of ICPs are polyaniline (PANI), polyacetylene (PA), polypyrrole (PPy), poly(3,4-ethylene dioxythiophene) (PEDOT), polyfuran (PF), poly(p-phenylenevinylene) (PPV), and polythiophene (PTh), etc.

5.1.1 Polyaniline (PANI)

PANI is known to be the most versatile member among all conducting polymer families. PANI belongs to a semi-flexible rod polymer family. It is easy in synthesis, cost-effective, and gives the high polymerization yield [22, 23]. Moreover, good environmental stability tunable electrical properties and occurrence in different form character of PANI make it excellent materials for potential applications in EMI shielding, batteries, sensors, controlled drug delivery, and tissue engineering. Generally, fully and partially oxidized PANI forms the pernigraniline and emeraldine base while fully reduced PANI forms leucoemaraldine base. It is noteworthy that emeraldine is most stable, and its conductivity varies from base to salt form in the limit of 10^{-10}–30 Scm^{-1}. However, the lack of solubility of emeraldine base due to the stiff polymer backbone and hydrogen bonding interaction between the nearest chain produces the processing related difficulties. PANI polymer and its composites have been shaped in the form of films, fabrics, nano-particles. PANI has been reported with excellent EMI capabilities from low frequency (MHz) to higher frequency (GHz) limits. In the lower frequency range (0.1–1 GHz), shielding effectiveness found to vary from 16 to 50 dB while at higher frequencies (1–18 GHz) limit, EMI SE was obtained from 31.9 to 81 dB [24]. Sasikumar and coworker made the camphor sulphonic acid doped Polyaniline (PANI-CSA) then made the film by using CSA as the dopant and m-cresol as the solvent. They achieved superior SE value of 45 dB in X band [25]. Jiang and colleagues reported the Co/C/polyaniline (Co/C/PA) nanocomposites fabricated by the arc-discharge process and in situ chemical oxidative polymerization method [26]. The Co/C/polyaniline nanocomposites showed strong absorption peaks exceeding −40 dB in the 3.5–5.5 GHz frequency range. In the above composites, the double-shelled structure was found to be advantageous than a single-shelled structure. Bora

Table 1 Some PANI polymer-based EMI shielding materials and microwave absorber

Materials	RL_{min}/SE_T*(dB)	Frequency (GHz)	Absorption width/frequency range (GHz)	Thickness (mm)	References
$Ni_{0.6}Zn_{0.4}Fe_2O_4$/PANI	−41	12.8	5	2.6	[29]
Graphene@Fe_3O_4@SiO_2/PANI	−40.7	12.5	10.5–16.3	2.5	[30]
RGO/SF/PANI	−45	16.08	5.48	1.5	[31]
PANI/$Li_{0.35}Zn_{0.3}Fe_{2.35}O_4$	−37	14.6	4.8	2	[32]
PANI/$Ba_{0.9}La_{0.1}Fe_{11.9}Ni_{0.1}O_{19}$/RGO	−49	14.08	12.4–16.72	1.9	[12]
Fe_3O_4/PANI microsphere	−24.3	18	15.52–18	1.5	[33]
RGO/PANI/Cu_2O	−52.8	2.7	13.2	2	[34]
PVB/PANI	−79	10.4	Undetected	2	[35]
PANI@HCNTs	−32.5	8.9	7.1–11.2	3.7	[16]
PANI/$CoFe_2O_4$/$Ba_3Co_2Fe_{24}O_{41}$	−30.2	10.5	3.8	3	[36]
$Li_{0.35}Zn_{0.3}Fe_{2.35}O_4$/PANI	−49.4	4.96	5.56	5.1	[37]
NiO@PANI	−32.8	10.1	11.12–15.76	2	[38]
Ag/CF/PANI	−13.2	9.3	Undetected	2	[39]
Co doped ZnNi ferrite/PANI	−54.3	15.1	6.02	6.8	[40]
PANI/PBC	−44.8	10.02	5.58–18	2	[41]
$BaFe_{12}O_{19}$@PANI	−65.35	17.28	Undetected	1.47	[42]
PANI@GE	−64.3	10.1	8.6–13.7	2.9	[43]
CNTs@PANI	−45.7	12	10.2–14.8	2.4	[44]
PANI/$NiFe_2O_4$:V	−42.17	7.02	Undetected	2	[45]
$ZnFe_2O_4$/PANI/graphene oxide	−58	9.5	7.80–11.71	3.29	[46]
Co_3O_4@PANI	−37.39	7.28	5.92–8.40	4	[47]
$Sm_2Co_{14}B$/PANI	−23.1	9.9	2.6	2.3	[48]
Fe_3O_4@PANI@MnO_2	14.7	15.76	4.75	3.5	[49]
Graphene/Ag@PANI	−44.5	10.5	Undetected	1.3	[50]
Nanoring-shaped PANI	−39.10	15.1	11.80–16.55	2	[51]

and coworker [27] made a coatable polyvinyl butyral (PVB)–polyaniline (PANI) nanocomposite and checked its absorption properties. In comparison with a pristine PANI-containing composite that shows absorption efficiency 53.5 dB GHz/mm, a high absorption efficiency 88.2 dB GHz/mm with large bandwidth was achieved for the PANI nanofiber-PVB nanocomposite in the frequency range 8.2–18 GHz. Li et al made the $Tb_{0.2}Dy_{0.8}(Fe_{0.8}Co_{0.2})_{1.93}$/polyaniline composites by arc-melting process, high-energy ball-milling and chemical oxidation polymerization methods. They achieved RL value −38.3 dB at 17.8 GHz at thickness of 1.6 mm which establish $Tb_{0.2}Dy_{0.8}(Fe_{0.8}Co_{0.2})_{1.93}$/polyaniline as excellent absorber against electromagnetic disturbance [28]. Some PANI based composites with their EMI performance are listed in Table 1.

5.1.2 Polypyrrole (PPy)

After the PANI polymer, polypyrrole (PPy) is another conducting polymer of particular interest due easy of synthesis, stability in oxidized form, good redox properties,

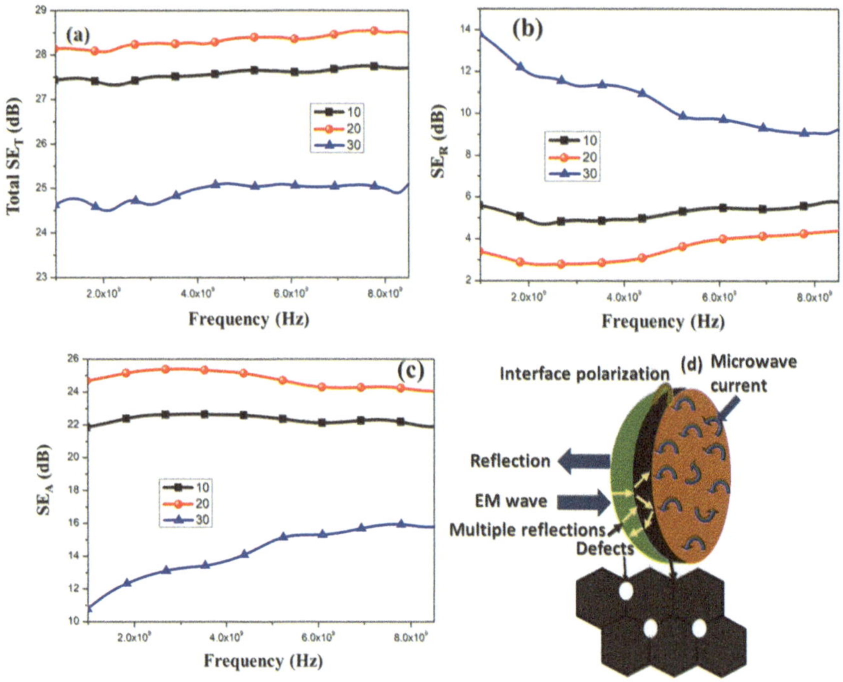

Fig. 7 **a** Plots of frequency versus SE_T, **b** SE_R, **c** SE_A of 10, 20, 30 composites and **d** schematic diagram of EMI attenuation mechanism [53] , reproduced by permission of the Springer

and tunable conductivity. The benefits of PPy with respect to other conducting polymers are the low-cost process, simple oxidation reaction, and better solubility than other ICPs. Therefore, PPy based composites are widely used in a supercapacitor, textile, and fabrics, drug delivery, corrosion protection, neutral tissue engineering. Even though PPy possesses good Vitro and Vivo bio-compatibility. Wu and coworkers synthesized the core-shell nanostructures of PPy shells coated on silicon carbide (SiC) nanowires. The rate of polymerization was chosen in such a way that thicknesses of the shells can be monitored [52]. The broad bandwidth 6.88 GHz, and the excellent reflection loss –58.6 dB were achieved in PPy@SiC composite. Figure 7a–c depicts the SE_T, SE_R and SE_A variation in Fe_3O_4/C/PPy-10, Fe_3O_4/C/PPy-20 and Fe_3O_4/C/PPy-30 composites. The Fe_3O_4/C/PPy-20 exhibits highest shielding efficiency 28.2 dB that result by the absorption loss (25 dB) rather than reflection loss. However, absorption loss decreases for Fe_3O_4/C/PPy-30 and hence reflection loss dominates for this composition. The probable mechanism of attenuation in pictorial representation through Fe_3O_4/C/PPy composites shown in Fig. 7d. In presence of EM wave, mobile charge carriers cloud get distorted around multi-interfaces/junctions that increases the Debye relaxation process. Meanwhile, Fe_3O_4/C/PPy transforms more and more EM wave energy into microcurrent that transmit through dual core-shell structure via carbon and PPy shells which work as an excellent electrical con-

Table 2 EMI shielding materials and microwave absorber

Materials	RL_{min}/SE_T (dB)	Frequency (GHz)	Absorption width/frequency range (GHz)	Thickness (mm)	References
S-PPy/RGO	−54.4	12.7	10.2–16.9	3	[54]
$BaFe_{12}O_{19}/Fe_3O_4$/PPy	−28	11.5	2.65	1.5	[55]
FeCo/PPy nanoplates	52.30	13.40	11.31–16.37	2.15	[56]
TiO_2@PPy	−60	11.3	Undetected	3	[57]
Fe_3O_4/PPy microsphere	−31.5	15.5	12.8–18	2.5	[58]
FeCo/RGO/PPy	−40.7	4.5	5.7	2.5	[59]
Carbon-sphere@$NiFe_2O_4$@PPy	−53	10.5	8.9–12.3	1.97	[60]
Co/C@PPy	−44.76	17.32	6.56	2	[61]
PPy Nanosphere/RGO	−59.2	5	14.7	3.8	[62]
Ag@PPy	−66	12.2	4.12	2.6	[63]
Fe/m-SiO_2/PPy	−51.24	7.44	4.16	4	[64]
Fe_3O_4@SiO_2/PPy	−56.9	10.41	10.53–16.91	3.7	[65]
Fe_3O_4@SiO_2/PPy/RGO film	32*	Undetected	8–12	0.27	[66]
PPy@SiC nanowires	−58.6	17.2	Undetected	2	[52]
γ-$Fe_2O_3/(SiO_2)_x$-SO_3H/PPy	−43.1	15.1	6.1	2	[67]
Fe/PPy/PMMA	−76.02	8.96	>3.4	3.2	[68]
RGO/PPy nanotube/Fe_3O_4	−49.2	11.8	9.8–15.9	3	[69]
SiCw-BDC/PPy	−52.4	11.4	9.0–17.1	2.26	[70]
PPy/ZIFs	−49	12.1	6.2	2.9	[71]
NiS_2@RGO/PPy	−58.7	16.44	7.04–18	2.03	[72]
Fe_3O_4@PPy/graphite	−49.3	9.92	12.47–16.55	1.5	[73]
$ZnFe_2O_4$@PPy	−42.31	30.24	9.66–37.86	2.5	[74]
Fe_3O_4@SiO_2/PPy	−40.9	6	11.12–18	5	[75]
Ni/PPy	−48	7	3.8	5	[76]

ductive pathways. Further, scattering of EM wave from residual defect sites of carbon layer or from voids in PPy and multiple reflections from multi-interfaces also influence the absorption process. Nevertheless, agglomeration of Fe_3O_4 nanoparticles and the PPy coating thickness may increase the surface-to-volume ratio of Fe_3O_4/C/PPy-30 composites that suppress the absorption process [53]. Table 2 shows the capacities of PPy based heterostructures for EMI application.

5.1.3 Poly(3,4-Ethylene Dioxythiophene) (PEDOT)

Poly (3,4-ethylenedioxythiophene) (PEDOT) and its complex with poly(styrene sulfonate) (PEDOT:PSS) is one of the most promising material among conductive polymers platform to protect the gigahertz devices against the the EMI pollution. PEDOT possess high electrical conductivity, superior electrochemical activity, low redox potential, low thermal conductivity, relatively larger Seebeck coefficient, excellent chemical and environmental stability [77–80]. The PEDOT and its derivative is extensively used in energy conversion and storage devices including solar cells, supercapacitor and thermoelectric devices [81–84]. Zhou et al. [77] studied the microwave absorption properties of core-shell Fe_3O_4 -PEDOT microspheres. The excellent microwave absorption properties with minimum RL~ -30 dB at 9.5 GHz was observed for 3, 4-ethylenedioxythiophene (EDOT)/Fe_3O_4 ratio of 20 at thickness of 4 mm, but for 50% volume fraction, minimum RL was obtained 27.6 dB at 13 GHz at 2 mm layer thickness. This shows that conductivity, volume fraction and layer thickness greatly influence to the microwave absorbing properties. On other hands, Wang and coworkers [80] employed the green approach to prepare the PEDOT:PSS-Fe_3O_4-RGO (P-GF) hybrid material by using molecular-atomic deposition routes. Figure 8a–c shows the microstructures of Fe_3O_4-RGO (GF). It is visible form 8(a, b,c) that Fe_3O_4 forms the small clusters that are uniformly distributed on RGO while enlarged image, as depicted in Fig. 8c, displays that Fe_3O_4 particles are successfully decorated on the RGO sheet. It is clear from Fig. 8d, e that PEDOT:PSS layers with multiple wrinkles are decorated on the GF hybrid. Figure 8f indicates the typical positions of Fe_3O_4, RGO, and PEDOT: PSS in P-GF materials. Figure 9a–l shows the real permittivity (ϵ'), permeability (μ'), and the imaginary permittivity (ϵ''), permeability (μ'') of the samples. Figure 9a–d demonstrates the complex permittivity of P-GF with different hybrid ratios at 50% filler loading. With increasing frequency, ϵ' decreases. Also, with decreasing the conductive PEDOT:PSS content, ϵ' fall off from 17.21 to 7.24. Furthermore, ϵ'' depicts the two relaxation peaks at $\sim$9 and $\sim$15 GHz. Figure 9e, f shows the Cole-Cole plots that endorse the fact that the material

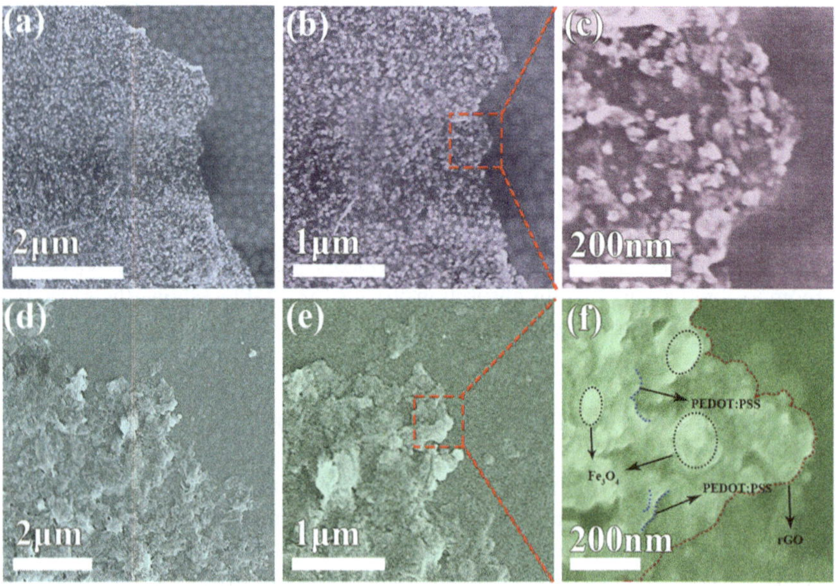

Fig. 8 SEM images at different magnifications for **a–c** GF nanohybrid, **d–f** P-GF nano-hybrid. The black line shows the Fe_3O_4 particles concentrated area, the blue line represents the PEDOT:PSS fold and the red line refers the graphene sheet [80], reproduced by permission of the American Chemical Society

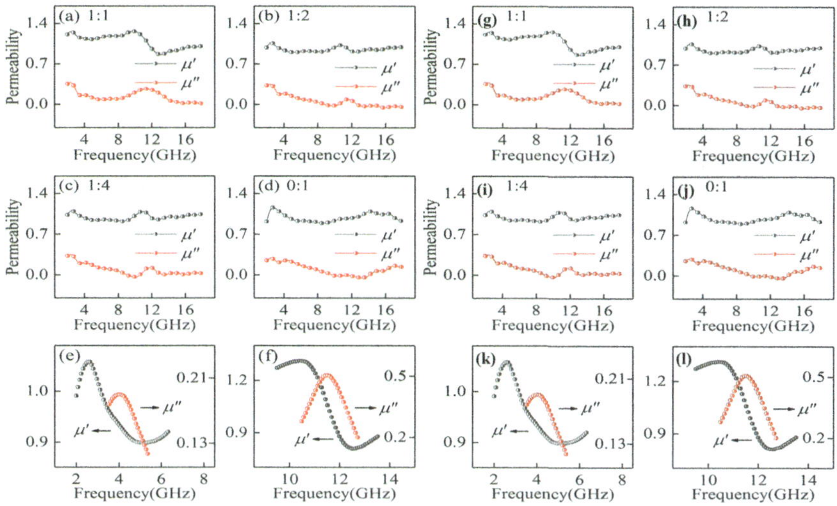

Fig. 9 **a–f** and **g–l** shows the ϵ' and ϵ'' of P-GF with different hybrid ratios **a** P:GF = 0:1, **b** P:GF = 1:4, **c** P:GF = 1:2, and **d** P:GF = 1:1 at the the 50% filler loading, **e**, **f** Cole–Cole plots of P:GF = 1:1 [80], reproduced by permission of the American Chemical Society

has multiple relaxations. It was predicted that might be the outcome of the dipole and interfacial polarization. Similarly, the permeability plot shows the multiple magnetic resonance peaks, as shown in Fig. 9g–j. It is well accepted that magnetic resonance losses are given by the domain-wall resonance loss, hysteresis loss, natural loss, and exchange interaction mechanism. Nevertheless, magnetic hysteresis loss and domain-wall resonance loss can be neglected in the GHz frequency limit because of the weak magnetic field. Thus, low-frequency peaks and high frequency can be attributed by natural resonance (Fig. 9e–f) and exchange resonance, respectively if the small size effect is taken into account. Figure 10a–e shows RL versus frequency plots of P:GF = 1:1 samples for variant filler loadings and thicknesses. It was observed that 50% filler loaded sample possess highest absorption peak and two peaks appear at a thickness of 1.81 mm, that indicate the larger absorption region of materials. Figure 10f shows a 3D bar graph of maximum RL at variant filler loadings and thicknesses for better visualization. The maximal RL was achieved ~–61.4 dB with effective absorption width~6.4 GHz at –10 dB. Moreover, the maximum RL of a 50% loading sample is much better than others, especially 3.86 mm thickness. The conductivity loss greatly influences the dielectric loss. It was anticipated that the deposition of conducting PEDOT:PSS on magnetic Fe_3O_4 and conducting graphene create the conductive networks for the aggregation-induced charge transport, giving rise the excellent dielectric loss. While Ferro-ferric oxide magnetic clusters lead the magnetic loss. Thus overall electromagnetic wave attenuation is great influences by all these factors. Table 3 demonstrates the capacities of PEDOT based heterostructures against the EMI pollution.

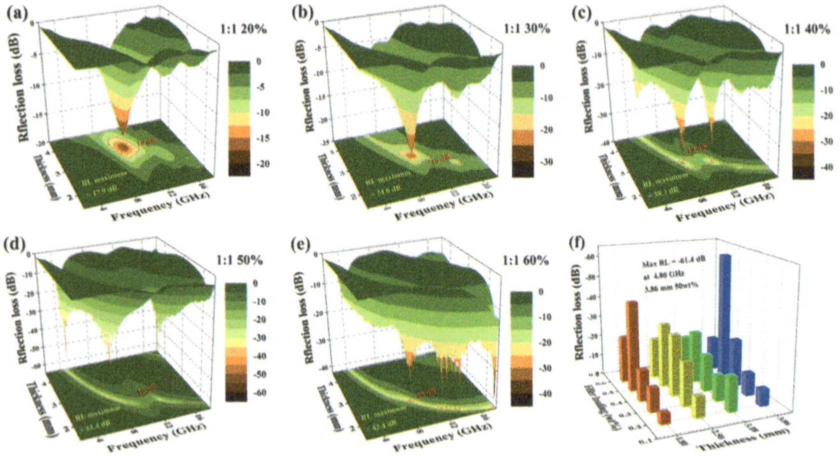

Fig. 10 **a–f** 3D RL plots of P:GF = 1:1 with varying filler loadings: **a** 20%, **b** 30%, **c** 40%, **d** 50%, and **e** 60%, **f** maximum RL plot for varying thicknesses and filler loadings [80], reproduced by permission of the American Chemical Society

Table 3 EMI shielding materials and microwave absorber

Materials	RL_{min}/SE_T* (dB)	Frequency (GHz)	Absorption width/frequency range (GHz)	Thickness (mm)	References
PEDOT/PCNT	58*	Undetected	12.4–18	2.8	[85]
GNP/PEDOT:PSS	30*	Undetected	8.2–12.4	0.8	[86]
Gn/PEDOT/$CoFe_2O_4$	−43.2	9.4	8.2–11.3	2.4	[87]
Leaflet like PEDOT-SWCNT	−43	9.4	>4	3	[88]
PEDOT/PSS-HNTs film	−16.3*	Undetected	2–13	4.5	[89]
PEDOT:PSS/WPU	70*	Undetected	8.2–12.4	0.15	[90]
PEDOT:PSS-PEO-PVA	28*	Undetected	12.4–18	0.042	[91]
PEDOT/RGO	42.7*	Undetected	12.4–18	2.5	[92]
Fe_3O_4-PEDOT nanospindle	−55	16.2	4.34	1.3	[93]
PEDOT:PSS/Fe_3O_4 film	46*	9.2	8–12.5	Undetected	[94]
γ-Fe_2O_3-SiO_2-PEDOT	27.5	13.8	12–16.1	2	[95]
PEDOT/RGO/$PbTiO_3$	51.94*	Undetected	12.4–18	2.5	[96]
PEDOT:PSS/Fe_3O_4-RGO	−61.4	4.8	6.4	3.86	[80]
γ-Fe_2O_3@PEDOT	44.7	12.9	10.8–15.1	2	[97]
PEDOT:PSS film	40*	Undetected	8.2–12.4	0.009	[98]
PVB-PEDOT:PSS-Gd_5Si_4	−16	Undetected	8.5–18	1	[78]
Th-GO-PEDOT	−47.5	15.5	4.9	1.5	[79]
PEDOT/RGO/$SrFe_{12}O_{19}$	42.29*	12.4	8.2–12.4	2.5	[99]

5.1.4 Other ICPs

Besides PANI, PPy and PEDOT polymers, many other conducting polymer e.g. polyacetylene (PA), polyfuran (PF) and polythiophene (PTh) have many potential applications including EMI shielding and microwave absorption. However, some drawbacks of these polymers restrict their broad application. For example the inert sulphur atom in thiophene enhances the oxidation potential of polythiophene (PTh) material, which makes the fabrication of polythiophene more complicated.

5.2 Coordination or Inorganic Conducting Polymers

These inorganic conducting polymers consist of charge transfer complexes and can be achieved by admixing metal atoms with chelating agents i.e., polydentate ligands. These are not useful for EMI applications because of complex synthesis procedures and less Eco-friendly nature.

5.3 Extrinsically Conductive Polymers

Extrinsically conducting polymers (CPCs) are insulating materials but owe conductivity by externally added conducting ingredients, thus named as extrinsically conductive polymers. There are two types of insulating polymer matrix: thermoplastics, thermoset, or elastomers. Some examples of thermoplastics are polypropylene, polystyrene, cellulose acetate, PTFE (Teflon), nylon, etc. Thermoplastic, also known as thermo-softening plastic, is the plastic polymer material. An elastomer is a polymer with visco-elasticity (i.e., both viscosity and elasticity) and has very weak inter-molecular forces, generally low Young's modulus and high failure strain compared with other materials. As the filler matrices, metal and their alloys and oxides, ceramics, and carbon-based materials are extensively used in an insulating matrix. Moreover, highly doped ICPs particles sometimes also can be used as conducting fillers instead of metal or carbon materials with insulating polymer matrix that does not only improve the electrical and magnetic properties but establish the complementary relation between permittivity and permeability. Some of host polymer matrices are:

5.3.1 Thermoplastic

The thermoplastic polymer includes the poly (vinylidene fluoride) (PVDF), polypropylene (PP), polyethylene (PE), liquid crystalline polymer (LCP), poly (phenylene sulfide) (PPS) matrices. Thermoplastic composites generally have lower electrical conductivity. Therefore, conductive fillers can be incorporated in the thermoplastic matrix with a proper conductive network structure. Some of thermoplastics are

Polyvinylidene fluoride (PVDF) PVDF is a semi-crystalline polymer with significant thermal stability, chemical resistance, and excellent pyroelectric and piezoelectric properties [100]. PVDF has much application in the architectural coating, EMI shielding, in piezoelectric films, and electrode separator in batteries due to the ease of processing and high elasticity and transparency properties. PVDF occurs in different crystalline phases $\alpha, \beta, \gamma, \delta$, and ϵ phases in which β phase can be considered the most active phase electrically. Table 3 demonstrates the PVDF based EMI shielding materials with their performance. Adding of conducting filler such as Ag, graphene, CNTs to a PVDF matrix give rise to the conductivity and hence the shielding properties in EMI application [101–104]. Kumaran et al. observed the Ag nanoparticle-induced EMI shielding in Ag nanoparticles incorporated graphite/poly-vinylidene diuoride (PVDF) composite films [105]. Compared to neat PVDF (conductivity ~2.70 S/cm at 1MHz) the conductivity of Ag-graphite/PVDF nanocomposites increased up to 2.5 times. The shielding effectiveness of the Ag-graphite/PVDF nanocomposites was obtained 29.1 dB at 12.4 GHz for the sample having 5wt.% Ag and 10wt.% graphite in PVDF matrix. Table 4 shows the EMI performance of PVDF based heterostructures.

Polypropylene (PP) PP is one of the desirable thermoplastic polymer for potential application. The low cost and low density, a balance of stiffness and impact strength, high chemical and thermal resistance, and excellent moisture barrier properties make it demandable in field of applied science. PP polymer composites show the attractive properties when combined with the filler materials that extend its utility in critical potential applications [107, 131–134]. Soares and coworkers used semi-biodegradable polypropylene/poly(lactic acid) (PLA) (70:30 wt.%) blend as the matrix and MCNTs and the noncovalently functionalized CNT with alkyl phosphonium-based ionic liquid (IL-CNT) were chosen as conducting fillers. Improved EMI SE was achieved by using IL-CNT filler (3 and 5 wt.%.) in the X-band region because more mobile charge carrier and free ions are stimulated by the interaction between the IL and CNT surface [135]. Some PP based composites with their performance are given in Table 5.

5.3.2 Elastomer

An elastomer is an amorphous polymer with the virtue of elasticity and flexibility. They have low Young's modulus and high yield strain. In general, elastomers are thermosets, but may also be thermoplastic. Elastomer is loosely cross-linked and soft polymers, deformable at ambient temperatures. Mondal et al. reported the high EMI shielding value of 33.9 dB in the X band for 20 wt% Ketjen carbon black (K-CB) loading of ethylene methyl acrylate copolymer (EMA) polymer [148]. It was proposed that interfacial interactions between Ketjen carbon black and EMA polymer

Table 4 Some PVDF polymer-based EMI shielding materials and microwave absorber

Materials	RL_{min}/SE_T*(dB)	Frequency (GHz)	Absorption width/ frequency range (GHz)	Thickness (mm)	References
Ag-graphite/PVDF	29.1*	12.4	8–12	Undetected	[105]
RGO@BT/PVDF	−45.3	5.44	4–18	4.5	[106]
PVDF/Ni-chain	35.4*	Undetected	18–26.5	0.5	[107]
Bimetallic-MWCNT–RGO/PVDF	28.5 *	12	8–12	0.5	[102]
RGO/$SrFe_{12}O_{19}$/PVDF	33*	Undetected	8.2–12.4	3	[108]
PVDF/CNT/Co chain	32.2	18	18–26.5	0.3	[109]
MWCNT/PVDF	32	10.3	8.2–18	0.1	[110]
PANI@PVDF	−24.4	12.16	10.16–14.24	2	[111]
MWCNT/graphite/PVDF	14.64	8.2	8–12	Undetected	[112]
PVDF-copper	28.75	8.62	8.2–12.4	Undetected	[113]
PVDF–carbonized charcoal	70.06*	8.2	8–13	2.8	[114]
MoO_3/MoS_3/PVDF	−38.5	8.7	3.03–11.02	2.2	[115]
$SrAl_4Fe_8O_{19}$/RGO/PVDF	42*	11.5	8.2–12.4	2	[116]
Cobalt sulfide/PVDF	−43	6.6	Undetected	4.5	[117]
Mo_2C@C/PVDF	−39	10.4	3.1	2	[118]
C-Fe_3O_4/PVDF	−43.9	15.13	2.45	3.5	[119]
K-Mno_2@PDA/PVDF	−49.4	12.56	10.96–14.8	2	[120]
Strontium hexaferrite-PVDF	−109.5	8.5	>15	.26	[121]
$Sr_3YCo_4O_{10}$/PVDF	50.2*	Undetected	8.2–18	2.5	[122]
CB/Fe_3O_4/PVDF	55.3*	Undetected	8.2–12.4	2	[123]
Barium hexaferrite-PVDF	−93.5*	8.63	8–12	.21	[124]
Mn-Zn-Cu-ferrite/PVDF	−33.5	11.5	Undetected	0.299	[125]
CEF NF/GE/PVDF	48.5	Undetected	0.03–1.5	0.188	[126]
PVDF/CNT/GnPs/Ni	43.7*	Undetected	18–26	0.3	[127]
Si-Modified RGO@Fe_3O_4/PVDF-co-HFP	-32.1	3.68	2.88	7	[128]
NiO@SiO_2/PVDF	21*	8.9	8–14	0.3	[103]
PVDF/Ni-chains foam	−26.8	Undetected	8.2–12.4	2	[129]
Fe_3O_4@PVDF	−62.7	16.9	9.2–18	3.5	[130]

matrix significantly improve the electrical conductivity and hence total EMI performance. Kuester and coworkers [149] fabricated the poly (styrene-b-ethylene-ran-butylene-b-styrene) (SEBS), graphene nanoplatelets (GnP), and carbon nanotubes (CNT) hybrid nanocomposites by melt compounding method. Decoration of GnP and CNT increases conductivity 17 orders of magnitude. Meanwhile, 36.47 dB SE value was found in the SEBS/GnP/CNT composites. Zhao and colleagues introduced 4.45 and 8.91 vol% graphene content into poly(ether-block-amide) (PEBAX) by melt blending and compression molding. They observed that SE of composite films could reach 16.6 and 30.7 dB with increasing the graphene content [127].

Table 5 Polypropylene based EMI shielding and microwave absorber

Materials	RL_{min}/SE_T*(dB)	Frequency(GHz)	Absorption width/ frequency range (GHz)	Thickness (mm)	References
CB/sisal fiber/polyamide/PP	40.7*	16	0.4–18	3	[136]
carbon nanotubes/PP	45.9*	0.8	0.03–1.5	3.2	[137]
Synthetic graphite/PP	34.7*	0.8	0.03–1.5	3.2	[137]
Carbon black/polypropylene	23.4*	0.8	0.03–1.5	3.2	[137]
PP/Ni-coated CF	48.4*	10	1–12.4	2	[138]
PP/poly(lactic acid)/CF/CNT	50.5*	Undetected	8.2–12.4	2	[139]
PP/EPDM/Ni coated glass fiber	22.2*	Undetected	8.2–12.4	2	[140]
CNT/PP	43.1	Undetected	8.2–12.4	2	[141]
Segregated CNT/PP	48.3*	Undetected	8.2–12.4	2.2	[142]
PP/PDA/Ag/PFDT	48.2*	Undetected	8.2–12.4	Undetected	[143]
PP/PET blends	60*	12.4	8.2–12.4	5	[144]
PDA/Ag/PP	26*	3	0–3	Undetected	[145]
$NiFe_2O_4$-RGO/PP	29.4*	5.8	5.8–8.2	2	[146]
Segregated carbon nanotube/PP	32*	Undetected	8.2–12.4	2	[147]

6 Techniques of Polymerization

The polymerization process takes place when comparatively small molecules (known as monomers) gathered through the chemical agent to produce a long chain-like structure, known as a polymer. Some important polymerization techniques are explained in later subsections.

6.1 Bulk or Mass Polymerization

This is the simplest method of polymerization that takes place in the absence of any solvent. The reaction mixture comprises only the monomer and monomer, soluble initiator. However, difficulties occur in the stirring process as the result of the high viscosity of high-molecular-weight polymers.

6.2 Solution Polymerization

To overcome the problem of high viscosity in mass polymerization, solution polymerization method is used. In this method, the presence of the solvent lowers the viscosity of the reaction, which is advantageous for heat transfer in exothermic reactions.

6.3 Suspension Polymerization

This method is similar to mass polymerization and widely used to fabricate the thermoplastic polymers, e. g. polystyrene thermoplastic polymer. The necessary condition for suspension polymerization is the insolubility of monomer, initiator and polymer in the suspension media such as water. Since all reaction is carried out in big droplets, suspended in an inert medium.

6.4 Emulsion Polymerization

Apart from suspension polymerization, Emulsion polymerization uses the soluble initiator and insoluble in the monomer in suspension media. Emulsion polymerization technique offers the different industrial and academic scale potential applications.

7 Popular Methods for Synthesis of ICPs and CPCs Composites

The some most common preparation methods of conducting and nonconducting polymer-based composites for EMI shielding applications are:

7.1 In-Situ Polymerization Methods

In situ polymerization method is the effective technique that offers the easy and uniform dispersion of filler into polymer matrix, followed by strong interaction between organic/non-organic filler and polymer polymer (intrinsic or extrinsic) matrix (Fig. 11). This method involve the mixing of filler (nanomaterial) in monomer/prepolymer into common solvent. Polymerization process takes place by adjusting the temperature and time. This method has been established to prepare the composites, having thermally unstable or insoluble polymer matrices. It has been seen that covalent linkages roll up between filler and the matrix when nanocomposite is fabricated by in situ polymerization methods. Nevertheless, in-situ polymerization can also be employed for non-covalent nano-composites [150, 151].

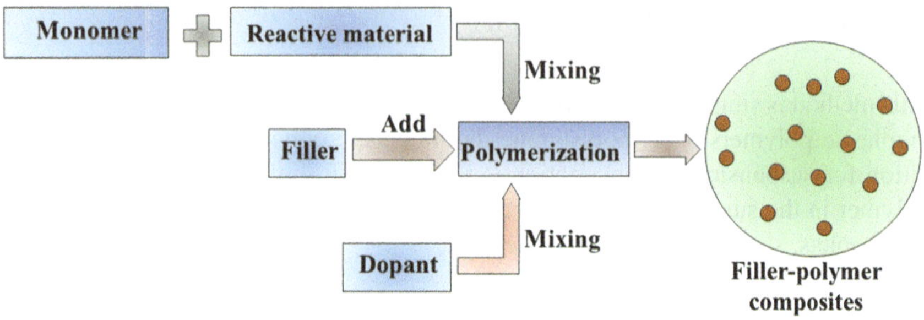

Fig. 11 A schematic representation of the in-situ polymerization method that can be used for intrinsic and extrinsic polymer composites

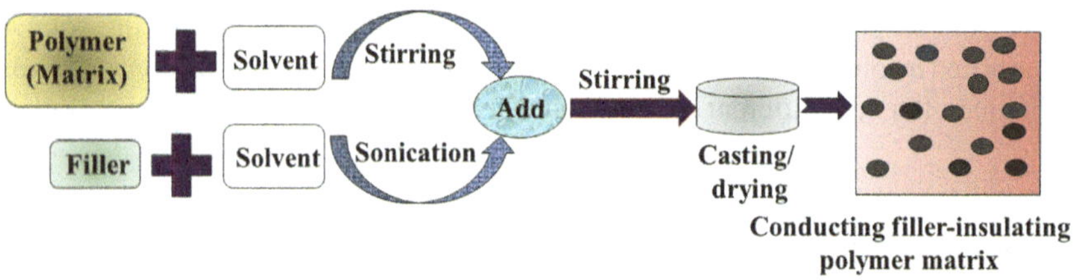

Fig. 12 A schematic representation of Solution intercalation methods for extrinsic polymer composites

7.2 *Solution Intercalation Methods*

The solution casting methods are supposed to be the simplest approach among all methods used for the making of extrinsic polymer composites. In this method, conductive fillers are dispersed in an appropriate solvent through the ultra-sonication process, as depicted in Fig. 12. Ultrasonication exploits the sound energy at high frequencies that disperse the particles by removing of particle agglomeration. This filler-solvent mixture is then mixed with the polymer-solvent mixture under the stirring process for a few hours. After this solvent can be evaporated either by thermal treatment or precipitation/distillation to form CPCs composite films. The solution casting method is commonly preferred for the synthesis of graphene or graphene oxide-based polymer composites that gives the benefits of homogeneous dispersion and lower viscosities etc. However, the requirement of a large number of solvents that hinder the fast evaporation may be the drawback of this method.

7.3 *Melt Blending Methods*

In comparison with above mentioned techniques, melt blending is widely used in industrial consumption owing to low cost, simple procedure that is desirable for large

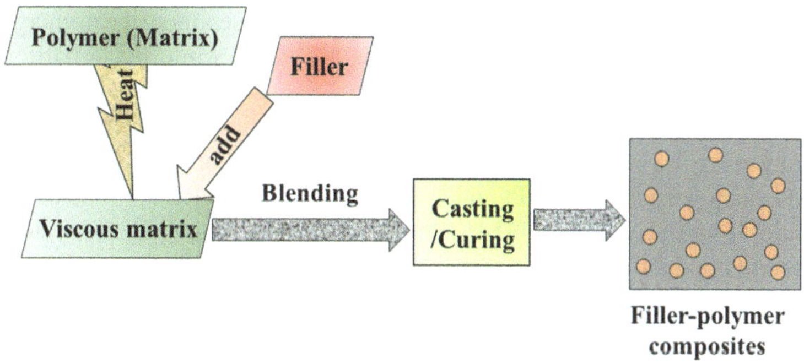

Fig. 13 A pictorial image of melt blending method

scale production of CPCs (Fig. 13). In this approach, the melting of polymers results from the viscous liquid, and high shear force leads to the dispersion of the nano-fillers with the polymer matrix. Therefore, there is no need for a solvent for dispersion of filler or polymers. In contrast to the solution casting method, low dispersion ability of the graphene in this making it less useful for synthesizing graphene-based polymer composites.

8 Metal, Alloys, Ceramics and Oxides Anchored Heterogeneous ICPs and CPCs Composites

In EMI terminology, reflection and absorption require material to have free charge carriers and electric and/or magnetic dipoles. Thus metallic fillers such as Ag, Cu, and magnetic fillers such as Ni, Fe, and their oxides are used with polymers matrices. Moreover, high dielectric constant materials such as ZnO, SiO_2, TiO_2 and $BaTiO_3$ and high magnetic permeability consisting materials Ni, Co, Fe, and Fe_3O_4 or combination of high permittivity and permeability materials are advantageous for absorption properties due to the balancing of these two factors. Thus the decoration of polymers with high permittivity and permeability materials are in trend to the development of effective EMI absorption materials. Hosseini and coworkers prepared PANI-$MnFe_2O_4$ core-shell nanocomposites by in situ polymerization method that showed the RL$\sim$ –15.3 dB values at frequency 10.4 GHz [152]. Wang et al. prepared FeCo@SnO_2@graphene@PANI quaternary nanocomposite via a three-step method. The maximum RL$\sim$ – 39.8 dB at 6.4 GHz at 3 mm thickness was found with effective bandwidth in 4.6–7.7 GHz range. Authors stated that high specific surface area, presence of functional groups, and residual defects in RGO act as polarized centers, which results in the polarization relaxation process along with multiple reflections. Additionally, multi-interfaces between FeCo, SnO_2, PANI lead the dipole and interfacial polarization. The eddy current may increase because of charge transfer between graphene and PANI that releases more heat energy. The SnO_2, graphene and PANI are dielectric loss absorbers while FeCo particles are proposed the magnetic loss absorbers in FeCo@SnO_2@graphene@PANI composites. Thus, the balance between dielectric and magnetic loss may be accountable for enhancing microwave absorption properties [153]. Long and colleagues modified the Fe_3O_4 nanoparticles with PPy made PANI and obtained RL~ -39.2 dB for 1: 4 mass ratio of aniline to Fe_3O_4 [154]. On the other hand, Yan et al. [155] decorated the different conducting polymers, polyaniline, polypyrrole, and poly (3,4-ethylene dioxythiophene) with $NiFe_2O_4$ coated on reduced graphene oxide sheets through the two-step method. Result revealed that maximum reflection loss −49.7 dB was obtained for RGO-PANI-$NiFe_2O_4$ (2.4 mm with absorption bandwidth 5.3 GHz) then −44.8 dB for RGO-PPy-$NiFe_2O_4$ (1.7 mm with absorption bandwidth 5.3 GHz) and lastly RGO-PEDOT-$NiFe_2O_4$ possess the RL value −45.4 dB (2 mm with absorption bandwidth 3.7 GHz). It was predicted that small $NiFe_2O_4$ particle size produced the dipole polarization, which can be contributed to the dielectric loss while conducting RGO can form a complete conductive network that also improves the dielectric loss. Meanwhile, conducting polymer coating layer on the RGO-$NiFe_2O_4$ enhances the Debye dipole relaxation as well as relaxation of RGO. Moreover, electronic polarization form the electron tunneling between conducting polymers and RGO sheet and tunnel effect from dielectric loss and magnetic loss also favors the excellent microwave absorption. The RGO and conducting polymers serve as dielectric loss absorber while $NiFe_2O_4$ particles are prominent magnetic loss absorber, the complementary relation between dielectric loss and magnetic absorber play the crucial role in impedance matching and enhancing the EM absorption properties. Recently, Thi and coworkers obtained the minimum RL of −40.6, −38.9, and −13.9 dB for of P(ANi-co-Py)/RGO@SiO_2, PANI/RGO@SiO_2, and PPy/RGO@SiO_2 composites [156]. The interfacial interaction between SiO_2 particles, rGO sheets, and conducting polymer chains can be considered for such improvement in absorption capabilities.

9 2D Filler Based Polymer Composites

With the advancement in technology, 2 dimensional (2D) fillers with polymer matrices have prompt the researcher's attention due to intriguing properties such as high electrical, thermal conductivity, and excellent mechanical stability of these fillers. After the discovery of the first 2D material known as graphene, the researcher explored the other 2D material such as MoS_2, metal carbides, and nitrides (MXene) for the possibility of searching the excellent material that can be tuned according to application in required fields. Apart for these 2D materials, different 1D and 3D carbonaceous filler and its derivative such as carbon nanotubes (CNTs), carbon black, graphite/expanded graphite, graphene oxide, reduced graphene oxide, and carbon fibers (CFs) innovate the EMI shielding materials in compare with traditional metallic filler based shielding materials for protection of next-generation technology devices.

9.1 Carbon Nanofiller Incorporated with ICPs and CPCs Nanocomposites

Carbons materials of different dimension such as 1D carbon nanotubes (CNTs), 2D graphene and its derivative GO, and RGO, carbon fiber (CFs) or expanded graphite (EG), 3D graphite have been widely studied for EMI shielding and microwave applications due to their low density and chemical resistance properties. Graphene demonstrates the extraordinary electronic, thermal stability, mechanical stability, optical properties, while Carbon nanotubes (CNTs) are attributed to the strongest and the stiffest material. The dispersion of graphene, CNTs, and poor magnetic properties with polymer matrix is always being the major problem for the material scientist [157]. Fortunately, the functionalization of the organic group on the surface of carbon could make its facile dispersion with solvents. Meanwhile, the magnetic properties of carbon material can be improved either by defect creation via heat treatment or irradiation or using the magnetic material along with these carbon materials. It is well known that a proper combination of permittivity and permeability is required for superior EMI performance. Thus, carbon materials usually used with the magnetic and/or polymers materials [43].

Graphene, graphene oxide, and reduced graphene oxide Graphene is the first experimentally obtained 2 dimensional sp^2 bonded carbon atom one layer sheet [158]. It has grown the interest of scientists due to extraordinary properties such as high charge carrier mobility, excellent mechanical properties, ultra-high specific surface area, and having high dielectric constant. Nevertheless, hydrophobic nature, poor magnetic properties, skin depth problems make it less useful in practice. Alternatively, graphene derivative such as graphene oxide (GO) and reduced graphene oxide (RGO or rGO) have been used for energy storage and conversion devices. When 3 dimensional (3D) graphite is oxidized with a strong oxidizing agent, then the intercalation of functional groups such as hydroxyl, carbonyl/carboxyl, epoxy increase the layer separation along c-axis of graphite. After the sonication of graphite, one can obtain the graphene oxide. The most interesting properties of GO is its hydrophilic behavior that makes its facile dispersion in both organic and inorganic type of solvent. Singh et al. [159] incorporated 30% of GO along with ferrofluid in the cement matrix and obtained the excellent SE value 46 dB. The decoration of GO and ferrofluid with a cement matrix give rise the strong polarization and magnetic losses. GO has a large surface area, various attached oxygen-containing groups that improve the interfacial properties between filler and matrix. Meanwhile, these functionalities lead the various relaxation phenomenon that improves the overall shielding performance. Chen et al. [160] introduced the GO onto the carbon fiber surface. Even low filler loading 0.4% GO-CF with cement matrix gave the superior SE ~34 dB in the X band frequency range. The main drawback of using GO is its poor electrical properties. GO is established as an insulating material due to the disruption of sp^2 network of carbon atoms. To recover the sp^2 bonding, several reducing agents such as sodium boro hydrate, hydrazine hydrate, sodium oxide, etc. can be used. The reduction of oxygen functionalities from graphene oxide gives the reduced graphene oxide that has high electrical conductivity, large surface area but lesser than graphene. The advantage of RGO over the graphene is it reduces the skin depth problem up to some extent, and remanent functional groups and defects lead the multiple relaxation processes and magnetic losses, which is favorable for absorption and reflection mechanism in EMI. That establishes the balance relation between permittivity and permeability. Therefore, graphene as filler with polyaniline (PANI), polypyrrole (PPy), polyvinyl alcohol (PVA) and poly(methyl methacrylate) (PMMA) polymers have the broad range of studies for EMI shielding materials. Yan and coworkers [161] prepared the insitu thermally reduced graphene/polyethylene composites and achieved the SE value in the range 28.3–32.4 dB at small filler loading of 0.66 vol% of graphene. On the other hand, other researchers group obtained the max RL~ -51.5 dB 11.2 GHz with absorption with 4 GHz at 2.5 mm thickness for covalently bonded PANI/amino-functionalized graphene sheets (AFG) composites [162].

Graphite (GR)/expanded graphite (EG) Graphite is a well known layered carbon structure in which carbon atoms are attached by weak Vander Waals forces. Graphite is a lightweight material lubricant and having a high aspect ratio with high mechanical and thermal stabilities. On the other hand, expanded graphite (EG) is the two-dimensional counterpart of graphite with small stacking of layers that can be achieved by the thermal treatment of graphite. Similar to graphite, EG is having the benefits of high mechanical stability, thermal stability, and significant electrical properties. Being a conducting filler, both GR and EG have been studied with polymer host matrices for EMI application. Cui and coworkers prepared polyethylene composites by structuring segregated carbon black (CB)/graphite (GR) network [163] that achieved the good electrical conductivity (33.9 S/m) and superior SE value 40.2 dB at low filler loading of 7.7 vol% in the composites. They revealed that the peculiar structure of segregated conductive networks and the synergetic effect of large and small-sized graphite flakes and CB nanoparticles might cause superior EMI performance. Chen et al. investigated the absorption properties of expanded graphite (EG)/polyaniline/$CoFe_2O_4$ ferrite and achieved RL value -19.13 dB at 13.28 GHz with 5.94 GHz bandwidth [164]. They proposed that EG/PANI/CF composites exhibit the dielectric loss properties with the bound or localized charges and the dipolar reorientation processes caused by PANI and EG. Likewise, magnetic losses come from the $CoFe_2O_4$. The spin-rotation resonance, interfacial polarization and relaxation effects among components gives rise the total absorbance properties of EG/PANI/$CoFe_2O_4$ composites.

Carbon nanotubes (CNTs) Carbon nanotubes (CNTs) are the rolled-up one-dimensional sheet of sp^2 hybridized carbon atoms. They may be single-walled Carbon nanotubes (SWCNTs) or multi-wall Carbon nanotubes (MWCNTs). The low density, small diameter, high conductivity, and large aspect ratio, easy percolation are some appealing properties of CNTs for making the modern energy conversion devices. Until, various CNTs-polymer-based composites have been reported with the superior EMI performances [44, 165]. Abraham et al. [166] decorated non-covalently functionalized multi-walled carbon nanotube (MWCNT) with ionic liquid into Styrene-butadiene rubber (SBR) matrix. The synergy between ionic liquid and

MWCNT makes it dispersible through cation-π interaction. Thus, better dispersion and microstructural development of MWCNTs within the SBR matrix accounts for superior shielding efficiency 35.06 dB at 18 GHz. Lee and colleagues obtained the >25 dB SE value for epoxy composites filled with multi-walled carbon nanotubes (CNTs), exfoliated graphite nanoplatelets (xGnPs) [167].

Carbon black (CB) Among the carbonaceous materials, nano-size conducting black (CB) is another promising filler owing to excellent electrical conductivity, lightweight, and low cost. Liu and coworkers prepared silicone rubber (SR)/polyolefin elastomer (POE) blends containing ionic liquids (IL) modified with CB-IL and MWCNT-IL by melt-blending and hot pressing method. Thc EMI SE was found to be 36.5 dB at 9.0 GHz for SR/POE/CB-CNT-IL composite due to the synergistic effect of ionic liquids (ILs)-modified CBs and CNTs [168]. Jani et al. prepared the carbon black/silicone rubber composites and investigated its applicability for EMI shielding in the frequency range of 8–18 GHz [169]. The 15 wt% of CB loading with silicone matrix shows –40 dB SE at thickness 2.8 mm.

Carbon fibres (CFs) Carbon Fibers (CFs) as conducting filler are widely used as reinforcement material in polymer composites. In comparison to other reinforcing fibers, CF presents the highest specific modulus and specific strength of all reinforcing fibers. Wang et al. synthesized the conductive filler nickel-coated carbon fiber (Ni-CF) via electroplating method and then dispersed with silicone rubber [170]. The randomly oriented and cross-linked Ni-CFs filler form a 3D conductive network with silicon rubber. At 80 phr Ni-CF filler content, the volume resistivity was 0.042 Ω cm that gives rise the SE value 80 dB in 30 MHz to 1200 MHz frequency range.

Table 6 depicts the EMI performance of some carbon filler such RGO, CNTs based composites.

Table 6 Summary of Carbon-polymer based composites against EMI pollution

Materials	RL_{min}/SE_T* (dB)	Frequency (GHz)	Absorption width/frequency range (GHz)	Thickness (mm)	References
S-PPy/RGO	−54.4	12.76	6.76	3	[54]
PANI/$NiFe_2O_4$/graphene	−50.5	12.5	11–16.3	2.5	[171]
S doped RGO/PS	24.5*	18	12–18	0.25	[172]
CFs/$CoFe_2O_4$/PANI	−38.2	12.7	Undetected	4.1	[173]
Graphene oxide/PPy	−58.1	12.4	9.2–15.4	2.96	[174]
PANI/Ni/CF	−12.4	8.8	Undetected	2	[175]
Graphene@Fe_3O_4@WO_3/PANI	−46.7	9.4	12.4–14.2	1.5	[176]
PANI/AFG	−51.5	11.2	9.6–13.6	2.5	[162]
TiO_2/PANI/GO	−51.74	9.67	3.91	3.12	[177]
FeCo@RGO@PPy	−40.7	4.5	3.1–6	2.5	[59]
PPy/Ni/RGO	−47.32	5.76	1.8	4	[178]
PANI/p-C/$NiFe_2O_4$	−37	10	3.9	2.5	[179]
MWCNTs/Fe_3O_4/PPy/C	−53.07	13.92	6.4	2.2	[180]
Fe_3O_4@Ppy/3D graphene	−40.53	6.32	11.12–16.24	2.5	[181]
PPy-graphene	33*	Undetected	8.2–12.4	2	[182]

Transition metal dichalcogenide (TMD) Transition metal dichalcogenide (TMD) can be represented by MX_2 where M refers the transition metal atom such as Tungsten (W), Molybdenum (Mo) and X represent the chalcogen atom such as Selenium (Se), Sulfur (S), Tellurium (Te), etc. TMDs are part of 2D materials, and thus these are thin semiconductors. Santoor et al. fabricated the liquid exfoliated molybdenum disulfide (MoS_2) nanosheets and polyaniline (PANI) nanoparticles and dispersed it in polystyrene (PS) matrix and studied the low frequency (100 Hz–5 MHz range) EMI performance. With the inclusion of MoS_2 nanosheets in the PS/PANI blends, the maximum estimated EMI SE reaches up to 92 dB at 100 Hz [183]. Memon and coworkers [193] studied the EMI shielding capabilities of dopamine functionalized polyurethane (PU) with multiwalled carbon nanotubes (MWNTs) and ferrite doped with molybdenum disulfide (Fe_3O_4@MoS_2). The as-prepared nanocomposite showed the outstanding EMI SE values −36.6 dB. Zhang et al. grown the PANI Nanoneedles on MoS_2 Nanosheets and obtained the RL values −44.8 dB of MoS_2/PANI-nanoneedles at 14.5 GHz with a thickness of 1.6 mm [186]. The author revealed that unique architecture MoS_2/PANI-nanoneedles play an important role in EM wave attenuation. The high interfacial polarization loss contributes to total absorption performance for the MoS_2/PANI. The electrons hooping takes place from MoS_2 interlayer to PANI. Additionally, the large surface area of molybdenum disulfide (MoS_2) nanosheets and densely packed PANI result from the more propagation paths that give rise to the multiple reflection loss and improve the total EM absorption performances. Thus, the investigation of other transition metal dichalcogenide (TMD) with polymers may help to search the new EMI shielding materials. Some reported EMI performance of TMD-polymer based composites against EMI pollution are listed in Table 7.

Table 7 Summary of some TMD-polymer based composites against EMI pollution

Materials	RL_{min}/SE_T* (dB)	Frequency (GHz)	Absorption width/frequency range (GHz)	Thickness (mm)	References
RGO@MoS_2/PVDF	−43.1	14.48	3.6–17.8	2	[184]
NiS_2@MoS_2/PVDF	−41.05	12.08	10.16–14.56	2.2	[185]
MoS_2/PANI nanoneedle	−44.8	14.5	13.5–15.9	1.6	[186]
MoS_2@Fe_3O_4@PANI	−40.97	10.06	8.2–12.2	2.3	[187]
3D nanoflower MoS_2/PANI	−50.57	5.04	4.24–6.32	5	[188]
MoS_2/PANI	-40.79	14.01	11.88-16.90	2	[189]
PPy@MoS_2	−49.1	6.1	11.5–17.5	2.5	[190]
MoS_2@PPy@Fe_3O_4	−32	15.9	4.3	2	[191]
3D sphere MoS_2/PANI	−59.78	8.08	6.56–9.68	3.44	[192]

9.2 2D Transition Metal Carbides (MXenes)

The 2D metal carbides and nitrides MXenes is currently the rising star in the 2D family of material that have the advantage of metal like conductivity (2–6*10^5 Sm^{-1}), high mechanical properties and hydrophilic surfaces. MXene is represented by formula $M_{n+1}X_nT_x$, where M is the transition metal such as titanium (Ti), Vanadium (V), Molybdenum (Mo), etc. while X refers the Carbon (C) and Nitrogen (N) (where n = 1, 2, or 3) and T_x describes the surface terminations such as hydroxyl (OH), oxygen (O), and/or fluorine (F) groups [194, 195]. Because of hydroxyl, oxygen, and fluorine organic groups on the surface, the MXene sheet can be easily dispersed in some organic solvent and water also. Along with applicability in EMI application, MXene has many attractive applications in supercapacitor, batteries, sensors, cellular imaging [196]. shahzad and coworkers obtained the EMI shielding effectiveness of 92 decibels for 45-μm thin Ti_2CT_x film [197]. Xu with colleagues made the lightweight porous MXene (Ti_2CT_x)/polyvinly alcohol composites foam by freezing the dying method. The specific effectiveness value 5136 $dBcm^2g^{-1}$ was obtained at a very low filler volume fraction of 0.15 vol%. Authors anticipated that internal reflection, multiple porous structures, dipole/interfacial polarization synergistically affect absorption efficiency. Although MXene has rather few studied, which shows the possibilities to combine it with intrinsic and extrinsic polymers. Recent studies on MXene-polymer based composites are shown in Table 8.

Table 8 Summary of MXene-polymer based composites materials against EMI pollution

Materials	RL_{min}/SE_T* (dB)	Absorption width (GHz)	Absorption width/frequency range	Thickness (mm)	References
$Ti_3C_2T_x$@PS	62*	12.4	8.2–12.4	2	[198]
$Ti_3C_2T_x$@PPy	−49.5	7.6	6.44–11.58	3.6	[199]
$Ti_3C_2T_x$/c-PANI	36*	Undetected	8.2–12.4	0.04	[200]
Annealed $Ti_3C_2T_x$/epoxy	41	Undetected	8.2–12.4	2	[201]
f-Ti_2CT_x/PVA	−18.7	8.2	Undetected	3.9	[202]
PPy/MXene/PET	90*	Undetected	8.2–12.4	1.3	[203]
MXene-PANI	23*	12.2	8.2–12.2	1.5	[204]
Ti_3C_2/Fe_3O_4/PANI	−40.3	15.3	12.8–18	1.9	[205]
$Ti_3C_2T_x$/PANI	−56.3	13.8	8–12.4	1.8	[206]
$Ti_3C_2T_x$/natural rubber	53.6	Undetected	8.2–12.4	0.251	[207]
Co-doped NiZn ferrite/PANI/$Ti_3C_2T_x$	−37.1	10.2	8.2–12.3	2.2	[208]
PDMS coated MXene	70.5*	Undetected	8.2–12.4	2	[209]
MXene/EPDM	52*	Undetected	12.4–18	0.3	[210]
RGO-MXene/epoxy	55*	Undetected	8.2–12.4	0.5	[211]
PEDOT:PSS/$Ti_3C_2T_x$	40.5	Undetected	8.2-12.4	0.0066	[212]
PVA/MXene	44.4*	12.4	8.2–12.4	0.027	[213]
MXene/graphene hybrid foam	50.7	12.4	8.2–12.4	1.5	[214]
MXene/PVDF	48.47±3.5*	Undetected	8.2–12.5	2	[215]
MXene/polyurethane	21*	Undetected	8–12	0.2	[216]

10 Conclusion

The conducting polymers based composites are highly in demand in field of energy storage and energy conversion. Due to attractive physical and chemical properties, conducting polymers present its candidacy for EMI shielding and microwave absorption application. Moreover, the integration of conducting polymers with two-dimensional family materials such as graphene, MXenes, and MoS_2 stimulates the requirement of these materials for the next generation's devices protection. This chapter present the brief introduction of EMI shielding mechanism and the materials that can be used to make a superior EMI shielding material and microwave absorber.

11 Challenges and Future Opportunities

The conducting polymers based composites have significantly prompted the research in the field of electromagnetic interference (EMI) shielding application. The use of two-dimensional family materials such as graphene, MXenes, and MoS_2 with these polymers stimulated the EMI shielding materials for the next generation's devices protection. An abbreviated introduction to the EMI shielding and microwave absorption materials have been summarized in the present chapter. Besides the great achievements in this field, some drawback is still present that could not be rectified. For instance, the use of carbonaceous materials such as graphene and CNTs have the skin depth problem due to high charge mobility, and poor magnetic properties suppress the absorption phenomenon due to impedance mismatching. Additionally, costly synthesis and their hydrophobic nature may require the multi-step preparation method. Moreover, surface modification or creation of defects may diminish the structural stability and physical properties of the material. On the other hand, insulating polymer-based conducting composites have poor thermal stability; hence above a certain temperature, they can not be operated. Apart from this, there are a lot of possibilities for future EMI materials. In general, people have studied the room temperature EMI or microwave absorption properties. Thus, it has the possibility of EMI shielding materials to protect the low temperature working electronic devices.

Acknowledgements V. Shukla is thankful to the Indian Institute of Technology Kharagpur for providing financial assistantship to carry out the research work.

References

1. Pawar, S.P., Marathe, D.A., Pattabhi, K., Bose, S.: Electromagnetic interference shielding through mwnt grafted fe 3 o 4 nanoparticles in pc/san blends. J. Mater. Chem. A **3**(2), 656–669 (2015)

2. Nasouri, K., Shoushtari, A.M.: Fabrication of magnetite nanoparticles/polyvinylpyrrolidone composite nanofibers and their application as electromagnetic interference shielding material. J. Thermoplastic Compos. Mater. **31**(4), 431–446 (2018)
3. Pawar, S.P., Gandi, M., Saraf, C., Bose, S.: Polycarbonate composites containing carbon encapsulated "brick-like" fe3o4 nanoparticles as efficient microwave absorbers with a large bandwidth. ChemistrySelect **1**(13), 3829–3838 (2016)
4. Mordina, B., Kumar, R., Tiwari, R.K., Setua, D.K., Sharma, A.: Fe3o4 nanoparticles embedded hollow mesoporous carbon nanofibers and polydimethylsiloxane-based nanocomposites as efficient microwave absorber. J. Phys. Chem. C **121**(14), 7810–7820 (2017)
5. Yablokov, M.Y., Shevchenko, V.G., Mukhortov, L.A., Ozerin, A.N.: Electromagnetic Interference Shielding of Carbon Nanotube-Fluoropolymer Elastomer Composites with Layered Structure, pp. 1–5. Nanotubes and Carbon Nanostructures, Fullerenes (2019)
6. Mehranvari, M., Katbab, A.A., Nayyeri, V., Nazokdast, H., Preparation of a radar absorbing material based on pp, epdm thermoplastic elastomer and carbon nanotube nanocomposite. In: Asia-Pacific Microwave Conference (APMC), pp. 1–4. IEEE (2016)
7. Chen, Y., Wang, Y., Zhang, H.-B., Li, X., Gui, C.-X., Yu, Z.-Z.: Enhanced electromagnetic interference shielding efficiency of polystyrene/graphene composites with magnetic fe3o4 nanoparticles. Carbon **82**, 67–76 (2015)
8. Al-Ghamdi, A., Al-Hartomy, O.A., Al-Salamy, F., Al-Ghamdi, A.A., El-Mossalamy, E., Abdel Daiem, A., El-Tantawy, F.: Novel electromagnetic interference shielding effectiveness in the microwave band of magnetic nitrile butadiene rubber/magnetite nanocomposites. J. Appl. Polym. Sci. **125**(4), 2604–2613 (2012)
9. Darwish, M.S., Bakry, A., Al-Harbi, L.M., Khowdiary, M.M., El-Henawy, A., Yoon, J.: Core/shell pa6@ fe3o4 nanofibers: magnetic and shielding behavior. J. Dispers. Sci. Technol. 1–9 (2019)
10. Lv, H., Ji, G., Liang, X., Zhang, H., Du, Y.: A novel rod-like mno 2@ fe loading on graphene giving excellent electromagnetic absorption properties. J. Mater. Chem. C **3**(19), 5056–5064 (2015)
11. Ribadeneyra, M.C., de Diego, J.P., González, M.G.: Emi Shielding Composites Based on Magnetic Nanoparticles and Nanocarbons. Ph.D. thesis, Ph. D. thesis, Universidad Carlos III de Madrid. http://hdl.handle.net (2014)
12. Luo, J., Zuo, Y., Shen, P., Yan, Z., Zhang, K.: Excellent microwave absorption properties by tuned electromagnetic parameters in polyaniline-coated ba 0.9 la 0.1 fe 11.9 ni 0.1 o 19/reduced graphene oxide nanocomposites. RSC Adv. **7**(58), 36433–36443 (2017)
13. Solafide, W., Murakami, R.-I.: Effect of moisture absorption and fiber orientation on electrical conductivity and electromagnetic interference shielding of carbon fiber reinforced bioplastic composites. Int. J. Mod. Phys. B **33**(10), 1950082 (2019)
14. Kim, H.G., Shin, H.J., Kim, G.-C., Park, H.J., Moon, H.J., Kwac, L.K.: Electromagnetic interference shielding characteristics for orientation angle and number of plies of carbon fiber reinforced plastic. Carbon letters **15**(4), 268–276 (2014)
15. Li, X.-H., Li, X., Liao, K.-N., Min, P., Liu, T., Dasari, A., Yu, Z.-Z.: Thermally annealed anisotropic graphene aerogels and their electrically conductive epoxy composites with excellent electromagnetic interference shielding efficiencies. ACS Appl. Mater. Interfaces **8**(48), 33230–33239 (2016)
16. Tian, X., Meng, F., Meng, F., Chen, X., Guo, Y., Wang, Y., Zhu, W., Zhou, Z.: Synergistic enhancement of microwave absorption using hybridized polyaniline@ helical cnts with dual chirality. ACS Appl. Mater. Interfaces **9**(18), 15711–15718 (2017)
17. Jafar, H., Husaen, S., Al-Ajaj, E.: Effect of water absorption on some electrical and dielectrical properties of epoxy resin reinforced with chopped carbon fibers. Al-Nahrain J. Sci. **14**(4), 87–94 (2011)
18. Liu, Y., Wei, S., Xu, B., Wang, Y., Tian, H., Tong, H.: Effect of heat treatment on microwave absorption properties of ni-zn-mg-la ferrite nanoparticles. J. Magnet. Magnet. Mater. **349**, 57–62 (2014)

19. Lv, H., Ji, G., Wang, M., Shang, C., Zhang, H., Du, Y.: Feco/zno composites with enhancing microwave absorbing properties: effect of hydrothermal temperature and time. RSC Adv. **4**(101), 57529–57533 (2014)
20. Hota, N., Karna, N., Dubey, K., Tripathy, D., Sahoo, B.: Effect of temperature and electron beam irradiation on the dielectric properties and electromagnetic interference shielding effectiveness of ethylene acrylic elastomer/millable polyurethane/swcnt nanocomposites. Eur. Polym. J. **112**, 754–765 (2019)
21. Tian, C., Du, Y., Xu, P., Qiang, R., Wang, Y., Ding, D., Xue, J., Ma, J., Zhao, H., Han, X.: Constructing uniform core-shell ppy@ pani composites with tunable shell thickness toward enhancement in microwave absorption. ACS Appl. Mater. Interfaces **7**(36), 20090–20099 (2015)
22. Qiu, M., Zhang, Y., Wen, B.: Facile synthesis of polyaniline nanostructures with effective electromagnetic interference shielding performance. J. Mater. Sci. Mater. Electron. **29**(12), 10437–10444 (2018)
23. Li, H., Lu, X., Yuan, D., Sun, J., Erden, F., Wang, F., He, C.: Lightweight flexible carbon nanotube/polyaniline films with outstanding emi shielding properties. J. Mater. Chem. C **5**(34), 8694–8698 (2017)
24. Kumar, P., Narayan Maiti, U., Sikdar, A., Kumar Das, T., Kumar, A., Sudarsan, V.: Recent advances in polymer and polymer composites for electromagnetic interference shielding: review and future prospects. Polym. Rev. **59**(4), 687–738 (2019)
25. Sasikumar, S.P., Libi Mol, V.A.H., George, D.M., Lindo, A.O., Pushkaran, N.K., John, H., Aanandan, C.K.: Electromagnetic interference shielding efficiency enhancement of the pani-csa films at broad band frequencies. Progr. Electromagnet. Res. **57**, 163–174 (2017)
26. Jiang, L., Wang, Z., Li, D., Geng, D., Wang, Y., An, J., He, J., Liu, W., Zhang, Z.: Excellent microwave-absorption performances by matched magnetic-dielectric properties in double-shelled co/c/polyaniline nanocomposites. RSC Adv. **5**(50), 40384–40392 (2015)
27. Bora, P.J., Azeem, I., Vinoy, K., Ramamurthy, P.C., Madras, G.: Polyvinylbutyral-polyaniline nanocomposite for high microwave absorption efficiency. ACS Omega **3**(12), 16542–16548 (2018)
28. Li, J., Jiang, L., Wu, A., Chen, X., Qiu, S., He, M., Chen, H.: Tb0. 2dy0. 8 (fe0. 8co0. 2) 1.93/polyaniline composites as an excellent absorber against electromagnetic pollution. Mater. Res. Exp. **6**(12), 126116 (2020)
29. Wang, M., Ji, G., Zhang, B., Tang, D., Yang, Y., Du, Y.: Controlled synthesis and microwave absorption properties of ni0. 6zn0. 4fe2o4/pani composite via an in-situ polymerization process. J. Magnet. Magn. Mater. **377**, 52–58 (2015)
30. Wang, L., Zhu, J., Yang, H., Wang, F., Qin, Y., Zhao, T., Zhang, P.: Fabrication of hierarchical graphene@ fe3o4@ sio2@ polyaniline quaternary composite and its improved electrochemical performance. J. Alloys Compounds **634**, 232–238 (2015)
31. Luo, J., Shen, P., Yao, W., Jiang, C., Xu, J.: Synthesis, characterization, and microwave absorption properties of reduced graphene oxide/strontium ferrite/polyaniline nanocomposites. Nanoscale Res. Lett. **11**(1), 141 (2016)
32. Du, M., Yao, Z., Zhou, J., Liu, P., Yao, T., Yao, R.: Design of efficient microwave absorbers based on multi-layered polyaniline nanofibers and polyaniline nanofibers/li0. 35zn0. 3fe2. 35o4 nanocomposite. Synth. Metals **223**, 49–57 (2017)
33. Hou, J., Zhang, L., Qiu, H., Duan, W., Wang, X., Wan, X., Du, X.: Fabrication and microwave absorption performances of hollow-structure fe 3 o 4/pani microspheres. J. Mater. Sci. Mater. Electron. **28**(13), 9279–9288 (2017)
34. Yan, P., Miao, J., Cao, J., Zhang, H., Wang, C., Xie, A., Shen, Y.: Facile synthesis and excellent electromagnetic wave absorption properties of flower-like porous rgo/pani/cu 2 o nanocomposites. J. Mater. Sci. **52**(22), 13078–13090 (2017)
35. Bora, P.J., Azeem, I., Vinoy, K., Ramamurthy, P.C., Madras, G., Microwave absorption property of pvb-polyaniline nanocomposite. In: IEEE Asia Pacific Microwave Conference (APMC), pp. 674–677. IEEE (2017)

36. Yang, H., Dai, J., Lin, Y., Han, N., Liu, X., Wang, J.: Synthesis and microwave absorption property of ternary composites: polyaniline/cofe 2 o 4/ba 3 co 2 fe 24 o 41. J. Mater. Sci. Mater. Electron. **28**(15), 10853–10861 (2017)
37. Zuo, Y., Yao, Z., Zhou, J., Zhang, X., Ning, Y.: Hierarchical structures based on li 0.35 zn 0.3 fe 2.35 o 4/polyaniline nanocomposites: synthesis and excellent microwave absorption properties. J. Mater. Sci. Mater. Electron. **29**(2), 922–926 (2018)
38. Liu, Y., Xing, H., Wang, L., Liu, Z., Wang, H., Jia, H.: Novel microwave absorption materials of porous flower-like nickel oxide@ polyaniline in the x-band. Nano **13**(06), 1850059 (2018)
39. Cheng, B., Wang, J., Zhang, F., Qi, S.: Preparation of silver/carbon fiber/polyaniline microwave absorption composite and its application in epoxy resin. Polym. Bull. **75**(1), 381–393 (2018)
40. Yao, Z., Lin, H., Zhou, J., Haidry, A.A., et al.: The effect of polymerization temperature and reaction time on microwave absorption properties of co-doped znni ferrite/polyaniline composites. RSC Adv. **8**(51), 29344–29355 (2018)
41. Yang, Q., Yang, W., Shi, Y., Yu, L., Li, X., Yu, L., Dong, Y., Zhu, Y., Fu, Y.: Aligned polyaniline/porous biomass carbon composites with superior microwave absorption properties. J. Mater. Sci. Mater. Electron. **30**(2), 1374–1382 (2019)
42. Wang, M., Lin, Y., Liu, Y., Yang, H.: Core-shell structure bafe 12 o 19@ pani composites with thin matching thickness and effective microwave absorption properties. J. Mater. Sci. Mater. Electron. **30**(15), 14344–14354 (2019)
43. Guo, Y., Li, J., Meng, F., Wei, W., Yang, Q., Li, Y., Wang, H., Peng, F., Zhou, Z.: Hybridization-induced polarization of graphene sheets by intercalation-polymerized polyaniline toward high performance of microwave absorption. ACS Appl. Mater. Interfaces **11**(18), 17100–17107 (2019)
44. Wang, H., Meng, F., Huang, F., Jing, C., Li, Y., Wei, W., Zhou, Z.: Interface modulating cnts@ pani hybrids by controlled unzipping of the walls of cnts to achieve tunable high-performance microwave absorption. ACS Appl. Mater. Interfaces **11**(12), 12142–12153 (2019)
45. Sahin, E.I., Paker, S., Kartal, M.: Microwave absorbing properties of polyaniline-nife2o4: V composites. J. Chem. Soc. Pak. **41**(2), 246–257 (2019)
46. Qiao, Y., Xiao, J., Jia, Q., Lu, L., Fan, H.: Preparation and microwave absorption properties of znfe2o4/polyaniline/graphene oxide composite. Results Phys. **13**, 102221 (2019)
47. Jia, H., Xing, H., Ji, X., Gao, S.: Synergistic effect of hexagonal flake co 3 o 4@ pani core-shell composites with excellent microwave-absorbing properties. J. Mater. Sci. Mater. Electron. **30**(4), 3386–3395 (2019)
48. Chen, P., Jiang, L., Li, J., Chen, H., He, J., Wang, Y.: Synthesis and microwave-absorption properties of organic-inorganic sm2co14b/polyaniline composites with embedded structure. J. Appl. Phys. **126**(8), 085501 (2019)
49. Yang, W., Yang, X., Li, X., Islam, M.Z., Dong, Y., Fu, Y., Liu, X., Fu, F., Zhu, Y.: Construction and microwave absorption properties of core@ double-shell structured fe3o4@ polyaniline@ mno2 nanospheres, vacuum 2050032, 2 (2020)
50. Guo, S., Wen, J., Song, L., Qu, J., He, W., Liu, S.: One-step fabrication and electromagnetic wave absorption of graphene/ag@ polyaniline ternary nanocomposites. Nanotechnology **31**(22), 225606 (2020)
51. Shi, Y., Chen, J., Dai, J., Qiu, J.: Frequency selective absorbing property of nanoring-shaped polyaniline with broadband absorption. J. Mater. Sci. Mater. Electron. **31**(4), 3622–3630 (2020)
52. Wu, F., Sun, M., Chen, C., Zhou, T., Xia, Y., Xie, A., Shang, Y.: Controllable coating of polypyrrole on silicon carbide nanowires as a core-shell nanostructure: a facile method to enhance attenuation characteristics against electromagnetic radiation. ACS Sustain. Chem. Eng. **7**(2), 2100–2106 (2018)
53. Shukla, V.: Role of spin disorder in magnetic and emi shielding properties of fe 3 o 4/c/ppy core/shell composites. J. Mater. Sci. **55**(7), 2826–2835 (2020)
54. Wu, F., Xie, A., Sun, M., Wang, Y., Wang, M.: Reduced graphene oxide (rgo) modified spongelike polypyrrole (ppy) aerogel for excellent electromagnetic absorption. J. Mater. Chem. A **3**(27), 14358–14369 (2015)

55. Hosseini, S., Asadnia, A., Moloudi, M.: Preparation and electromagnetic wave absorption hard-soft ba ferrite/polypyrrole core-shell nanocomposites. Mater. Res. Innov. **19**(2), 107–112 (2015)
56. Wang, H., Yan, Z., An, J., He, J., Hou, Y., Yu, H., Ma, N., Yu, G., Sun, D.: Iron cobalt/polypyrrole nanoplates with tunable broadband electromagnetic wave absorption. RSC Adv. **6**(95), 92152–92158 (2016)
57. Jiang, W., Wang, Y., Xie, A., Wu, F.: Microwave absorption of a tio2@ ppy hybrid and its nonlinear dielectric resonant attenuation mechanism. J. Phys. D Appl. Phys. **49**(38), 385502 (2016)
58. Qiao, M., Lei, X., Ma, Y., Tian, L., Su, K., Zhang, Q.: Well-defined core-shell fe3o4@ polypyrrole composite microspheres with tunable shell thickness: synthesis and their superior microwave absorption performance in the ku band. Indus. Eng. Chem. Res. **55**(22), 6263–6275 (2016)
59. Wang, Y., Wu, X., Zhang, W., Luo, C., Li, J.: Synthesis of ferromagnetic sandwich feco@ graphene@ ppy and enhanced electromagnetic wave absorption properties. J. Magnet. Magn. Mater. **443**, 358–365 (2017)
60. Wang, X., Yang, M., Yan, H., Qi, S.: The characterization and preparation of core-shell structure particles of carbon-sphere@ nife 2 o 4@ ppy as microwave absorbing materials in x band. J. Mater. Sci. Mater. Electron. **28**(20), 14988–14995 (2017)
61. Sun, X., Lv, X., Sui, M., Weng, X., Li, X., Wang, J.: Decorating mof-derived nanoporous co/c in chain-like polypyrrole (ppy) aerogel: A lightweight material with excellent electromagnetic absorption. Materials **11**(5), 781 (2018)
62. Wang, Y., Du, Y., Wu, B., Han, B., Dong, S., Han, X., Xu, P.: Fabrication of ppy nanosphere/rgo composites via a facile self-assembly strategy for durable microwave absorption. Polymers **10**(9), 998 (2018)
63. Xie, A., Zhang, K., Sun, M., Xia, Y., Wu, F.: Facile growth of coaxial ag@ polypyrrole nanowires for highly tunable electromagnetic waves absorption. Mater. Des. **154**, 192–202 (2018)
64. Li, C., Ji, S., Jiang, X., Waterhouse, G.I., Zhang, Z., Yu, L.: Microwave absorption by watermelon-like microspheres composed of γ-fe 2 o 3, microporous silica and polypyrrole. J. Mater. Sci. **53**(13), 9635–9649 (2018)
65. Sun, X., Lv, X., Li, X., Yuan, X., Li, L., Gu, G.: Fe3o4@ sio2 nanoparticles wrapped with polypyrrole (ppy) aerogel: a highly performance material as excellent electromagnetic absorber. Mater. Lett. **221**, 93–96 (2018)
66. Yuan, Y., Yin, W., Yang, M., Xu, F., Zhao, X., Li, J., Peng, Q., He, X., Du, S., Li, Y.: Lightweight, flexible and strong core-shell non-woven fabrics covered by reduced graphene oxide for high-performance electromagnetic interference shielding. Carbon **130**, 59–68 (2018)
67. Li, C., Zhang, Y., Ji, S., Jiang, X., Zhang, Z., Yu, L.: Microwave absorption properties of γ-fe 2 o 3/(sio 2) x-so 3 h/polypyrrole core/shell/shell microspheres. J. Mater. Sci. **53**(7), 5270–5286 (2018)
68. Peymanfar, R., Norouzi, F., Javanshir, S.: A novel approach to prepare one-pot fe/ppy nanocomposite and evaluation of its microwave, magnetic, and optical performance. Mater. Res. Exp. **6**(3), 035024 (2018)
69. Zhang, C., Chen, Y., Li, H., Tian, R., Liu, H.: Facile fabrication of three-dimensional lightweight rgo/ppy nanotube/fe3o4 aerogel with excellent electromagnetic wave absorption properties. ACS Omega **3**(5), 5735–5743 (2018)
70. Dong, S., Zhang, X., Hu, P., Zhang, W., Han, J., Hu, P.: Biomass-derived carbon and polypyrrole addition on sic whiskers for enhancement of electromagnetic wave absorption. Chem. Eng. J. **359**, 882–893 (2019)
71. Jiao, Y., Li, J., Xie, A., Wu, F., Zhang, K., Dong, W., Zhu, X.: Confined polymerization strategy to construct polypyrrole/zeolitic imidazolate frameworks (ppy/zifs) nanocomposites for tunable electrical conductivity and excellent electromagnetic absorption. Compos. Sci. Technol. **174**, 232–240 (2019)

72. Zhang, Z., Lv, Q., Chen, Y., Yu, H., Liu, H., Cui, G., Sun, X., Li, L.: Nis2@ rgo nanosheet wrapped with ppy aerogel: A sandwich-like structured composite for excellent microwave absorption. Nanomaterials **9**(6), 833 (2019)
73. Su, X., Wang, J., Zhang, X., Zhang, B., Wu, Q., Dai, W., Zou, Y., Shao, C.: Synthesis of core-shell fe3o4@ ppy/graphite nanosheets composites with enhanced microwave absorption performance. Mater. Lett. **239**, 136–139 (2019)
74. Ge, Y., Li, C., Jiang, X., Waterhouse, G.I., Zhang, L., Zhang, Z., Yu, L.: Znfe2o4@polypyrrole nanocomposites as an efficient broadband electromagnetic wave absorber at 2–40 ghz. Ceram. Int. **45**(11), 13883–13893 (2019)
75. Liu, T., Liu, N., Zhai, S., Gao, S., Xiao, Z., An, Q., Yang, D.: Tailor-made core/shell/shell-like fe3o4@ sio2@ ppy composites with prominent microwave absorption performance. J. Alloys Compounds **779**, 831–843 (2019)
76. Zhang, N., Li, J., Men, X., Li, Z., Zhao, H.: Cleaning synthesis of core-shell structured ni@ ppy composite as excellent lightweight electromagnetic wave absorber. J. Mater. Sci. Mater. Electron. **31**(2), 1483–1490 (2020)
77. Zhou, W., Hu, X., Bai, X., Zhou, S., Sun, C., Yan, J., Chen, P.: Synthesis and electromagnetic, microwave absorbing properties of core-shell fe3o4-poly (3, 4-ethylenedioxythiophene) microspheres. ACS Appl. Mater. Interfaces **3**(10), 3839–3845 (2011)
78. Bora, P.J., Gupta, S., Pecharsky, V.K., Vinoy, K., Ramamurthy, P.C., Hadimani, R.L.: Enhancement of microwave absorption bandwidth of polymer blend using ferromagnetic gadolinium silicide nanoparticles. Mater. Lett. **252**, 178–181 (2019)
79. Yan, J., Huang, Y., Zhou, S., Han, X., Liu, P.: Preparation and microwave absorption properties of nanomesh poly (3, 4-ethylenedioxythiophene) covalently functionalized graphene oxide. J. Mater. Sci. Mater. Electron. **30**(5), 5273–5283 (2019)
80. Wang, X., Shu, J.-C., He, X.-M., Zhang, M., Wang, X.-X., Gao, C., Yuan, J., Cao, M.-S.: Green approach to conductive pedot: Pss decorating magnetic-graphene to recover conductivity for highly efficient absorption. ACS Sustain. Chem. Eng. **6**(11), 14017–14025 (2018)
81. Zhang, F., Johansson, M., Andersson, M.R., Hummelen, J.C., Inganäs, O.: Polymer photovoltaic cells with conducting polymer anodes. Adv. Mater. **14**(9), 662–665 (2002)
82. Sun, K., Zhang, S., Li, P., Xia, Y., Zhang, X., Du, D., Isikgor, F.H., Ouyang, J.: Review on application of pedots and pedot: Pss in energy conversion and storage devices. J. Mater. Sci. Mater. Electron. **26**(7), 4438–4462 (2015)
83. Du, Y., Li, H., Jia, X., Dou, Y., Xu, J., Eklund, P.: Preparation and thermoelectric properties of graphite/poly (3, 4-ethyenedioxythiophene) nanocomposites. Energies **11**(10), 2849 (2018)
84. Gao, X., Zu, L., Cai, X., Li, C., Lian, H., Liu, Y., Wang, X., Cui, X.: High performance of supercapacitor from pedot: Pss electrode and redox iodide ion electrolyte. Nanomaterials **8**(5), 335 (2018)
85. Farukh, M., Singh, A.P., Dhawan, S.: Enhanced electromagnetic shielding behavior of multi-walled carbon nanotube entrenched poly (3, 4-ethylenedioxythiophene) nanocomposites. Compos. Sci. Technol. **114**, 94–102 (2015)
86. Agnihotri, N., Chakrabarti, K., De, A.: Highly efficient electromagnetic interference shielding using graphite nanoplatelet/poly (3, 4-ethylenedioxythiophene)-poly (styrenesulfonate) composites with enhanced thermal conductivity. RSC Adv. **5**(54), 43765–43771 (2015)
87. Liu, P., Huang, Y., Zhang, X.: Preparation and excellent microwave absorption properties of ferromagnetic graphene/poly (3, 4-ethylenedioxythiophene)/cofe2o4 nanocomposites. Powder Technology **276**, 112–117 (2015)
88. Bai, X., Hu, X., Zhou, S., Li, L., Rohwerder, M.: Controllable synthesis of leaflet-like poly (3, 4-ethylenedioxythiophene)/single-walled carbon nanotube composites with microwave absorbing property. Compos. Sci. Technol. **110**, 166–175 (2015)
89. Luo, S.-J., Zhang, P., Mei, Y.-A., Chang, J.-B., Yan, H.: Electromagnetic interference shielding properties of pedot/pss–halloysite nanotube (hnts) hybrid films. J. Appl. Polym. Sci. 133 (47)
90. Li, P., Du, D., Guo, L., Guo, Y., Ouyang, J.: Stretchable and conductive polymer films for high-performance electromagnetic interference shielding. J. Mater. Chem. C **4**(27), 6525–6532 (2016)

91. K. K. Khanum, P. J. Bora, K. Vinoy, P. C. Ramamurthy, Evaluation of electromagnetic interference shielding using poly (3, 4-ethylenedioxythiophene) polystyrene sulfonate blend. In: 2016 3rd International Conference on Emerging Electronics (ICEE), pp. 1–4. IEEE (2016)
92. Dalal, J., Gupta, A., Lather, S., Singh, K., Dhawan, S., Ohlan, A.: Poly (3, 4-ethylene dioxythiophene) laminated reduced graphene oxide composites for effective electromagnetic interference shielding. J. Alloys Compounds **682**, 52–60 (2016)
93. Yan, L., Wang, X., Zhao, S., Li, Y., Gao, Z., Zhang, B., Cao, M., Qin, Y.: Highly efficient microwave absorption of magnetic nanospindle-conductive polymer hybrids by molecular layer deposition. ACS Appl. Mater. Interfaces **9**(12), 11116–11125 (2017)
94. Xia, Y., Fang, J., Li, P., Zhang, B., Yao, H., Chen, J., Ding, J., Ouyang, J.: Solution-processed highly superparamagnetic and conductive pedot: Pss/fe3o4 nanocomposite films with high transparency and high mechanical flexibility. ACS Appl. Mater. Interfaces **9**(22), 19001–19010 (2017)
95. Ji, S., Zhang, Z., Jiang, X., Yu, L.: Synthesis and microwave absorbing properties of γ-fe 2 o 3-sio 2-poly (3, 4-ethylenedioxythiophene) core-shell-shell nanocomposites. J. Mater. Sci. **52**(20), 12358–12369 (2017)
96. Dalal, J., Lather, S., Gupta, A., Dahiya, S., Maan, A., Singh, K., Dhawan, S., Ohlan, A.: Emi shielding properties of laminated graphene and pbtio3 reinforced poly (3, 4-ethylenedioxythiophene) nanocomposites. Compos. Sci. Technol. **165**, 222–230 (2018)
97. Ji, S., Li, C., Zhang, Z., Jiang, X., Yu, L.: Hollow γ-fe2o3@ poly (3, 4-ethylenedioxythiophene) versus γ-fe2o3@ sio2@ poly (3, 4-ethylenedioxythiophene) core-shell structures for highly effective microwave absorption. Synth. Metals **239**, 59–65 (2018)
98. Bora, P.J., Anil, A.G., Vinoy, K.J., Ramamurthy, P.C.: Outstanding absolute electromagnetic interference shielding effectiveness of cross-linked pedot: Pss film. Adv. Mater. Interfaces **6**(22), 1901353 (2019)
99. Dalal, J., Lather, S., Gupta, A., Tripathi, R., Maan, A.S., Singh, K., Ohlan, A.: Reduced graphene oxide functionalized strontium ferrite in poly (3, 4-ethylenedioxythiophene) conducting network: a high-performance emi shielding material. Adv. Mater. Technol. **4**(7), 1900023 (2019)
100. Shukla, V.: Review of electromagnetic interference shielding materials fabricated by iron ingredients. Nanoscale Adv. **1**(5), 1640–1671 (2019)
101. Raagulan, K., Braveenth, R., Jang, H.J., Seon Lee, Y., Yang, C.-M., Mi Kim, B., Moon, J.J., Chai, K.Y.: Electromagnetic shielding by mxene-graphene-pvdf composite with hydrophobic, lightweight and flexible graphene coated fabric. Materials **11**(10), 1803 (2018)
102. Rengaswamy, K., Sakthivel, D.K., Muthukaruppan, A., Natesan, B., Venkatachalam, S., Kannaiyan, D.: Electromagnetic interference (emi) shielding performance of lightweight metal decorated carbon nanostructures dispersed in flexible polyvinylidene fluoride films. New J. Chem. **42**(15), 12945–12953 (2018)
103. Dutta, B., Kar, E., Sen, G., Bose, N., Mukherjee, S.: Lightweight, flexible nio@ sio2/pvdf nanocomposite film for uv protection and emi shielding application. Mater. Res. Bull. **124**, 110746 (2020)
104. Pawar, S.P., Kumar, S., Jain, S., Gandi, M., Chatterjee, K., Bose, S.: Synergistic interactions between silver decorated graphene and carbon nanotubes yield flexible composites to attenuate electromagnetic radiation. Nanotechnology **28**(2), 025201 (2016)
105. Kumaran, R., Alagar, M., Dinesh Kumar, S., Subramanian, V., Dinakaran, K.: Ag induced electromagnetic interference shielding of ag-graphite/pvdf flexible nanocomposites thinfilms. Appl. Phys. Lett. **107**(11), 113107 (2015)
106. Guo, A.-P., Zhang, X.-J., Qu, J.-K., Wang, S.-W., Zhu, J.-Q., Wang, G.-S., Guo, L.: Improved microwave absorption and electromagnetic interference shielding properties based on graphene-barium titanate and polyvinylidene fluoride with varying content. Mater. Chem. Front. **1**(12), 2519–2526 (2017)
107. Zhao, B., Park, C.B.: Tunable electromagnetic shielding properties of conductive poly (vinylidene fluoride)/ni chain composite films with negative permittivity. J. Mater. Chem. C **5**(28), 6954–6961 (2017)

108. Acharya, S., Gopinath, C.S., Alegaonkar, P., Datar, S.: Enhanced microwave absorption property of reduced graphene oxide. Diamond Related Mater. **89**, 28–34 (2018)
109. Li, X., Zeng, S., Liang, S.E.L., Bai, Z., Zhou, Y., Zhao, B., Zhang, R.: Quick heat dissipation in absorption-dominated microwave shielding properties of flexible poly (vinylidene fluoride)/carbon nanotube/co composite films with anisotropy-shaped co (flowers or chains). ACS Appl. Mater. Interfaces **10**(47), 40789–40799 (2018)
110. Kumar, G.S., Patro, T.U.: Efficient electromagnetic interference shielding and radar absorbing properties of ultrathin and flexible polymer-carbon nanotube composite films. Mater. Res. Exp. **5**(11), 115304 (2018)
111. Wang, L., Li, Z., Wan, J.-Q., Lin, C.-H., Cai, Y.-L., Wang, G.-S.: Pure organic polymer composite with enhanced microwave absorption performance based on polyaniline and polyvinylidene fluoride. Sci. Adv. Mater. **10**(8), 1125–1132 (2018)
112. Halder, K.K., Tomar, M., Sachdev, V., Gupta, V.: To study the effect of mwcnt incorporated into pvdf-graphite composites for emi shielding applications. Mater. Today Proc. **5**(7), 15348–15353 (2018)
113. Kamal Halder, K., Sonker, R.K., Sachdev, V., Tomar, M., Gupta, V.: Study of electrical, dielectric and emi shielding behavior of copper metal, copper ferrite and pvdf composite. Integr. Ferroelectr. **194**(1), 80–87 (2018)
114. Halder, K.K., Tomar, M., Sachdev, V., Gupta, V.: Carbonized charcoal-loaded pvdf polymer composite: A promising emi shielding material. Arab. J. Sci. Eng. 1–10 (2019)
115. Li, C.-Q., Shen, X., Ding, R.-C., Wang, G.-S.: Controllable synthesis of one-dimensional moo3/mos2 hybrid composites with their enhanced efficient electromagnetic wave absorption properties. ChemPlusChem **84**(2), 226 232 (2019)
116. Acharya, S., Alegaonkar, P., Datar, S.: Effect of formation of heterostructure of sral4fe8o19/rgo/pvdf on the microwave absorption properties of the composite. Chem. Eng. J. **374**, 144–154 (2019)
117. Wu, Q., Wu, J., Wang, G.-S., Zhang, H.-Z., Gao, H.-J., Shui, W.-J.: Enhanced wave absorption and mechanical properties of cobalt sulfide/pvdf composite materials. Sci. Rep. **9**(1), 1–12 (2019)
118. Li, C.-Q., Shen, X., Ding, R.-C., Wang, G.-S.: Excellent microwave absorption properties based on a composite of one dimensional mo 2 c@ c nanorods and a pvdf matrix. RSC Adv. **9**(37), 21243–21248 (2019)
119. Adebayo, L.L., Soleimani, H., Yahya, N., Abbas, Z., Ridwan, A.T., Wahaab, F.A.: Investigation of the broadband microwave absorption of citric acid coated fe3o4/pvdf composite using finite element method. Appl. Sci. **9**(18), 3877 (2019)
120. Lv, S.-Q., Xu, W., Huang, W., Lv, G.-C., Wang, G.-S.: Microwave absorption enhancement by adjusting reactant ratios and filler contents based on 1d k-mno 2@ pda and poly (vinylidene fluoride) matrix. RSC Adv. **9**(23), 13088–13095 (2019)
121. Sutradhar, S., Saha, P., Chowdhury, A., Das, S.: Reduction of electromagnetic pollution by the enhancement of microwave absorption of strontium hexaferrite functionalized poly (vinylidene fluoride) composite film. Mater. Res. Exp. **6**(8), 086424 (2019)
122. V. Lalan, A. Puthiyedath Narayanan, K. P. Surendran, S. Ganesanpotti, Room-temperature ferromagnetic sr3yco4o10+ δ and carbon black-reinforced polyvinylidenefluoride composites toward high-performance electromagnetic interference shielding. ACS Omega 4 (5) (2019) 8196–8206
123. Lalan, V., Ganesanpotti, S.: Broadband electromagnetic response and enhanced microwave absorption in carbon black and magnetic fe 3 o 4 nanoparticles reinforced polyvinylidenefluoride composites. J. Electron. Mater. 1–11 (2019)
124. Sutradhar, S., Saha, S., Javed, S.: Shielding effectiveness study of barium hexaferrite-incorporated, β-phase-improved poly (vinylidene fluoride) composite film: a metamaterial useful for the reduction of electromagnetic pollution. ACS Appl. Mater. Interfaces **11**(26), 23701–23713 (2019)
125. Saha, P., Debnath, T., Das, S., Chatterjee, S., Sutradhar, S.: β-phase improved mn-zn-cu-ferrite-pvdf nanocomposite film: a metamaterial for enhanced microwave absorption. Mater. Sci. Eng. B **245**, 17–29 (2019)

126. Mei, X., Lu, L., Xie, Y., Wang, W., Tang, Y., Teh, K.S.: An ultra-thin carbon-fabric/graphene/poly (vinylidene fluoride) film for enhanced electromagnetic interference shielding. Nanoscale **11**(28), 13587–13599 (2019)
127. Zhao, B., Zhang, X., Deng, J., Zhang, C., Li, Y., Guo, X., Zhang, R.: Flexible pebax/graphene electromagnetic shielding composite films with a negative pressure effect of resistance for pressure sensors applications. RSC Adv. **10**(3), 1535–1543 (2020)
128. Li, Y., Duan, Y., Wang, C.: Enhanced microwave absorption and electromagnetic properties of si-modified rgo@ fe3o4/pvdf-co-hfp composites. Materials **13**(4), 933 (2020)
129. Zhang, H., Zhang, G., Gao, Q., Tang, M., Ma, Z., Qin, J., Wang, M., Kim, J.-K.: Multifunctional microcellular pvdf/ni-chains composite foams with enhanced electromagnetic interference shielding and superior thermal insulation performance. Chem. Eng. J. **379**, 122304 (2020)
130. Adebayo, L.L., Soleimani, H., Yahya, N., Abbas, Z., Sabet, M., Wahaab, F.A., Ayinla, R.T.: Facile preparation and enhanced electromagnetic wave absorption properties of fe3o4@ pvdf nanocomposite. J. Mater. Res. Technol. **9**(2), 2513–2521 (2020)
131. Zoughi, R., Arias-Monje, P.J., Gallion, J., Sarkar, S., Wang, P.-H., Gulgunje, P., Verghese, N., Kumar, S.: Microwave dielectric properties and targeted heating of polypropylene nanocomposites containing carbon nanotubes and carbon black. Polymer **179**, 121658 (2019)
132. He, Q., Yuan, T., Zhang, X., Yan, X., Guo, J., Ding, D., Khan, M.A., Young, D.P., Khasanov, A., Luo, Z., et al.: Electromagnetic field absorbing polypropylene nanocomposites with tuned permittivity and permeability by nanoiron and carbon nanotubes. J. Phys. Chem. C **118**(42), 24784–24796 (2014)
133. Rezania, J., Rahimi, H.: Investigating the carbon materials' microwave absorption and its effects on the mechanical and physical properties of carbon fiber and carbon black/polypropylene composites. J. Compos. Mater. **51**(16), 2263–2276 (2017)
134. Cheng, C., Jiang, Y., Sun, X., Shen, J., Wang, T., Fan, G., Fan, R.: Tunable negative permittivity behavior and electromagnetic shielding performance of silver/silicon nitride metacomposites. Compos. Part A Appl. Sci. Manuf. **130**, 105753 (2020)
135. Soares, B.G., Cordeiro, E., Maia, J., Pereira, E.C., Silva, A.A.: The effect of the noncovalent functionalization of cnt by ionic liquid on electrical conductivity and electromagnetic interference shielding effectiveness of semi-biodegradable polypropylene/poly (lactic acid) composites. Polym. Compos. **41**(1), 82–93 (2020)
136. He, H., Cheng, S., Lian, Y., Xing, Y., He, G., Huang, Z., Wu, M.: Electrical conductivity and electromagnetic interference shielding effectiveness of carbon black/sisal fiber/polyamide/polypropylene composites. J. Appl. Polym. Sci. **132**(46)
137. King, J.A., Pisani, W.A., Klimek-McDonald, D.R., Perger, W.F., Odegard, G.M., Turpeinen, D.G.: Shielding effectiveness of carbon-filled polypropylene composites. J. Compos. Mater. **50**(16), 2177–2189 (2016)
138. Lee, S.H., Kim, J.Y., Koo, C.M., Kim, W.N.: Effects of processing methods on the electrical conductivity, electromagnetic parameters, and emi shielding effectiveness of polypropylene/nickel-coated carbon fiber composites. Macromolec. Res. **25**(9), 936–943 (2017)
139. Lee, S.H., Lee, Y., Jang, M.G., Han, C., Kim, W.N.: Comparative study of emi shielding effectiveness for carbon fiber pultruded polypropylene/poly (lactic acid)/multiwall cnt composites prepared by injection molding versus screw extrusion. J. Appl. Polym. Sci. **134**(34), 45222 (2017)
140. Duan, H., Zhao, M., Yang, Y., Zhao, G., Liu, Y.: Flexible and conductive pp/epdm/ni coated glass fiber composite for efficient electromagnetic interference shielding. J. Mater. Sci. Mater. Electron. **29**(12), 10329–10336 (2018)
141. Wu, H.-Y., Zhang, Y.-P., Jia, L.-C., Yan, D.-X., Gao, J.-F., Li, Z.-M.: Injection molded segregated carbon nanotube/polypropylene composite for efficient electromagnetic interference shielding. Indust. Eng. Chem. Res. **57**(37), 12378–12385 (2018)
142. Wu, H.-Y., Jia, L.-C., Yan, D.-X., Gao, J.-F., Zhang, X.-P., Ren, P.-G., Li, Z.-M.: Simultaneously improved electromagnetic interference shielding and mechanical performance of

segregated carbon nanotube/polypropylene composite via solid phase molding. Compos. Sci. Technol. **156**, 87–94 (2018)
143. Gao, J., Luo, J., Wang, L., Huang, X., Wang, H., Song, X., Hu, M., Tang, L.-C., Xue, H.: Flexible, superhydrophobic and highly conductive composite based on non-woven polypropylene fabric for electromagnetic interference shielding. Chem. Eng. J. **364**, 493–502 (2019)
144. Sadeghi, A., Moeini, R., Yeganeh, J.K.: Highly conductive pp/pet polymer blends with high electromagnetic interference shielding performances in the presence of thermally reduced graphene nanosheets prepared through melt compounding. Polym. Compos. **40**(S2), E1461–E1469 (2019)
145. Liu, C., Liu, J., Ning, X., Chen, S., Liu, Z., Jiang, S., Miao, D.: The effect of polydopamine on an ag-coated polypropylene nonwoven fabric. Polymers **11**(4), 627 (2019)
146. Yadav, R.S., Kuřitka, I., Vilčáková, J., Machovský, M., Škoda, D., Urbánek, P., Masař, M., Gořalik, M., Urbánek, M., Kalina, L., et al.: Polypropylene nanocomposite filled with spinel ferrite nife2o4 nanoparticles and in-situ thermally-reduced graphene oxide for electromagnetic interference shielding application. Nanomaterials **9**(4), 621 (2019)
147. Zhang, Y.-P., Zhou, C.-G., Sun, W.-J., Wang, T., Jia, L.-C., Yan, D.-X., Li, Z.-M.: Injection molding of segregated carbon nanotube/polypropylene composite with enhanced electromagnetic interference shielding and mechanical performance. Compos. Sci. Technol. **197**, 108253 (2020)
148. Mondal, S., Ganguly, S., Das, P., Khastgir, D., Das, N.C.: Low percolation threshold and electromagnetic shielding effectiveness of nano-structured carbon based ethylene methyl acrylate nanocomposites. Compos. Part B Eng. **119**, 41–56 (2017)
149. Kuester, S., Demarquette, N.R., Ferreira Jr., J.C., Soares, B.G., Barra, G.M.: Hybrid nanocomposites of thermoplastic elastomer and carbon nanoadditives for electromagnetic shielding. Eur. Polym. J. **88**, 328–339 (2017)
150. Sankaran, S., Deshmukh, K., Ahamed, M.B., Pasha, S.K.: Recent advances in electromagnetic interference shielding properties of metal and carbon filler reinforced flexible polymer composites: a review. Compos. Part A Appl. Sci. Manuf. **114**, 49–71 (2018)
151. Sharma, M., Sharma, S., Abraham, J., Thomas, S., Madras, G., Bose, S.: Flexible emi shielding materials derived by melt blending pvdf and ionic liquid modified mwnts. Mater. Res. Exp. **1**(3), 035003 (2014)
152. Hosseini, S.H., Mohseni, S., Asadnia, A., Kerdari, H.: Synthesis and microwave absorbing properties of polyaniline/mnfe2o4 nanocomposite. J. Alloys Compounds **509**(14), 4682–4687 (2011)
153. Wang, Y., Zhang, W., Luo, C., Wu, X., Yan, G.: Superparamagnetic feco@ sno2 nanoparticles on graphene-polyaniline: synthesis and enhanced electromagnetic wave absorption properties. Ceram. Int. **42**(10), 12496–12502 (2016)
154. Long, Z., Xiaona, W., Wenjing, D., Hu, Q., Jieqiong, H., Xiaorui, W., Hui, L., Xueyan, D.: Synthesis and microwave absorption performance of fe3o4@ ppy@ pani composites. Chem. J. Chin. Univers.-Chin. **39**(1), 185–192 (2018)
155. Yan, J., Huang, Y., Chen, X., Wei, C.: Conducting polymers-nife2o4 coated on reduced graphene oxide sheets as electromagnetic (em) wave absorption materials. Synth. Metals **221**, 291–298 (2016)
156. Thi, Q.V., Lim, S., Jang, E., Kim, J., Van Khoi, N., Tung, N.T., Sohn, D.: Silica particles wrapped with poly (aniline-co-pyrrole) and reduced graphene oxide for advanced microwave absorption. Mater. Chem. Phys. **122691** (2020)
157. Bora, P.J., Vinoy, K., Ramamurthy, P.C., Madras, G.: Electromagnetic interference shielding efficiency of mno2 nanorod doped polyaniline film. Mater. Res. Exp. **4**(2), 025013 (2017)
158. Shukla, V.: Observation of critical magnetic behavior in 2d carbon based composites. Nanoscale Adv. **2**, 962–990 (2020)
159. Singh, A.P., Mishra, M., Chandra, A., Dhawan, S.: Graphene oxide/ferrofluid/cement composites for electromagnetic interference shielding application. Nanotechnology **22**(46), 465701 (2011)

160. Chen, J., Zhao, D., Ge, H., Wang, J.: Graphene oxide-deposited carbon fiber/cement composites for electromagnetic interference shielding application. Constr. Build. Mater. **84**, 66–72 (2015)
161. Yan, D.-X., Pang, H., Xu, L., Bao, Y., Ren, P.-G., Lei, J., Li, Z.-M.: Electromagnetic interference shielding of segregated polymer composite with an ultralow loading of in situ thermally reduced graphene oxide. Nanotechnology **25**(14), 145705 (2014)
162. Yan, J., Huang, Y., Wei, C., Zhang, N., Liu, P.: Covalently bonded polyaniline/graphene composites as high-performance electromagnetic (em) wave absorption materials. Compos. Part Appl. Sci. Manuf. **99**, 121–128 (2017)
163. Cui, C.-H., Yan, D.-X., Pang, H., Jia, L.-C., Bao, Y., Jiang, X., Li, Z.-M.: Towards efficient electromagnetic interference shielding performance for polyethylene composites by structuring segregated carbon black/graphite networks. Chin. J. Polym. Sci. **34**(12), 1490–1499 (2016)
164. Chen, K., Xiang, C., Li, L., Qian, H., Xiao, Q., Xu, F.: A novel ternary composite: fabrication, performance and application of expanded graphite/polyaniline/cofe 2 o 4 ferrite. J. Mater. Chem. **22**(13), 6449–6455 (2012)
165. Liu, Z., Bai, G., Huang, Y., Li, F., Ma, Y., Guo, T., He, X., Lin, X., Gao, H., Chen, Y.: Microwave absorption of single-walled carbon nanotubes/soluble cross-linked polyurethane composites. J. Phys. Chem. C **111**(37), 13696–13700 (2007)
166. Abraham, J., Xavier, P., Bose, S., George, S.C., Kalarikkal, N., Thomas, S., et al.: Investigation into dielectric behaviour and electromagnetic interference shielding effectiveness of conducting styrene butadiene rubber composites containing ionic liquid modified mwcnt. Polymer **112**, 102–115 (2017)
167. Lee, Y.S., Park, Y.H., Yoon, K.H.: Flexural, electrical, thermal and electromagnetic interference shielding properties of xgnp and carbon nanotube filled epoxy hybrid nanocomposites. Carbon Lett. (Carbon Lett.) **24**, 41–46 (2017)
168. Liu, C., Yu, C., Sang, G., Xu, P., Ding, Y.: Improvement in emi shielding properties of silicone rubber/poe blends containing ils modified with carbon black and mwcnts. Appl. Sci. **9**(9), 1774 (2019)
169. Jani, R.K., Patra, M.K., Saini, L., Shukla, A., Singh, C.P., Vadera, S.R.: Tuning of microwave absorption properties and electromagnetic interference (emi) shielding effectiveness of nanosize conducting black-silicone rubber composites over 8–18 ghz. Progr. Electromagnet. Res. **58**, 193–204 (2017)
170. Wang, R., Yang, H., Wang, J., Li, F.: The electromagnetic interference shielding of silicone rubber filled with nickel coated carbon fiber. Polym. Test. **38**, 53–56 (2014)
171. Liu, P., Huang, Y., Zhang, X.: Cubic nife2o4 particles on graphene-polyaniline and their enhanced microwave absorption properties. Compos. Sci. Technol. **107**, 54–60 (2015)
172. Shahzad, F., Yu, S., Kumar, P., Lee, J.-W., Kim, Y.-H., Hong, S.M., Koo, C.M.: Sulfur doped graphene/polystyrene nanocomposites for electromagnetic interference shielding. Compos. Struct. **133**, 1267–1275 (2015)
173. Fang, J., Chen, Z., Wei, W., Li, Y., Liu, T., Liu, Z., Yue, X., Jiang, Z.: A carbon fiber based three-phase heterostructure composite cf/co 0.2 fe 2.8 o 4/pani as an efficient electromagnetic wave absorber in the k u band, RSC Adv. **5**(62) (2015) 50024–50032
174. Chen, X., Chen, J., Meng, F., Shan, L., Jiang, M., Xu, X., Lu, J., Wang, Y., Zhou, Z.: Hierarchical composites of polypyrrole/graphene oxide synthesized by in situ intercalation polymerization for high efficiency and broadband responses of electromagnetic absorption. Compos. Sci. Technol. **127**, 71–78 (2016)
175. Chen, X., Wang, X., Li, L., Qi, S.: Preparation and microwave absorbing properties of nickel-coated carbon fiber with polyaniline via in situ polymerization. J. Mater. Sci. Mater. Electron. **27**(6), 5607–5612 (2016)
176. Wang, Y., Wu, X., Zhang, W., Luo, C., Li, J., Wang, Q.: 3d heterostructure of graphene@ fe3o4@ wo3@ pani: preparation and excellent microwave absorption performance. Synth. Metals **231**, 7–14 (2017)

177. Jia, Q., Wang, W., Zhao, J., Xiao, J., Lu, L., Fan, H.: Synthesis and characterization of tio2/polyaniline/graphene oxide bouquet-like composites for enhanced microwave absorption performance. J. Alloys Compounds **710**, 717–724 (2017)
178. Han, S., Wang, S., Li, W., Lai, Y., Zhang, N., Yang, N., Wang, Q., Jiang, W.: Synthesis of ppy/ni/rgo and enhancement on its electromagnetic wave absorption performance. Ceram. Int. **44**(9), 10352–10361 (2018)
179. Chen, X., Zhong, K., Shi, T., Meng, X., Wu, G., Lu, Y.: Urchin-like polyaniline/magnetic carbon sphere hybrid with excellent electromagnetic wave absorption performance. Synth. Metals **248**, 59–67 (2019)
180. Zhang, K., Chen, X., Gao, X., Chen, L., Ma, S., Xie, C., Zhang, X., Lu, W.: Preparation and microwave absorption properties of carbon nanotubes/iron oxide/polypyrrole/carbon composites. Synth. Metals **260**, 116282 (2020)
181. Li, J., Ji, H., Xu, Y., Zhang, J., Yan, Y.: Three-dimensional graphene supported fe3o4 coated by polypyrrole toward enhanced stability and microwave absorbing properties. J. Mater. Res. Technol. **9**(1), 762–772 (2020)
182. Gill, N., Gupta, V., Tomar, M., Sharma, A.L., Pandey, O., Singh, D.P.: Improved electromagnetic shielding behaviour of graphene encapsulated polypyrrole-graphene nanocomposite in x-band. Compos. Sci. Technol. **192**, 108113 (2020)
183. Saboor, A., Khalid, S.M., Jan, R., Khan, A.N., Zia, T., Farooq, M.U., Afridi, S., Sadiq, M., Arif, M.: Ps/pani/mos2 hybrid polymer composites with high dielectric behavior and electrical conductivity for emi shielding effectiveness. Materials **12**(17), 2690 (2019)
184. Guo, A.-P., Zhang, X.-J., Wang, S.-W., Zhu, J.-Q., Yang, L., Wang, G.-S.: Excellent microwave absorption and electromagnetic interference shielding based on reduced graphene oxide@mos_2/poly (vinylidene fluoride) composites. ChemPlusChem **81**(12), 1305 (2016)
185. Zhang, X.-J., Wang, S.-W., Wang, G.-S., Li, Z., Guo, A.-P., Zhu, J.-Q., Liu, D.-P., Yin, P.-G.: Facile synthesis of nis 2@ mos 2 core-shell nanospheres for effective enhancement in microwave absorption. RSC Adv. **7**(36), 22454–22460 (2017)
186. Zhang, W.L., Jiang, D., Wang, X., Hao, B.N., Liu, Y.D., Liu, J.: Growth of polyaniline nanoneedles on mos2 nanosheets, tunable electroresponse, and electromagnetic wave attenuation analysis. J. Phys. Chem. C **121**(9), 4989–4998 (2017)
187. Zhang, W., Li, L., Zhu, W., Yan, H., Qi, S.: Preparation and microwave absorbing performance of mos 2@ fe 3 o 4@ pani composites. J. Mater. Sci. Mater. Electron. **28**(20), 15488–15494 (2017)
188. An, D., Bai, L., Cheng, S., Zhang, Z., Duan, X., Sun, Z., Liu, Y.: Synthesis and electromagnetic wave absorption properties of three-dimensional nano-flower structure of mos 2/polyaniline nanocomposites. J. Mater. Sci. Mater. Electron. **30**(15), 13948–13956 (2019)
189. Yang, J., Ye, M., Han, A., Zhang, Y., Zhang, K.: Preparation and electromagnetic attenuation properties of mos 2-pani composites: a promising broadband absorbing material. J. Mater. Sci. Mater. Electron. **30**(1), 292–301 (2019)
190. Gai, L., Zhao, Y., Song, G., An, Q., Xiao, Z., Zhai, S., Li, Z.: Construction of core-shell ppy@ mos2 with nanotube-like heterostructures for electromagnetic wave absorption: assembly and enhanced mechanism. Applied Science and Manufacturing, Composites Part A, p. 105965 (2020)
191. Chen, X., Shi, T., Wu, G., Lu, Y.: Design of molybdenum disulfide@ polypyrrole compsite decorated with fe3o4 and superior electromagnetic wave absorption performance. J. Colloid Interf. Sci. **572**, 227–235 (2020)
192. Ma, J., Ren, H., Liu, Z., Zhou, J., Wang, Y., Hu, B., Liu, Y., Kong, L.B., Zhang, T.: Embedded mos2-pani nanocomposites with advanced microwave absorption performance. Compos. Sci. Technol. **108239** (2020)
193. Menon, A.V., Madras, G., Bose, S.: Mussel-inspired self-healing polyurethane with "flower-like" magnetic mos2 as efficient microwave absorbers. ACS Appl. Polym. Mater. **1**(9), 2417–2429 (2019)
194. Li, X., Yin, X., Xu, H., Han, M., Li, M., Liang, S., Cheng, L., Zhang, L.: Ultralight mxene-coated, interconnected sicnws three-dimensional lamellar foams for efficient microwave absorption in the x-band. ACS Appl. Mater. Interfaces **10**(40), 34524–34533 (2018)

195. Cao, M.-S., Cai, Y.-Z., He, P., Shu, J.-C., Cao, W.-Q., Yuan, J.: 2d mxenes: electromagnetic property for microwave absorption and electromagnetic interference shielding. Chem. Eng. J. **359**, 1265–1302 (2019)
196. Song, D., Li, X., Li, X.-P., Jia, X., Min, P., Yu, Z.-Z.: Hollow-structured mxene-pdms composites as flexible, wearable and highly bendable sensors with wide working range. J. Colloid Interface Sci. **555**, 751–758 (2019)
197. Shahzad, F., Alhabeb, M., Hatter, C.B., Anasori, B., Hong, S.M., Koo, C.M., Gogotsi, Y.: Electromagnetic interference shielding with 2d transition metal carbides (mxenes). Science **353**(6304), 1137–1140 (2016)
198. Sun, R., Zhang, H.-B., Liu, J., Xie, X., Yang, R., Li, Y., Hong, S., Yu, Z.-Z.: Highly conductive transition metal carbide/carbonitride (mxene)@ polystyrene nanocomposites fabricated by electrostatic assembly for highly efficient electromagnetic interference shielding. Adv. Funct. Mater. **27**(45), 1702807 (2017)
199. Liu, T., Liu, N., An, Q., Xiao, Z., Zhai, S., Li, Z.: Designed construction of ti3c2tx@ ppy composites with enhanced microwave absorption performance. J. Alloys Compounds **802**(25), 445–457 (2019)
200. Zhang, Y., Wang, L., Zhang, J., Song, P., Xiao, Z., Liang, C., Qiu, H., Kong, J., Gu, J.: Fabrication and investigation on the ultra-thin and flexible ti3c2tx/co-doped polyaniline electromagnetic interference shielding composite films. Compos. Sci. Technol. **183**, 107833 (2019)
201. Wang, L., Chen, L., Song, P., Liang, C., Lu, Y., Qiu, H., Zhang, Y., Kong, J., Gu, J.: Fabrication on the annealed ti3c2tx mxene/epoxy nanocomposites for electromagnetic interference shielding application. Compos. Part B: Eng. **171**, 111–118 (2019)
202. Xu, H., Yin, X., Li, X., Li, M., Liang, S., Zhang, L., Cheng, L.: Lightweight ti2ct x mxene/poly (vinyl alcohol) composite foams for electromagnetic wave shielding with absorption-dominated feature. ACS Appl. Mater. Interfaces **11**(10), 10198–10207 (2019)
203. Wang, Q.-W., Zhang, H.-B., Liu, J., Zhao, S., Xie, X., Liu, L., Yang, R., Koratkar, N., Yu, Z.-Z.: Multifunctional and water-resistant mxene-decorated polyester textiles with outstanding electromagnetic interference shielding and joule heating performances. Adv. Funct. Mater. **29**(7), 1806819 (2019)
204. Kumar, S., Kumar, P., Singh, N., Verma, V., et al.: Steady microwave absorption behavior of two-dimensional metal carbide mxene and polyaniline composite in x-band. J. Magnet. Magn. Mater. **488**, 165364 (2019)
205. Wang, Y., Han, X., Xu, P., Liu, D., Cui, L., Zhao, H., Du, Y.: Synthesis of pomegranate-like mo2c@ c nanospheres for highly efficient microwave absorption. Chem. Eng. J. **372**, 312–320 (2019)
206. Wei, H., Dong, J., Fang, X., Zheng, W., Sun, Y., Qian, Y., Jiang, Z., Huang, Y.: Ti3c2tx mxene/polyaniline (pani) sandwich intercalation structure composites constructed for microwave absorption. Compos. Sci. Technol. **169**, 52–59 (2019)
207. Luo, J.-Q., Zhao, S., Zhang, H.-B., Deng, Z., Li, L., Yu, Z.-Z.: Flexible, stretchable and electrically conductive mxene/natural rubber nanocomposite films for efficient electromagnetic interference shielding. Compos. Sci. Technol. **182**, 107754 (2019)
208. Lei, Y., Yao, Z., Li, S., Zhou, J., Haidry, A.A., Liu, P.: Broadband high-performance electromagnetic wave absorption of co-doped nizn ferrite/polyaniline on mxenes. Ceram. Int. **46**(8), 10006–10015 (2020)
209. Wu, X., Han, B., Zhang, H.-B., Xie, X., Tu, T., Zhang, Y., Dai, Y., Yang, R., Yu, Z.-Z.: Compressible, durable and conductive polydimethylsiloxane-coated mxene foams for high-performance electromagnetic interference shielding. Chem. Eng. J. **381**, 122622 (2020)
210. Lu, S., Li, B., Ma, K., Wang, S., Liu, X., Ma, Z., Lin, L., Zhou, G., Zhang, D.: Flexible mxene/epdm rubber with excellent thermal conductivity and electromagnetic interference performance. Appl. Phys. A **126**(7), 1–12 (2020)
211. Song, P., Qiu, H., Wang, L., Liu, X., Zhang, Y., Zhang, J., Kong, J., Gu, J.: Honeycomb structural rgo-mxene/epoxy nanocomposites for superior electromagnetic interference shielding performance. Sustain. Mater. Technol. **24**, e00153 (2020)

212. Wan, Y.-J., Li, X.-M., Zhu, P.-L., Sun, R., Wong, C.-P., Liao, W.-H.: Lightweight, flexible mxene/polymer film with simultaneously excellent mechanical property and high-performance electromagnetic interference shielding. Compos. Part A Appl. Sci. Manuf. **130**, 105764 (2020)
213. Jin, X., Wang, J., Dai, L., Liu, X., Li, L., Yang, Y., Cao, Y., Wang, W., Wu, H., Guo, S.: Flame-retardant poly (vinyl alcohol)/mxene multilayered films with outstanding electromagnetic interference shielding and thermal conductive performances. Chem. Eng. J. **380**, 122475 (2020)
214. Fan, Z., Wang, D., Yuan, Y., Wang, Y., Cheng, Z., Liu, Y., Xie, Z.: A lightweight and conductive mxene/graphene hybrid foam for superior electromagnetic interference shielding. Chem. Eng. J. **381**, 122696 (2020)
215. Rajavel, K., Luo, S., Wan, Y., Yu, X., Hu, Y., Zhu, P., Sun, R., Wong, C.: 2d ti3c2tx mxene/polyvinylidene fluoride (pvdf) nanocomposites for attenuation of electromagnetic radiation with excellent heat dissipation. Compos. Part A Appl. Sci. Manuf. **129**, 105693 (2020)
216. Yuan, W., Yang, J., Yin, F., Li, Y., Yuan, Y.: Flexible and stretchable mxene/polyurethane fabrics with delicate wrinkle structure design for effective electromagnetic interference shielding at a dynamic stretching process. Compos. Commun. **19**, 90–98 (2020)

Advanced Hybrid Conducting Polymers: Tissue Engineering Aspects

Suresh Sagadevan, Mohd. Rafie Johan, Md Enamul Hoque, J. Anita Lett, Kamrun Nahar Fatema, and Nanthini Sridewi

Abstract Conducting polymers (CPs) consist of appropriate intrinsic monomers intertwined like biomolecules. These stable CPs exhibit enhanced conductivity and sensitivity at the cellular level that finds various applications in the manufacturing of biomedical devices. Indeed, they can be synthesized using different techniques rapidly and cost-effectively with typical biocompatibility and biodegradability behavior. The affinity of CPs towards cells at the polymer-tissue interface extends their applications in the biomedical and clinical fields such as artificial nerves, biosensors, neural implants, drug delivery devices, etc. Thus, taking advantage of the critical role played by the CPs in the biomedical sector, this chapter focuses on the CPs outlook towards tissue engineering aspects with a special emphasis on the synthesis and fine-tuning of the surface properties providing enhanced physicochemical characteristics required in the biomedical field.

Keywords Conducting polymer · Biofabrication · Tissue engineering · Biomedicine · Biocompatible · Hybrid polymer

S. Sagadevan (✉) · Mohd. R. Johan
Nanotechnology & Catalysis Research Centre, University of Malaya, Kuala Lumpur, Malaysia
e-mail: sureshsagadevan@gmail.com

M. E. Hoque (✉)
Department of Biomedical Engineering, Military Institute of Science and Technology (MIST), Dhaka, Bangladesh
e-mail: enamul1973@gmail.com

J. A. Lett
Department of Physics, Sathyabama Institute of Science and Technology, Chennai, India

K. N. Fatema
Department of Advanced Materials Science and Engineering, Hanseo University, Seosan-si, Chungnam 356-706, Korea

N. Sridewi
Department of Maritime Science and Technology, Faculty of Defence Science and Technology, National Defence University of Malaysia, Kuala Lumpur, Malaysia

© Springer Nature Switzerland AG 2021

S. Shahabuddin et al. (eds.), *Advances in Hybrid Conducting Polymer Technology*, Engineering Materials, https://doi.org/10.1007/978-3-030-62090-5_10

1 Introduction

Conducting polymers (CPs) are smart electroactive biomaterials that permit immediate haulage of electrical, electrochemical, and mechanical stimuli across the living cells. For instance, materials such as electrets or piezoelectric and photovoltaic systems allow limited electrical stimuli across the cells without utilizing an external power source [43, 65]. The CPs are a unique form of polymeric materials with a microporous architecture possessing both ionic and electronic conductivity that can be utilized in wet as well as dry state [66, 86]. Moreover, nonlinear defects viz. solitons, polarons or bipolarons owing to doping/polymerization induce conductivity in CPs [77]. Secondly, the conjugated two-fold bonds present in the protected structure like CPs in the form of alternate single and double bonds between the atoms (carbon) also induce electrical conductivity. In brief, each single and double bond of the backbone is localized with a sturdy 'sigma' (s) bond and weak 'pi' (p) bonds, respectively [31]. To achieve successful conductivity in CPs, a dopant is introduced into the polymer in its oxidized state to modulate the electrical, optical, and structural properties. Alongside, the charge and motility of the dopant introduced can be used to determine the stability of polymer with improved applications [65].

Polyacetylene (PA), was first discovered as a CP by Hideki Shirakawa in the 1970s and since then, it has been widely exploited and found that this noncyclic polyene is unstable in air and thus making it difficult to be processed [45]. Consequently, the solution-processable aromatic CPs, for instance, polypyrrole (PPy), polythiophene (PT), poly(p-phenylene vinylene) (PPV), polyaniline (PANI) and poly(3,4-ethylene dioxythiophene) (PEDOT) have gained widespread interest due to the outstanding thermal conductivity and stability owing to their phenyl or pentyl groups [63]. The CPs can be synthesized using physical and chemical methods such as electrospinning, soft chemical and hard physical template guided synthesis viz. interfacial polymerization, dilute, reverse emulsion polymerization, template-free method, and lithography techniques [64]. The physical, chemical and electrical properties of CPs can easily be adapted by incorporating antibodies, antigenic moieties, and various enzymes [41, 79]. Furthermore, these properties can be tainted and elicited from the beginning to the end and designed with the electrical stimulus [1, 16, 17].

The CPs have been comprehensively studied materials for biomedical applications that include majorly the tissue engineering sector [2, 92, 26, 34, 71]. The electroactive biomaterials unite numerous functions such as biocompatibility, biodegradation/absorption in physiological media, generation of non-toxic products, and sterilization [59, 83]. The CPs play significant roles in the area of tissue engineering through the stimulation of cells via providing electrical current, delivering drugs with the help of controllable and reversible redox reactions, exerting antimicrobial effects through conduction of electricity, and resembling muscles by behaving as electromechanical actuators through an electromechanical response [62].

The need for engineered and replete biomaterials that mimick biochemical and physicochemical characteristics for the successful replacement of biological origin has led to the emergence of tissue engineering. Here, the biomaterials used for engineering tissues include natural (e.g. chitosan, gelatin, collagen, alginate) and synthetic polymers (e.g. poly(lactic-co-glycolic acid) [PLGA], polylactide [PLA], polycaprolactone [PCL], polyglycerol sebacate, and polyurethane [PU]) [13, 27, 32, 33, 35]. These biomaterials have a significant role in repairing the damaged tissues utilizing a scaffold (matrix) that enables the formation of viable tissue architecture [16, 17]. In recent years, the CNTs-based conducting biomaterials have been used in the biosensor and bone tissue engineering sector for its plausible conductivity and tensile strength [21]. Conversely, non-biodegradability and polydispersity of such conducting materials in a composite system remains a mainstay for long-term in vivo toxicity. However, CPs like PANI, PPy, polythiophene and their derivatives/composites exhibits attractive properties viz. biocompatibility, rapid synthesis, easy manipulation and electronic control over physical and chemical properties using surface functionalization of dopant for its extensive application in biomedical and clinical field [47]. The scheme of CPs application and strategies towards tissue engineering in biomedicine is illustrated in Fig. 1.

The biocompatible CPs help greatly in promoting cellular metabolism augmenting cell adhesion, proliferation, and differentiation irrespective of electrical stimuli [5]. This was demonstrated using different cell lines that are electrically sensitive namely fibroblasts (L929), myoblasts (C2C12), cardiac (H9c2), and mesenchymal stem cells [38, 72]. According to reports, nearly half of the medical expenses in the US correspond to organ failure and tissue loss in addition to surgical procedures (~8 billion) and length of treatment (~90 million hospital days/year) [53]. To reduce medical costs, improve healthcare standards, and a predetermined length of treatment regime,

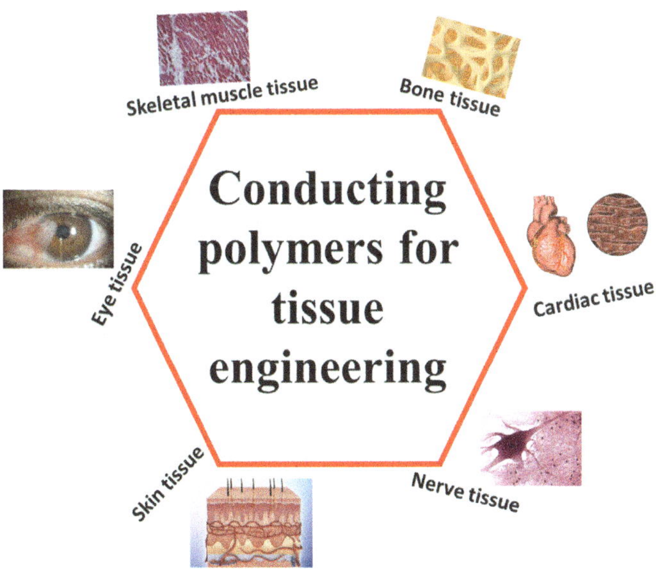

Fig. 1 Schematic presentation of different types of CPs applicable in various tissue engineering applications

regenerative medicine or bioengineering of tissue is needed [2, 44, 78]. Therefore, in this review, an overview of CPs as a next-generation smart system in biomedical applications is provided, in addition to highlighting the fabrication, precision, and processability in terms of a tissue engineering perspective.

2 History and Specific Behavior of CPs

CPs constitute a special class of electronically conducting behavior with high conductivity to weight ratio. They are suitable for biomedical applications since they are biocompatible, biodegradable, porous in nature and neither the materials nor degradation products show toxicity [50]. The structural and mechanical compatibility of organic CPs with cells, tissues, and organs make them a better candidate over conventional ones [2, 92, 18, 71]. At present, there are around 19 different types of CPs available. Figure 2 represents some well-researched monomeric structures of CPs. The available CP structures include, PPy, PANI, PEDOT, PTh, PPV, polyfuran (PFu), polyazulene (PAZ), polyisoprene (PIP), polybutadiene (PBD), poly(a-naphthylamine) (PNA), poly(isothianaphthene) (PITN), polyacetylene (PAc), poly(p-phenylene) (PPP), Poly(3-octylthiophnene-3-methyl thiophene) (POTMT), poly(2,5-thienylenevinylene) (PTV), poly(3-alkyl thiophene) (PAT), poly-p-phenylene-sulphide (PPS), poly(p-phenylene-terephthalamide) (PPTA), and polythiophene-vinylene (PTh-V). The advantages of conducting polymers over the

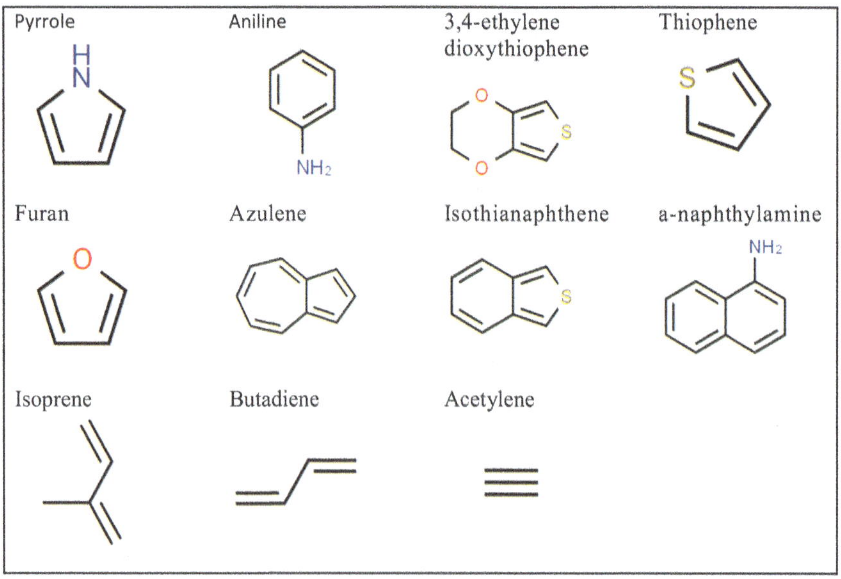

Fig. 2 Chemical structures of monomers for some representative conducting polymers (CPs)

Table 1 The conductivity values of some CPs

Polymer	Conductivity (s cm^{-1})
Polythiophene (PT)	$10–10^3$
Poly(p-phenylene) (PPP)	$10^2–10^3$
Polyaniline (PANI)	30–200
Polypyrrole (PPy)	$10^{2–7}.5 \times 10^3$
Poly(3,4-ethylenedioxythiophene) (PEDT, PEDOT)	$3 \times 10^{5–5} \times 10^5$
Polyacetylene (PAc)	$10^3–1.7 \times 10^5$
Poly(p-phenylenevinylene) (PPV)	$3–5 \times 10^3$

conventional polymers like ionic transport, redox nature, chemical and photocatalytic activity, electrochemical behavior, and junctional effects where all these characteristics are possible to maintain in a single polymeric structure. Also, the flexibility, lightweight, and non-metallic surface nature allow for the enhanced mechanical properties that specifically matches for the tissue engineering purposes.

The accidental discovery of an increase in the conductivity after 'doping' (by 10 million-fold) with iodine vapor in the preparation of PAc has drawn much attention in this new field of research [12, 26, 54]. Once doped, the electrical conductivity reaches the metallic conducting regime. In this case, PAc was very difficult to synthesize due to its unstable nature, so the hunt for CPs with better electrical conductivity has begun [90, 91]. Furthermore, polyheterocycles have emerged as the most sought-after polymers for their stability and high electronic conductivity [3, 88] as shown in Table 1.

3 Fabrication of Conducting Biomaterials

3.1 Pure CP Films

Pure CP films can be obtained by the electrochemical deposition method. The synthesized CPs usually get deposited on the electrode under the influence of electric current across the electrodes dipped in an electrolytic solution containing the monomeric units of the polymer, for its use in cell culture [55]. Here a uniform layer of monomer gets deposited as a thin film upon oxidation resulting in the formation of insoluble polymer chains. Several polymers are already employed in medicinal chemistry viz. PPy, PANI, PT, etc. possessing the best characteristics that have been reported so far. They indeed act as a 3D supporting matrix in the process of adhesion, proliferation, and differentiation of a cell [6, 22, 69]. For instance, the PPy membrane at the nanoscale (62 nm) could be electrically deposited on an indium-tin-oxide glass slide. The amount of bioactive molecule deposition of CPs in situ formed by the electrochemical synthesis which gets doped with the polymers used as the solvent and has certain restrictions due to the electrode [82, 80]. For any desired connectivity achieving support, the geometry and surface area of the electrode plays a pivotal role.

3.2 Conducting Blends or Composites

As the surface area and geometry of the electrode remain a mainstay, the chemical synthesis methods have been adapted in the fabrication of composites by combining PPy with polymers such as PU, poly(methylmethacrylate), polyvinylchloride, polystyrene and poly(D, L-lactide) (PDLLA) [85]. Since the CPs are extremely delicate which makes their fabrication complex [8] and this shortcoming of CPs can be resolved by blending them with other polymers to combine their properties for achieving good results [19]. This blending is commonly known as the composite formation and here, both natural and synthetic polymers (e.g. chitosan, PCL, PLGA, PLA and silk fibroin) are fused with CPs such as PPy and PANI [11] for enhanced conductivity. Initially, CP film is immersed into a pyrrole deionized water and polystyrene sulfonic acid blend to form a conductive PPy/PCL film. Then, an oxidant viz. $FeCl_3$ is added to the blend resulting in the interpenetrating polymeric network of PPy encapsulated PCL conductive film. The developed film demonstrated the resistivity of 1.0 ± 0.4 kΩ cm similar to biologic cardiac tissue. To achieve polymeric conductivity, the dip-coating and emulsion polymerization methods [82, 80] are used for the formation of PDLLA composite films and nerve conduits. Similarly, the characteristic segmented-block architecture of PU has been harnessed for tuning its mechanical, physical, and biological functionalities to attain biocompatibility [7], and in that way, the CPs-blended-biodegradable polymers find their application in tissue engineering. Figure 3 demonstrates the chemical structure of some representative conducting polymers that could provide various properties including high conductivity for different real-world needs.

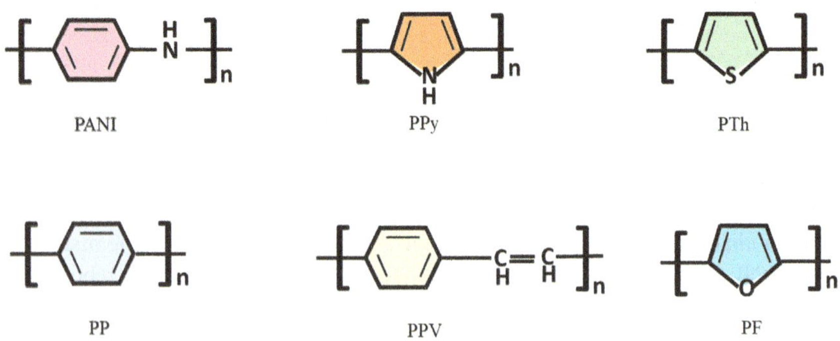

Fig. 3 Chemical structure of some representative conducting polymers

3.3 Conducting Copolymer Films

The CP blended copolymers can be synthesized in the form of a nano or microfiber using the electrospinning method. In brief, a high-voltage electrostatic field is applied to PLA and PANI/CSA in hexafluoroisopropanol that draws charged threads in the form of intertwined fiber. The electrical conductivity of the resulting electrospun fiber would be enhanced by varying the PANI content from 0 to 3 wt% in the PLA polymer. The resulting scaffold would be conductive, flexible, and biocompatible for use in neural and cardiac tissue generation [15, 29, 49]. The phenomenon was systematically examined by a detailed analysis of electrospun and solvent membrane casts. The mechanical properties and plasticity of electrospun polymer nanofibers are affected by the manufacturing process as shown in Fig. 4.

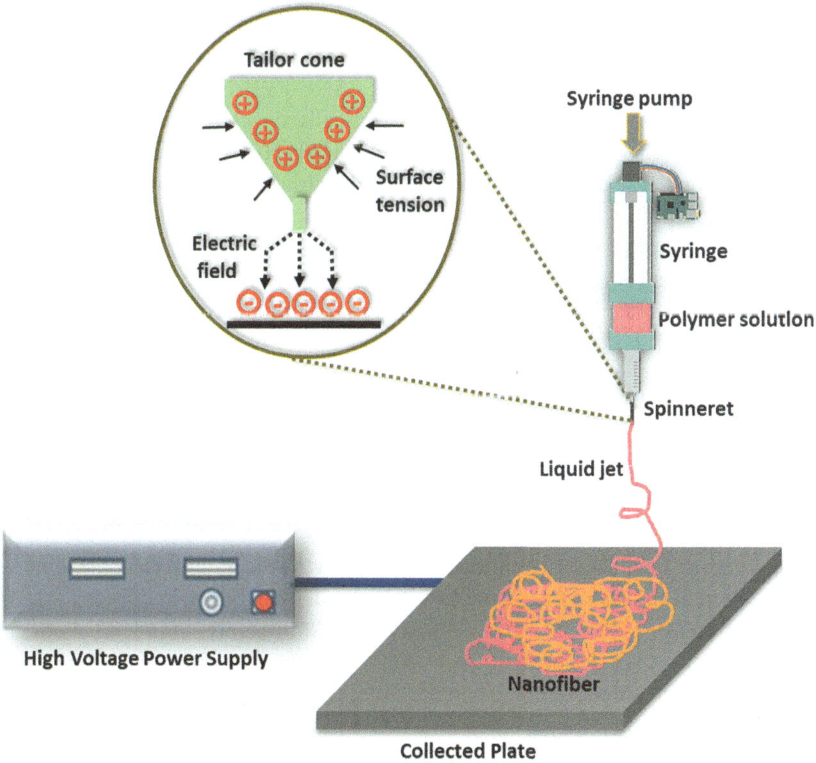

Fig. 4 Schematic representation of the electrospinning process for nanofiber fabrication

3.4 Hydrogels

CPs have also shown to be polymerized successfully inside hydrogel networks. The hydrogels are a 3D network of hydrophilic polymers and they are spongy, porous and their high-water absorbing capacity resembles natural living tissues [28, 40]. Their controlled pore morphology and conductive properties greatly support cell signaling during tissue regeneration pertaining to its application in drug release devices, implantable biosensors and deep-brain stimulators. The conductive hydrogel biomaterials such as PANI, PEDOT, PPy, PPV PT and so on are utilized to improve cell attachment, proliferation, and differentiation [64, 79]. The conductive hydrogel can provide both physical support and electrical cues for the attached cells that find great use in biomedical applications [41, 43]. The CP hydrogel matrix can simulate the physiological microenvironment of electroactive tissues and also enhance the communication of electrical signals between cells. The CPs are currently used in both forms of customized 3D printable and in situ injectable hydrogels.

Some commonly used composite electroactive hydrogels include, PPy/PANI–polyacrylamide, PANI–polyacrylamide and PANI–PVP hydrogels [52, 75]. One such polymer, PPy has been also electrochemically developed with the support of hydrogels viz. poly(2-hydroxyethyl methacrylate) p(HEMA) [25] or turned into a hydrogel through use of a PMAS dopant [68]. Their high surface/volume ratio confers the PPy hydrogel much lower impedance than the naturally occurring PPy film. A hydrogel composite comprising PPy-mucopolysaccharides was shown to release the surface-bound proteins under the influence of the electric field that has been implied for sustained drug release and tissue engineering [10, 25]. Table 2 describes some selected CP-based composite hydrogels that have been successfully studied for various biomedical applications.

4 Tissue Engineering Applications

Tissue Engineering (TE) is a multidisciplinary area that deals with the replacement or regeneration of tissues/organs using generally biodegradable 3D scaffolds, bioactive molecules and/or cells. Therefore, TE can be able to resolve problems associated with tissue injury, in the present treatment with transplantation, mechanical tools, or surgical reconstruction. For example, organ transplants have major drawbacks, such as transplant rejection and lack of a donor to meet all global demand. Mechanical devices are not capable of performing all tissue-related functions and cannot prevent progressive patient deterioration. In the end, surgical reconstruction can lead to long-term problems. Therefore, TE emerges from the need to provide more concrete solutions for tissue repair in clinics and aims to achieve this objective by developing in vitro tools to repair in vivo damaged tissue. The innumerable properties of CPs viz. large surface area to volume ratio, conductivity, electroactivities, reversible oxidation, biocompatibility, surface topography, 3D geometry etc. make it

Table 2 CPs-based composite hydrogels used in tissue engineering applications

CP type	Preparation method	Advantages	Reference
PANI-PEG conductive hydrogel	PANI is precipitated in a polyethylene glycol diacrylate (PEGDA) solution then, crosslinked the polymer chains under UV irradiation	Enhanced conductivity with its hydrophilic nature	[23]
Gelatin-g-PANI (GP) hydrogel	At body temperature, PANI was incorporated into the gelatin backbone by genipin and then GP was crosslinked using genipin producing the hydrogel	Increased the conductivity of the hydrogel with the increased amount of PANI and also demonstrated non-cytotoxicity with C2C12 cells and BMSCs	[49]
Chitosan-g-PANI (QCSP) hydrogel	PANI was incorporated into the QCS backbone. Then QCSP was crosslinked using oxidized dextran (Odex) via Schiff base reaction producing conductive hydrogels which was injectable	Demonstrated self-healing and antibacterial properties, and provided the conductivity (~10^{-3} s cm^{-1}) similar to the biologic cardiac tissue. The hydrogel also improved biocompatibility with H9C2 cardiac cells and C2C12 myoblast cells	[15]
PEDOT-PSS hydrogel	PEDOT-PSS immersed into photo-cross-linkable gelatin methacryloyl (GelMA) hydrogels	The doped PEDOT-PSS enhanced conductivity through adjustment of the bandgap and in the doping state, contributed strong stability. Besides, the PEDOT: PSS film is stable and compatible with most organic solvents for fabrication process	[74]

a suitable candidate for tissue engineering. As an example, the PGA, PANI, PLCL, PLA-PPy, PCL, and PLGA nanofibers exhibit the above-mentioned properties. The roles and responsibilities of various CPs in a different type of tissue engineering are discussed as follows.

4.1 Bone Tissue Engineering

To form new bone tissue, a material is implanted, and cells need to colonize the material. The adhesive and cohesive properties in cells like mesenchymal stem cells can be improved by using conducting biomaterials [56]. Ice-templating, a novel route to template porosity in biomaterials, has been applied to electroconductive PEDOT-PSS scaffold (50 μm) capable of guiding MC3T3-E1 cells to penetrate and get along with the matrix. It was also studied that the PEDOT-PSS scaffold could improve the level of expression of genes enhances mineralization in the extracellular matrix (ECM) and promotes deposition of osteocalcin in MC3T3-E1 cells. Such an electro-conductive scaffold can also induce differentiation of osteogenic precursor cells to osteoblasts in MC3T3-E1 cells [48]. Moreover, 3D conductive scaffolds based PEDOT, PSS, gelatin, and bioglass at nanoscale show ~60% porosity with pore size in the range 110–160 μm. The pore size can be varied by varying the concentration of CPs in scaffold whereby the cell viability is increased as evident in human mesenchymal stem cells [42]. Similarly, an adequate proliferation of cells supported by electroactive shape memory polymeric (ESMP) films with a six-arm branched polylactide aniline trimer was obtained. Whereas no such proliferation was observed when aniline trimer-free SMP films were used.

Likewise, ESMP films were also found to improve the osteogenic differentiation of C2C12 myoblast cells as determined using ALP enzyme activity, immunofluorescence, and relative gene expression. Indeed, 3D conductive PLA/PANI cytocompatible nanocomposite scaffolds were used for the regeneration of bone using bone marrow-derived mesenchymal stem cells (BMSCs). Once under the influence of electric potential, these conducting scaffolds improved osteogenic differentiation of BMSCs associated with the upregulation of ALP, OCN, Runx2 genes, and mineralization attributing for bone tissue engineering applications [57]. Another benefit of CPs as scaffolds is that they could produce an electrical stimulus on the substrates. For instance, heparin doped PPy/PLLA in the process of adhesion, proliferation, and osteogenic differentiation, when an electrical stimulus (ES) is elicited, the osteoblast adhesion and growth process heightened, with a significant augmentation of calcium and phosphate levels and mineralization. Alongside, ES also upregulated the osteoblast-specific marker expression viz. Runx2, BMP2 and ALP [4]. Figure 5 demonstrates the schematic representation of polymer-based bone tissue engineering approach.

4.2 Skeletal Muscle Tissue Engineering

Skeletal muscles are abundant, highly specialized and an organized tissue comprising of bundles of myofibers (multinucleated muscle fibers) meddled with blood vessels, nerves, and extracellular connective tissue. These muscles are metabolically active showing a characteristic regenerative potential [37, 61]. In the case of volumetric muscle loss (VML), this regeneration potential gets affected demanding interventional support. One such promising technology to mimic the structural and functional constructs of such complex tissues in vitro and in vivo is the skeletal muscle tissue engineering (SMTE) [9, 87]. The cytocompatibility of poly-N-isopropylacrylamide (PNIPAM) polymer significantly enhanced the production of vascularized tissue in vitro by modifying the hydrophilic property of cellular adhesion, thereby

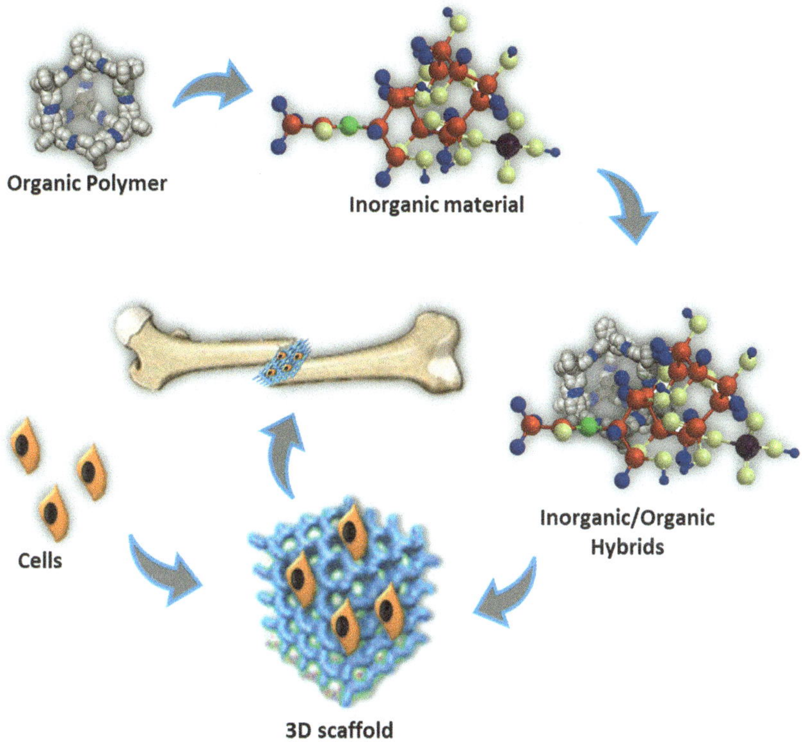

Fig. 5 Schematic representation of polymer-based bone tissue generation

enhancing cell survival. In this line, PLCL/PANI fibers also exhibited compatibility and viability by enhancing the expression of myogenin. The upregulation of myosin heavy chain gene and troponin T inferred PLCL/PANI CPs considerably modulated the induction of myotube formation [70]. Gelatin, yet another natural polymer when blended with CSA and PANI fibers permitted the cell attachment, interaction, and myogenic differentiation of C2C12 cells on par with pure gelatin or CSA doped nanofibers. Also, conductive nanofibers augment colocalization of receptors (ryanodine and dihydropyridine), intracellular organization, regulate gene expression to accommodate contractile functionality with an appropriate elasticity [67].

To achieve a replica of native tissue, bioengineered muscle tissue must be fabricated into an appropriate scaffold. The scaffolding support must provide a solid framework for attachment, proliferation, migration, and differentiation into myotubes. The most important factor is biocompatibility maintaining the cell viability and without inducing any inflammatory responses. The electrospun scaffolds should act as a temporary matrix where the neo-generated matrix would be deposited and biodegrade gradually to accommodate the newly formed tissue. For example, PCL/PANI nanofibers have been used to increase myotube maturation than

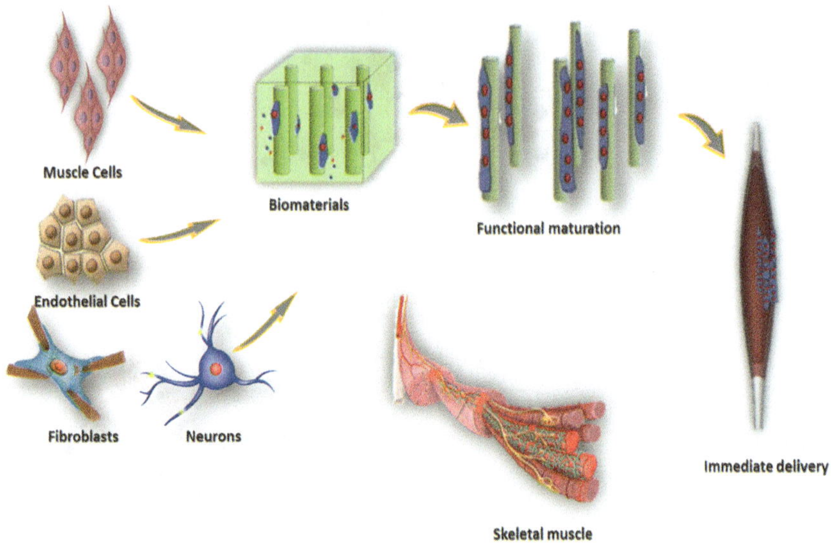

Fig. 6 Skeletal muscle tissue generation represented schematically

that of non-conductive PCL fibers [20]. Schematic representation of Skeletal muscle tissue engineering as shown in Fig. 6.

4.3 Nerve Tissue Engineering

One of the complex biological phenomena is the generation of neuron/nerve tissue. Neurons can regenerate in the peripheral nervous system during injuries, but success varies with the intensity of the damage. To ease the repair process, bioengineering of nerve graft for the peripheral nervous system remains to be an alternative strategy for surgery. In this study, researchers have fabricated PANI and PPy as conductive polymer scaffolds to support the growth and regeneration of nerve tissues [46, 51]. These conductive nanocomposites of PCL/PPy exhibit superior cytocompatibility that supports the growth, maturation, and differentiation of cells [14]. Alongside, electroactive nanofibers were fabricated in situ using tobacco mosaic virus TMV/PANI/PSS augmenting the growth of neuronal cells linked with an increase in the length of neurites. This was quite contrasted with the number of cells cultured in TMV-derived nonconductive nanofibers. It is noteworthy to mention that the electroactive TMV/PANI/PSS nanofibers could act as a template in the neurites' outgrowth leading to a bipolar cellular morphology as evident from electroactivity and topographical cues. It is inferred that the electrical stimulus endured as an effective approach for neuronal function and regeneration [24, 82, 80]. Due to the effective enhancement of ion exchange linked to the recording sites of adjoining tissues, neural probes, and implantable electrodes have been much sought after [84]. For example, PANI-coated platinum (Pt) electrodes were prepared using in situ polymerization which is used to improve the functionalities of neural probes. The PANI-coated Pt electrode tends to aggregate retinal fragments on their surface supporting the maturation and differentiation for long-term implantation as a therapeutic device. Alongside this, this electrode coated with PANI remained stable for 6 months under electrical stimulation [73]. Lacour's group developed a microfabricated conformable electrode array with electrode sites of 100 μm diameter coated with PEDOT: PSS, that conferred high charge injection properties and stimulated the auditory system with small stimulation sites. This demonstrates the synergistic potential of PEDOT: PSS nanofibers for efficient while electrochemically safe stimulation of the central auditory system [81].

Natural and synthetic polymers have been tested substantially as a remedy for restoring function in malfunctioning neural tissues. Polymers deliver a wide variety of flexibility e.g. forms, processability, mechanical and biological characteristics that sin compared to other biomaterials like ceramics and metals. Numerous studies have demonstrated that polymers can be formed into appropriate support structures, like scaffolds, nerve conduits and electrospun matrices that facilitate the regeneration of damaged/diseased neural tissues. In general, natural polymers offer the advantage of improved biocompatibility and bioactivity, while synthetic or non-natural polymers have improved mechanical properties and structural stability. Sometimes, combinations of the two allow for the creation of polymeric conduits capable of imitating the native physiological environment of healthy neural tissues, thereby controlling cell activity and promoting the regeneration of damaged nerve tissues. Schematic representation of peripheral Nerve tissue engineering as shown in Fig. 7.

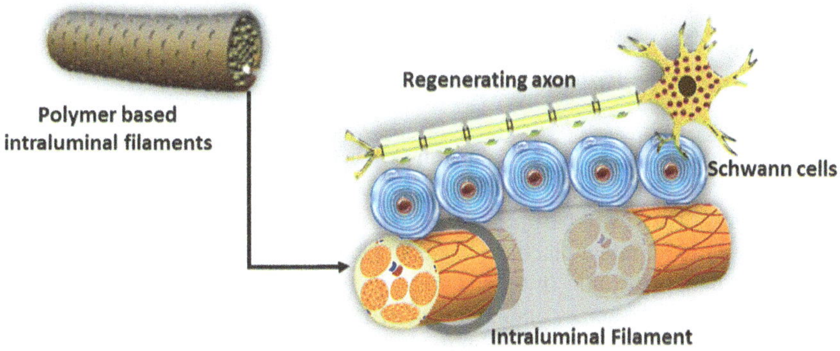

Fig. 7 Schematic representation of peripheral nerve tissue generation

4.4 Cardiac Tissue Engineering

Excitation-contraction coupling in the heart is related to the transmission of electrical signals via the cardiac cells in a synchronized fashion [39]. The process that drives electrical excitation to contraction is by mobilization of calcium. To understand the underlying mechanism, a PPy-PCL conductive film has been used to promulgate the calcium dynamics via velocity, wave propagation, and transient time of the cardiomyocyte monolayers [15].

One of the therapeutic challenges faced all around the world is myocardial infarction (MI) [36]. It generally denotes an event of a heart attack due to the formation of plaques in the arterial walls, reduced blood flow, and lack of oxygen supply [76]. In order to tackle the situation, a hydrogel is constructed consisting of a self-healing conductive chitosan-graft-aniline tetramer and dibenzaldehyde-terminated poly(ethylene glycol). They are known for their biocompatibility and a targeted delivery system as well in the treatment regime. These encapsulated C2C12 cells hydrogel shows a controlled release profile and biocompatibility in vivo suggesting its potential as a cell delivery vehicle for MI [30]. Together, conductive hydrogels showed an electroconductive correspondent as native myocardium which when loaded with plasmid DNA (encoding eNOs nanocomplexes and stem cells) could be used for treating MI. When a hydrogel was injected into an infarcted myocardium of SD rats, eNOs in myocardial tissue expression upregulated followed by the enhanced expression of myocardium-related mRNA and proangiogenic growth factors. The histological analysis and electrocardiography (ECG) evidenced the shortened interval, clear increase in ejection fraction and smaller infarction size, which indicates that for the treatment of MI, the conductive hydrogel incorporated with stem cells could be a robust therapeutic approach [89].

4.5 Skin Tissue Engineering

The most important purpose of the human skin is to protect us from microbial invasion. Every year several million deaths occur due to wounds caused by thermal burn [58]. Despite advances in burn care including resuscitation, early wound excision, respiratory support, infection control, etc. a viable permanent substitute for biological skin is under research. In this context, bioengineering of skin substitutes like biopolymer scaffolds, and a combination of cells-scaffolds are used in skin cell therapies [60]. The CPs such as PANI with antibacterial activity have been proven to promote cellular activities in skin cells namely fibroblasts and keratinocytes [90, 91]. Figure 8 represents schematically the biopolymer scaffolds used in skin cell therapy.

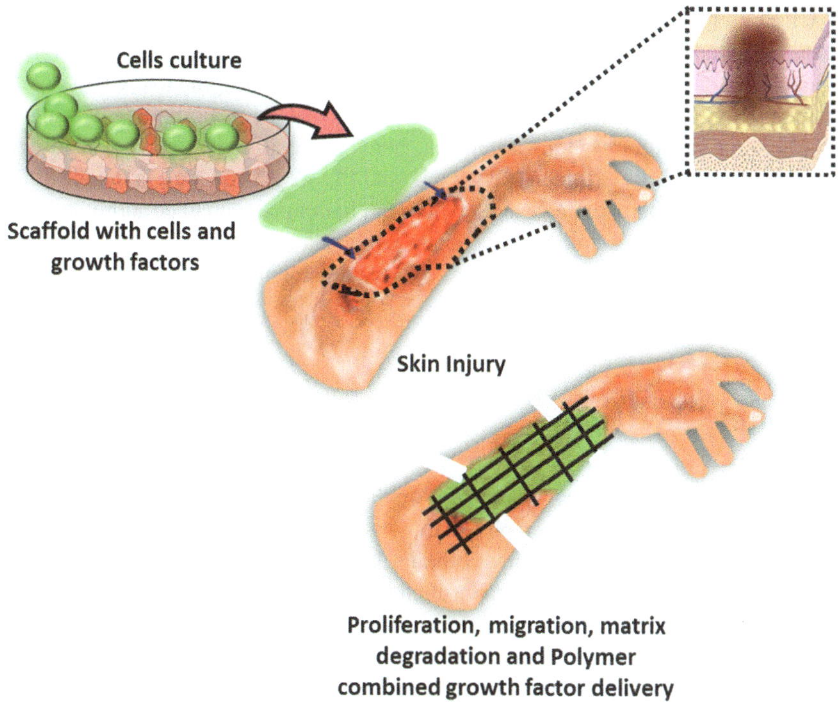

Fig. 8 Schematic representation of biopolymer scaffolds used in skin cell therapy

5 Conclusions

The tissue engineering community's extremely desired objective is to develop stimulus-responsive biomaterials with easy-to-tailor characteristics. One such material promises to become a novel type of electroactive biomaterial, the CP. The CPs are already being used in fuel cells, computer displays, and microsurgical instruments and are now finding biomaterials applications. These versatile polymers can be mixed into composites or electrospun into microfibers by themselves as hydrogels. To be biocompatible and biodegradable, they can be developed. By binding biologically significant molecules into the polymer using one of the many techniques available for their functionalization, their physical characteristics can be easily optimized for a specific application. Their conductive nature enables stimulation of the cells or tissue growth on them, to influence the own physical characteristics of the polymers after synthesis, and to release the drugs contained in them by applying an electrical signal. Therefore, it is not surprising that these polymers become very significant materials for biosensors, neural implants, drug delivery systems, and tissue engineering scaffolds. Since the versatile nature of CPs attracts researchers from many countries due to their cost-effectiveness, stability, lightness, better capacity, resistance to decomposition, and better electrical conductivity. The fabrication method of CPs by electrochemical, composite, electrospun, and hydrogel techniques has been presented in this paper with suitable examples and figures. There is a continuous effort for the development of biocompatible and biodegradable CPs towards biomedical applications. The specific application of CPs in tissue engineering as implantation substituting human cell viz. bone, skeletal, nerve, cardiac, and skin using PGA, PANI, PLCL, PLA-PPY, PCL, and PLGA is discussed in detail. In the development of a therapeutic device, numerous challenges need to be addressed with alternative strategies. A collaborative effort remains a mandate to explore the excellent properties of CPs to the maximum for its implementation in multidisciplinary areas.

6 Future Directions

In this chapter, most of the biomedical applications that involve CPs have been covered taking advantage of their physical characteristics. To explore future advanced applications of CPs, the biocompatible and biodegradable nature of the CPs can further be enhanced and/or modified through various methods. Applying an electrical signal (e.g. through electrical or electromechanical stimulus) to the CP substrate, the behavior of cells or tissues cultured onto the CP substrate could be influenced. The modification before and/or after synthesis provides great degrees of control over the properties of the CPs that no other material might provide. Hence, it is to see many more real-world examples of the applications of this smart material in wider tissue engineering areas, and it is hoped that this chapter has persuaded the reader for the same.

Acknowledgements The authors would like to acknowledge the financial support provided by research university grant number (RU001-2018) NANOCAT, University of Malaya, Malaysia.

Competing interests The authors declare no conflict of interest.

References

1. Aoki, T., Tanino, M., Sanui, K., Ogata, N., Kumakura, K.: Secretory function of adrenal chromaffin cells cultured on polypyrrole films. Biomaterials **17**, 1971–1974 (1996)
2. Ateh, D.D., Navsaria, H.A., Vadgama, P.: Polypyrrole-based conducting polymers and interactions with biological tissues. J. R. Soc. Interface **3**, 741–752 (2006)
3. Balint, R., Cassidy, N.J., Cartmell, S.H.: Conductive polymers: towards a smart biomaterial for tissue engineering. Acta Biomater. **10**, 2341–2353 (2014)
4. Baniasadi, H., Ramazani, S.A.A., Mashayekhan, S.: Fabrication, and characterization of conductive chitosan/gelatin-based scaffolds for nerve tissue engineering. Int. J. Biol. Macromol. **74**, 360–366 (2015)
5. Bettinger, C.J., Bruggeman, J.P., Misra, A., Borenstein, J.T., Langer, R.: Biocompatibility of biodegradable semiconducting melanin films for nerve tissue engineering. Biomaterials **30**, 3050–3057 (2009)

6. Brahim, S., Guiseppi-Elie, A.: Electroconductive hydrogels: electrical and electrochemical properties of polypyrrole-poly(HEMA) composites. Electroanalysis **17**, 556–570 (2005)
7. Caló, E., Khutoryanskiy, V.V.: Biomedical applications of hydrogels: a review of patents and commercial products. Eur. Polym. J. **65**, 252–267 (2015)
8. Chattopadhyay, D.K., Raju, K.: Structural engineering of polyurethane coatings for high-performance applications. Prog. Polym. Sci. **32**, 352–418 (2007)
9. Chen, M.-C., Sun, Y.-C., Chen, Y.-H.: Electrically conductive nanofibers with highly oriented structures and their potential application in skeletal muscle tissue engineering. Acta Biomater. **9**, 5562–5572 (2013)
10. Chen, J., Yu, M., Guo, B., Ma, P.X., Yin, Z.: Conductive nanofibrous composite scaffolds based on In-situ formed polyaniline nanoparticle and polylactide for bone regeneration. J. Colloid Interface Sci. **514**, 517–527 (2018)
11. Chronakis, I.S., Grapenson, S., Jakob, A.: Conductive polypyrrole nanofibers via electrospinning: electrical and morphological properties. Polymer **47**(5), 1597–1603 (2006)
12. Dai, L.: Conducting polymers. In: Dai, L. (ed.) Intelligent Macromolecules for Smart Devices: From Materials Synthesis to Device Applications, pp. 41–80. Springer, London (2004)
13. Denis, P., Wrzecionek, M., Gadomska-Gajadhur, A., Sajkiewicz, P.: Poly(glycerol sebacate)-poly(l-lactide) nonwovens. Towards attractive electrospun material for tissue engineering. Polymers **11**(12), 2113 (2019)
14. Di, L., Wang, L.-P., Lu, Y.-N., He, L., Lin, Z.-X., Wu, K.-J., Ren, Q.-S., Wang, J.-Y.: Protein adsorption and peroxidation of rat retinas under stimulation of a neural probe coated with polyaniline. Acta Biomater. **7**(10), 3738–3745 (2011)
15. Dong, R., Zhao, X., Guo, B., Ma, P.X.: Self-healing conductive injectable hydrogels with antibacterial activity as cell delivery carrier for cardiac cell therapy. ACS Appl. Mater. Interfaces **8**, 17138–17150 (2016)
16. Garner, B., Georgevich, A., Hodgson, A.J., Liu, L., Wallace, G.G.: Polypyrrole–heparin composites as stimulus-responsive substrates for endothelial cell growth. Inc. J Biomed Mater Res **44**, 121–129 (1999)
17. Garner, B., Hodgson, A.J., Wallace, G.G., Underwood, P.A.: Human endothelial cell attachment to and growth on polypyrrole–heparin is fibronectin dependent. J. Mater. Sci. Mater. Med. **10**, 19–27 (1999)
18. George, P.M., LaVan, D.A., Burdick, J.A., Chen, C.Y., Liang, E., Langer, R.: Electrically controlled drug delivery from biotin-doped conductive polypyrrole. Adv. Mater. **18**, 577–581 (2006)
19. Ghasemi-Mobarakeh, L., et al.: Application of conductive polymers, scaffolds and electrical stimulation for nerve tissue engineering. J. Tissue Eng. Regen. Med. **5**, e17–e35 (2011)
20. Ghasemi-Mobarakeh, L., Prabhakaran, M.P., Morshed, M., Nasr-Esfahani, M.H., Ramakrishna, S.: Electrical stimulation of nerve cells using conductive nanofibrous scaffolds for nerve tissue engineering. Tissue Eng. Part A **15**, 3605–3619 (2009)
21. Gomez, N., Schmidt, C.E.: Nerve growth factor-immobilized polypyrrole: bioactive electrically conducting polymer for enhanced neurite extension. J. Biomed. Mater. Res. A **81**(1), 135–149 (2007)
22. Green, R.A., Lovell, N.H., Wallace, G.G., Poole-Warren, L.A.: Conducting polymers for neural interfaces: challenges in developing an effective long-term implant. Biomaterials **29**, 3393–3399 (2008)
23. Guarino, V., Alvarez-Perez, M.A., Borriello, A., Napolitano, T., Ambrosio, L.: Conductive PANi/PEGDA macroporous hydrogels for nerve regeneration. Adv. Healthc. Mater. **2**(1), 218–227 (2013)
24. Guex, A.A., Vachicouras, N., Hight, A.E., Brown, M.C., Lee, D.J., Lacour, S.P.: Conducting polymer electrodes for auditory brainstem implants. J. Mater. Chem. B **3**, 5021–5027 (2015)
25. Guex, A.G., Puetzer, J.L., Armgarth, A., Littmann, E., Stav inidou, E., Giannelis, E.P., Malliaras, G.G., Stevens, M.M.: Highly porous scaffolds of PEDOT: PSS for bone tissue engineering. Acta Biomater. **62**, 91–101 (2017)

26. Guimard, N.K., Gomez, N., Schmidt, C.E.: Conducting polymers in biomedical engineering. Prog. Polym. Sci. **32**, 876–921 (2007)
27. Guimard, N.K.E., Sessler, J.L., Schmidt, C.E.: Toward a biocompatible and biodegradable copolymer incorporating electroactive oligothiophene units. Macromolecules **42**, 502–511 (2009)
28. Guo, B.L., Finne-Wistrand, A., Albertsson, A.C.: Degradable and electroactive hydrogels with tunable electrical conductivity and swelling behavior. Chem. Mater. **23**, 1254–1262 (2011)
29. Guo, B.L., Yuan, J.F., Gao, Q.Y.: Preparation and characterization of temperature and pH-sensitive chitosan material and its controlled release on coenzyme A. Colloids Surf. B **58**, 151–156 (2007)
30. Guo, B.L., Sun, Y., Finne-Wistrand, A., Mustafa, K., Albertsson, A.C.: Electroactive porous tubular scaffolds with degradability and non-cytotoxicity for neural tissue regeneration. Acta Biomater. **8**(1), 144–153 (2012). Biomacromolecules Review. https://doi.org/10.1021/acs.biomac.8b00276
31. Heeger, A.J., MacDiarmid, A.G., Shirakawa, H.: The Nobel Prize in Chemistry, 2000: Conductive Polymers. Royal Swedish Academy of Sciences, Stockholm, Sweden (2000)
32. Hoque, M.E., Feng, W., Wong, Y.S., Hutmacher, D.W., Li, S., Huang, M.H., Vert, M.: Scaffolds designed and fabricated with elastic biomaterials applying CAD-CAM technique. Tissue Eng. Part A **14**(5), 907–908 (2008)
33. Hoque, M.E., Meng, T.T.H., Chuan, Y.L., Chowdhury, M., Prasad, R.: Fabrication and characterization of hybrid PCL/PEG 3D scaffolds for potential tissue engineering applications. Mater. Lett. **131**, 255–258 (2014)
34. Hoque, M.E., Prasad, R.G.S.V., Sapuan, S.M.: Conducting polymers: drug delivery and tissue engineering. In: Munmaya, M. (ed.) Encyclopedia of Biomedical Polymers and Polymeric Biomaterials, pp. 2011–2023. Taylor and Francis (2015)
35. Hutmacher, D.W., Hoque, M.E., Wong, Y.S.: Design, fabrication and physical characterization of scaffolds made from biodegradable synthetic polymers in combination with RP systems based on melt extrusion. In: Bopaya, B., Paulo, B. (eds.) Virtual Prototyping & Bio Manufacturing in Medical Applications, pp. 261–291. Springer (2008)
36. Jahromi, M.A.M., Zangabad, P.S., Basri, S.M.M., Zangabad, K.S., Ghamarypour, A., Aref, A.R., Karimi, M., Hamblin, M.R.: Nanomedicine and advanced technologies for burns: preventing infection and facilitating wound healing. Adv. Drug Deliv. Rev. **123**, 33–64 (2018)
37. Jun, I., Jeong, S., Shin, H.: The stimulation of myoblast differentiation by electrically conductive sub-micron fibers. Biomaterials **30**, 2038–2047 (2009)
38. Karp, J.M., Langer, R.: Development and therapeutic applications of advanced biomaterials. Curr. Opin. Biotechnol. **18**, 454–459 (2007)
39. Kashi, M., Baghbani, F., Moztarzadeh, F., Mobasheri, H., Kowsari, E.: Green synthesis of degradable conductive thermosensitive oligopyrrole/chitosan hydrogel intended for cartilage tissue engineering. Int. J. Biol. Macromol. **107**, 1567–1575 (2018)
40. Khatun, M.M., Hoque, M.E., Wadud, S.I., Shawon, Z.B.K.: Recent developments in nanocellulose and nanohydrogel matrices—towards stem cell research and development. In: Mohammad, F., Al-Lohedan, H.A., Jawaid, M. (eds.) Sustainable Nanocellulose And Nanohydrogels From Natural Sources, pp 315–328. Elsevier (2020)
41. Kim, D.H., Richardson-Burns, S.M., Hendricks, J.L., Sequera, C., Martin, D.C.: Effect of immobilized nerve growth factor on conductive polymers: electrical properties and cellular response. Adv. Funct. Mater. **17**, 79–86 (2007)
42. Koning, M., Harmsen, M.C., van Luyn, M.J., Werker, P.: Current opportunities and challenges in skeletal muscle tissue engineering. J. Tissue Eng. Regener. Med. **3**, 407–415 (2009)
43. Kotwal, A., Schmidt, C.E.: Electrical stimulation alters protein adsorption and nerve cell interactions with electrically conducting biomaterials. Biomaterials **22**, 1055–1064 (2001)
44. Lakard, B., Ploux, L., Anselme, K., Lallemand, F., Lakard, S., Nardin, M., et al.: Effect of ultrasounds on the electrochemical synthesis of polypyrrole, application to the adhesion and growth of biological cells. Bioelectrochemistry **75**, 148–157 (2009)

45. Le, T.H., Kim, Y., Yoon, H.: Electrical and electrochemical properties of conducting polymers. Polymers, **9**(4), 150 (2017)
46. Lee, J.Y., Bashur, C.A., Goldstein, A.S., Schmidt, C.E.: Polypyrrole-coated electrospun PLGA nanofibers for neural tissue applications. Biomaterials **30**, 4325–4335 (2009)
47. Lee, J.W., Serna, F., Nickels, J., Schmidt, C.E.: Carboxylic acid-functionalized conductive polypyrrole as a bioactive platform for cell adhesion. Biomacromolecules **7**, 1692–1695 (2006)
48. Levenberg, S., Rouwkema, J., Macdonald, M., Garfein, E.S., Kohane, D.S., Darland, D.C., Marini, R., Van Blitterswijk, C.A., Mulligan, R.C., D'Amore, P.A., Langer, R.: Engineering vascularized skeletal muscle tissue. Nat. Biotechnol. **23**(7), 879–884 (2005)
49. Li, L., Ge, J., Guo, B., Ma, P.X.: In situ forming biodegradable electroactive hydrogels. Polym. Chem. **5**, 2880–2890 (2014)
50. Li, Y., Neoh, K.G., Kang, E.T.: Plasma protein adsorption and thrombus formation on surface functionalized polypyrrole with and without electrical stimulation. J. Colloid Interface Sci. **275**, 488–495 (2004)
51. Li, D.-F., Wang, W., Wang, H.-J., Jia, X.-S., Wang, J.-Y.: Polyaniline films with nanostructure used as neural probe coating surfaces. Appl. Surf. Sci. **255**, 581–584 (2008)
52. Li, L.C., Yu, M., Ma, P.X., Guo, B.L.: Electroactive degradable copolymers enhancing osteogenic differentiation from bone marrow-derived mesenchymal stem cells. J. Mater. Chem. B **4**, 471–481 (2016)
53. Makris, E.A., Gomoll, A.H., Malizos, K.N., Hu, J.C., Athanasiou, K.A.: Repair and tissue engineering techniques for articular cartilage. Nat. Rev. Rheumatol. **11**, 21–34 (2015)
54. Martins, N.C.T., Moura e Silva, T., Montemora, M.F., Fernandes, J.C.S., Ferreira, M.G.S.: Electrodeposition and characterization of polypyrrole films on aluminum alloy 6061-T6. Electrochim. Acta **53**, 4754–4763 (2008)
55. Meng, S., Rouabhia, M., Shi, G., Zhang, Z.: Heparin dopant increases the electrical stability, cell adhesion, and growth of conducting polypyrrole/poly(L, L-lactide) composites. J. Biomed. Mater. Res. **87A**, 332–3344 (2008)
56. Meng, S., Zhang, Z., Rouabhia, M.: Accelerated osteoblast mineralization on a conductive substrate by multiple electrical stimulation. J. Bone Miner. Metab. **29**, 535–544 (2011)
57. Mohamadali, M., Irani, S., Soleimani, M., Hosseinzadeh, S.: PANi/PAN copolymer as scaffolds for the muscle cell-like differentiation of mesenchymal stem cells. Polym. Adv. Technol. **28**, 1078–1087 (2017)
58. Moutsatsou, P., Coopman, K., Georgiadou, S.: Biocompatibility assessment of conducting PANI/chitosan nanofibers for wound healing applications. Polymers **9**, 687 (2017)
59. Nezakati, T., Seifalian, A., Tan, A., Seifalian, A.M.: Conductive polymers: opportunities and challenges in biomedical applications. Chem. Rev. **118**, 6766–6843 (2018)
60. Nuge, T., Tshai, K.Y., Lim, S.S., Nordin, N., Hoque, M.E.: Characterization and optimization of the mechanical properties of electrospun gelatin nanofibrous scaffolds. World J. Eng. **17**(1), 12–20 (2020)
61. Ostrovidov, S., Ebrahimi, M., Bae, H., Nguyen, H.K., Salehi, S., Kim, S.B., Kumatani, A., Matsue, T., Shi, X., Nakajima, K., Hidema, S., Osanai, M., Khademhosseini, A.: Gelatin–polyaniline composite nanofibers enhanced excitation-contraction coupling system maturation in myotubes. ACS Appl. Mater. Interfaces, **9**, 42444–42458 (2017)
62. Palza, H., Zapata, P.A., Angulo-Pineda, C.: Electroactive smart polymers for biomedical applications. Materials, **12**(2), 27712(2) (2019)
63. Park, Y., Jung, J., Chang, M.: Research progress on conducting polymer-based biomedical applications. Appl. Sci. (Switzerland), **9**(6), 1–20 (2019)
64. Ravichandran, R., Sundarrajan, S., Venugopal, J.R., Mukherjee, S., Ramakrishna, S.: Applications of conducting polymers and their issues in biomedical engineering. J. R. Soc. Interface, **7**, S559–S579 (2010)
65. Rivers, T.J., Hudson, T.W., Schmidt, C.E.: Synthesis of a novel, biodegradable electrically conducting polymer for biomedical applications. Adv Funct Mater **12**, 33–37 (2002)
66. Schultze, J.W., Karabulut, H.: 2005 Application potential of conducting polymers. Electrochim. Acta **50**, 1739–1745 (2005)

67. Shafei, S., Foroughi, J., Stevens, L., Wong, C.S., Zabihi, O., Naebe, M.: Electroactive nanostructured scaffold produced by controlled deposition of PPy on electrospun PCL fibres. Res. Chem. Intermed. **43**, 1235–1251 (2017)
68. Shahini, A., Yazdimamaghani, M., Walker, K.J., Eastman, M.A., Hatami-Marbini, H., Smith, B.J., Ricci, J.L., Madihally, S.V., Vashaee, D., Tayebi, L.: 3D conductive nanocomposite scaffold for bone tissue engineering. Int. J. Nanomed. **9**, 167 (2013)
69. Shi, G., Rouabhia, M., Wang, Z., Dao, L.H., Zhang, Z.: A novel electrically conductive and biodegradable composite made of polypyrrole nanoparticles and polylactide. Biomaterials **25**, 2477–2488 (2004)
70. Shi, Z., Gao, H., Feng, J., Ding, B., Cao, X., Kuga, S., Wang, Y., Zhang, L., Cai, J.: In situ synthesis of robust conductive cellulose/ polypyrrole composite aerogels and their potential application in nerve regeneration. Angew. Chem., Int. Ed. **53**, 5380–5384 (2014)
71. Shirakawa, H., Louis, E.J., MacDiarmid, A.G., Chiang, C.K., Heeger, A.J.: Synthesis of electrically conducting organic polymers: halogen derivatives of polyacetylene, (CH)x. J. Chem. Soc. Chem. Commun. **16**, 578–580 (1977)
72. Smith, L.A., Liu, X., Ma, P.X.: Tissue engineering with nanofibrous scaffolds. Soft Matter. **4**, 2144–2149 (2008)
73. Spearman, B.S., Hodge, A.J., Porter, J.L., Hardy, J.G., Davis, Z.D., Xu, T., Zhang, X., Schmidt, C.E., Hamilton, M.C., Lipke, E.A.: Conductive interpenetrating networks of polypyrrole and polycaprolactone encourage electrophysiological development of cardiac cells. Acta Biomater. **28**, 109–120 (2015)
74. Spencer, A.R., Primbetova, A., Koppes, A.N., Koppes, R.A., Fenniri, H., Annabi, N.: Electroconductive gelatin methacryloyl-PEDOT:PSS composite hydrogels: design, synthesis, and properties. ACS Biomater. Sci. Eng. **4**, 1558–1567 (2018)
75. Sun, B., Long, Y.Z., Zhang, H.D., Li, M.M., Duvail, J.L., Jiang, X.Y., Yin, H.L.: Advances in three-dimensional nanofibrous macrostructures via electrospinning. Prog. Polym. Sci. **39**, 862–890 (2014)
76. Tandon, B., Magaz, A., Balint, R., Blaker, J.J., Cartmell, S.H.: Electroactive biomaterials: vehicles for controlled delivery of therapeutic agents for drug delivery and tissue regeneration. Adv. Drug Deliv. Rev. **2018**(129), 148–168 (2017)
77. Tschmelak, J., et al.: Automated water analyser computer supported system (AWACSS). Part II. Intelligent, remote-controlled, cost-effective, on-line, water-monitoring measurement system. Biosens. Bioelectron. **20**, 1509–1519 (2005)
78. Vacanti, J.P., Vacanti, C.A.: The history and scope of tissue engineering. In: Principles of Tissue Engineering, 4th edn, pp 3–8. Elsevier (2014)
79. Wallace, G.G., Smyth, M., Zhao, H.: Conducting electroactive polymer-based biosensors. Trends Analyt. Chem. **18**, 245–251 (1999)
80. Wang, S., Sun, C., Guan, S., Li, W., Xu, J., Ge, D., Zhuang, M., Liu, T., Ma, X.: Chitosan/gelatin porous scaffolds assembled with conductive poly (3, 4-ethylene dioxythiophene) nanoparticles for neural tissue engineering. J. Mater. Chem. B **5**, 4774–4788 (2017)
81. Wang, W., Tan, B., Chen, J., Bao, R., Zhang, X., Liang, S., Shang, Y., Liang, W., Cui, Y., Fan, G., Jia, H., Liu, W.: An injectable conductive hydrogel encapsulating plasmid DNA-eNOs and ADSCs for treating myocardial infarction. Biomaterials **160**, 69–81 (2018)
82. Wang, L., Wu, Y.B., Hu, T.L., Guo, B.L., Ma, P.X.: Electrospun conductive nanofibrous scaffolds for engineering cardiac tissue and 3D bioactuators. Acta Biomater. **59**, 68–81 (2017)
83. Woodard, L.N., Grunlan, M.A.: Hydrolytic degradation, and erosion of polyester biomaterials. ACS Macro Lett. **7**, 976–982 (2018)
84. Wu, Y.B., Wang, L., Guo, B.L., Ma, P.X.: Interwoven aligned conductive nanofiber yarn/hydrogel composite scaffolds for engineered 3D cardiac anisotropy. ACS Nano **11**, 5646–5659 (2017)
85. Xu, H., Holzwarth, J.M., Yan, Y., Xu, P., Zheng, H., Yin, Y., Li, S., Ma, P.X.: Conductive PPY/PDLLA conduit for peripheral nerve regeneration. Biomaterials **35**, 225–235 (2014)
86. Xu, L.B., Chen, W., Mulchandani, A., Yan, Y.: Reversible conversion of conducting polymer films from superhydrophobic to superhydrophilic. Angew. Chem. Int. Ed. **44**, 6009–6012 (2005)

87. Zhang, Z., Rouabhia, M., Wang, Z., Roberge, C., Shi, G., Roche, P., Li, J., Dao, L.H.: Electrically conductive biodegradable polymer composite for nerve regeneration: electricity-stimulated neurite outgrowth and axon regeneration. Artif. Organs **31**, 13–22 (2007)
88. Zhang, Z., Roy, R., Dugré, F.J., Tessier, D., Dao, L.H.: In vitro biocompatibility study of electrically conductive polypyrrole-coated polyester fabrics. J. Biomed. Mater. Res. **57**, 63–71 (2001)
89. Zhao, X., Li, P., Guo, B.L., Ma, P.X.: Antibacterial and conductive injectable hydrogels based on quaternized chitosan-graftpolyaniline/oxidized dextran for tissue engineering. Acta Biomater. **26**, 236–248 (2015)
90. Zhao, F., Shi, Y., Pan, L.J., Yu, G.H.: Multifunctional nanostructured conductive polymer gels: synthesis, properties, and applications. Acc. Chem. Res. **50**, 1734–1743 (2017)
91. Zhao, X., Wu, H., Guo, B., Dong, R., Qiu, Y., Ma, P.X.: Antibacterial anti-oxidant electroactive injectable hydrogel as self-healing wound dressing with hemostasis and adhesiveness for cutaneous wound healing. Biomaterials **122**, 34–47 (2017)
92. Zhou, D.D., Cui, X.T., Hines, A., Greenberg, R.J.: Conducting polymers in neural stimulation applications. In: Zhou, D.D., Greenbaum, E. (eds.) Implantable Neural Prostheses, vol. 2, pp. 217–5215. Springer (2010)

Conducting Polymer-Based Nanocomposites Against Pathogenic Bacteria

Sumayah Abdelnasir, Areeba Anwar, and Ayaz Anwar

Abstract Conducting polymers are polymers with a network of free electrons and hence, can conduct electricity. Nanomaterials doped conducting polymers are generally referred to as conducting polymer nanocomposites. These nanocomposites have recently emerged as pivotal alternatives in the drug delivery and therapy of infectious diseases caused by pathogenic bacteria. Bacterial infections are one of the leading contributors in high rate of morbidity and mortality worldwide. Furthermore, the emergence of multi-drug resistance in pathogenic bacteria has worsened the disease burden. Albeit the advances in antimicrobial therapy, the success in development of new antibiotics have been limited. Recently, a variety of nanocomposites based on nitrogen-containing conducting polymers including polyaniline (PANI), polypyrrole (PPy) and polycarbazole (PC) as well as sulphur-containing conducting polymers including polythiophene (PT) and poly(3,4-ethylenedioxythiophene) (PEDOT), have been investigated as antibacterial platforms against pathogenic bacteria. Among nanoparticles, the most widely used metals include gold, silver, palladium, copper, titanium oxide, zinc oxide and iron oxide. However, the exact mechanism of their antibacterial activity is not completely understood. It is evident from some studies that conducting polymers-based nanocomposites are biocompatible materials making them ideal candidates for drug development against infectious diseases. Moreover, conducting polymers and their nanocomposites have also been used to study antibacterial activity. All in all, at present, the use of conducting polymer-based nanocomposites against bacterial infections has not been studied extensively. In this chapter we will summarize the recent reports of uses of conducting polymer-based nanocomposites for the therapy of pathogenic bacteria and will also identify the research gaps in this field. Furthermore, we will also propose a future perspective to overcome existing limitations and suggest some possible solutions based on our expertise. This chapter in our opinion, will benefit the researchers working in the field of nanomedicine as well as antimicrobials, and will be of high value to students, clinicians and pharmaceutical sectors.

S. Abdelnasir · A. Anwar · A. Anwar (✉)
Department of Biological Sciences, School of Science and Technology, Sunway University, Petaling Jaya 47500, Selangor, Malaysia
e-mail: ayazanwarkk@yahoo.com

© Springer Nature Switzerland AG 2021

S. Shahabuddin et al. (eds.), *Advances in Hybrid Conducting Polymer Technology*, Engineering Materials, https://doi.org/10.1007/978-3-030-62090-5_11

Keywords Conducting polymers · Nanocomposites · Nanomaterials · Antibacterial · Multi-drug resistance

1 Conducting Polymers

A polymer is a large molecule composed of many repeating subunits. Traditionally, polymers are utilised as insulators in electrical applications [61]. Their thermal stability, high resistivity, flexibility, mouldability, durability and low cost processability make them advantageous materials for wide daily use as such [65]. First successfully produced in the 1970s, conducting polymers (CP) have since gained much attention due to their ability to conduct electricity while retaining the feasible properties of conventional polymers such as flexibility. Today there are over 25 conducting polymer systems [3, 78].

Research on conducting polymers can be dated back to the 1960s when Giulio Natta polymerised acetylene gas using Zeigler-Natta catalyst [63]. The polyacetylene produced as a result was later measured to be semi-conducting at best, with a conductivity of about 10^{-7} S/cm^2. This work was picked up a decade later by Shirakawa, MacDiarmid and Heeger who synthesised polyacetylene as a thin shiny film and discovered that its conductivity can be increased up to 10^3 S/cm^2 when oxidised with iodine vapour [69]. This process by which a polymer is either reduced or oxidised to increase its conductivity was termed 'doping' [3].

Inherently, a pattern of alternating single and double bonds exists along the backbone of polymers and is known as conjugation. Conjugated polymers exhibit semiconductivity due to the presence of delocalised electrons that are able to move along the polymer chain [78]. In brief, in conjugated polymers, strong σ bonds are formed from 3 of the 4 valence electrons of each carbon atom through sp^2 hybridisation. The fourth valence electron remains in a p_z orbital and forms a π bond when it overlaps with a neighbouring p_z orbital. When the π electrons of the conjugated p_z orbitals overlap, an extended p_z orbital is created, in which electrons can move freely [19]. This delocalisation of π electrons gives conjugated polymers their minimal conductivity. However, doping is essential in creating polymers that are highly electrically conductive or conducting polymers. Movement of delocalised electrons in undoped, conjugated polymers is limited because the orbitals are occupied by electrons. The removal or addition of electrons through oxidation or reduction in doping allows more free movement and thus dramatically increases conductivity [19].

This early work of Shirakawa and team on conducting polymers triggered intense research on the topic which resulted in the discovery of many more novel conducting polymer systems including polyheterocycles; a family of conductive polymers that are stable and retain high conductivity. The antibacterial activity of π-conjugated polymers, polyaniline (PANI) and polypyrrole (Ppy), was first reported in 2005 by Seshadri and Bhat in textile applications [68]. The team prepared PANI fabrics that elicited a 95 and 85% reduction in the bacterial activity of gram positive (*Staphylococcus aureus*) as well as gram negative (*Escherichia coli*) respectively. Since,

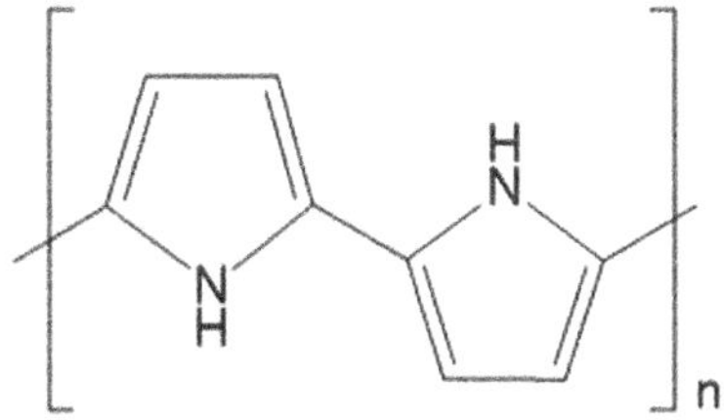

Fig. 1 Chemical structure of polypyrrole

these conducting polymers have been further explored for these purposes and studies expanded to incorporate other polyheterocycles. Current research on conducting polymers revolves mainly around the polyheterocycles [31]: polythiophene (PT), polypyrrole (PPy), poly(3,4-ethylenedioxythiophene) (PEDOT), polyaniline (PANI) and polycarbazole (PC). Generally, conducting polymers can be classified into two groups namely nitrogen-containing CPs and sulphur containing CPs.

1.1 Polypyrrole (PPy)

Polypyrrole is arguably the most extensively studied conducting polymer and has been utilised in a number of applications including as a biomaterial in tissue engineering [8], blood conduits and neural probes [29]. This conducting polymer is a polyheterocycle Fig. 1 and has also been made use of as protection from corrosion, in microsurgical tools and drug delivery systems. It has been found to harbour numerous advantageous properties [5]. For example, polypyrrole is notably stable in air and water, biocompatible in vitro as well as in vivo, highly conductive under physiological conditions and reasonably soluble in water [16]. Using a wide range of solvents, polypyrrole can be synthesised with large surface areas, varying porosities, and in large quantities. Moreover, polypyrrole allows easy incorporation of bioactive molecules to increase suitability of polypyrrole for use in biomedical applications [2]. However, after synthesis, the molecular structure of polypyrrole gives it a rigid, brittle, non-thermoplastic and insoluble nature making further processing of the polymer difficult. For these qualities, including its conductivity Table 1, polypyrrole has also been extensively used in the production of electrochemical biosensors and batteries [28]. As an antibacterial agent, Ppy has been relatively thoroughly explored to be effective especially in a composite of nanomaterials or another conducting polymer such as PANI, which is further discussed in succeeding sections [43].

1.2 Polyaniline (PANI)

Polyaniline, also known as aniline black, has a unique structure of alternating nitrogen atoms and benzene rings [28] Fig. 2. In the sp^2 hybridised state, the nitrogen atoms

Table 1 Reviewed conducting polymers and their abbreviation, conductivity and chemical structure

Conducting polymer	Abbreviation	Conductivity (S/cm)	Structure
Polypyrrole	Ppy	2–100	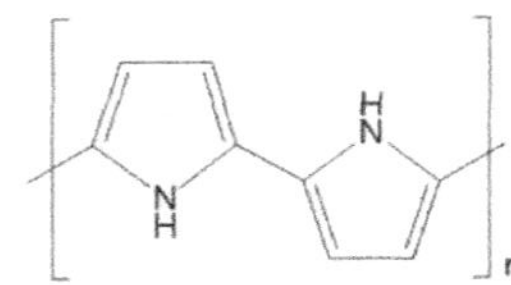
Polyaniline	PANI	~200	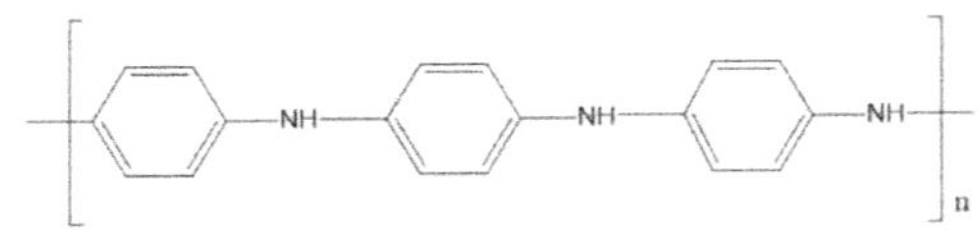
			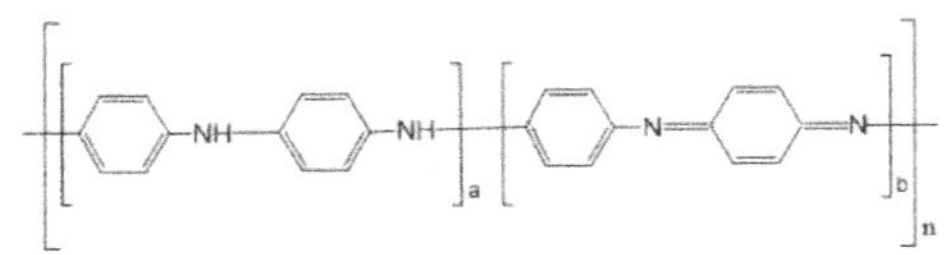
			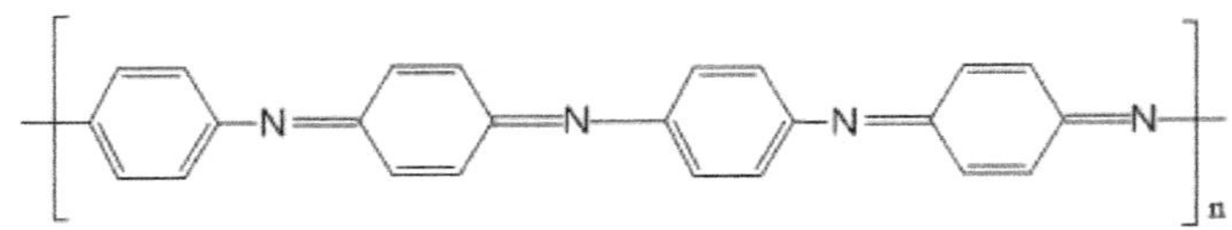

(continued)

Table 1 (continued)

Conducting polymer	Abbreviation	Conductivity (S/cm)	Structure
Polythiophene	PTh	>100	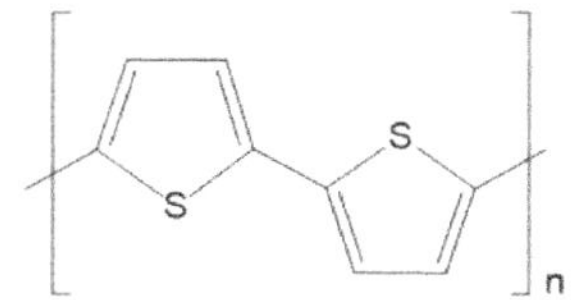
Poly(3,4-ethyldioxythiophene)	PEDOT	Up to 80	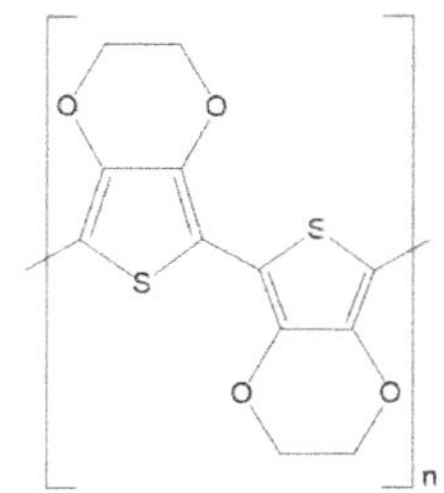
Polycarbazole	PCz	10^{-3}	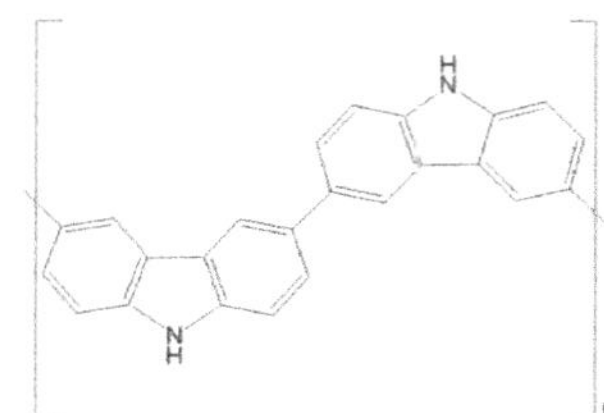

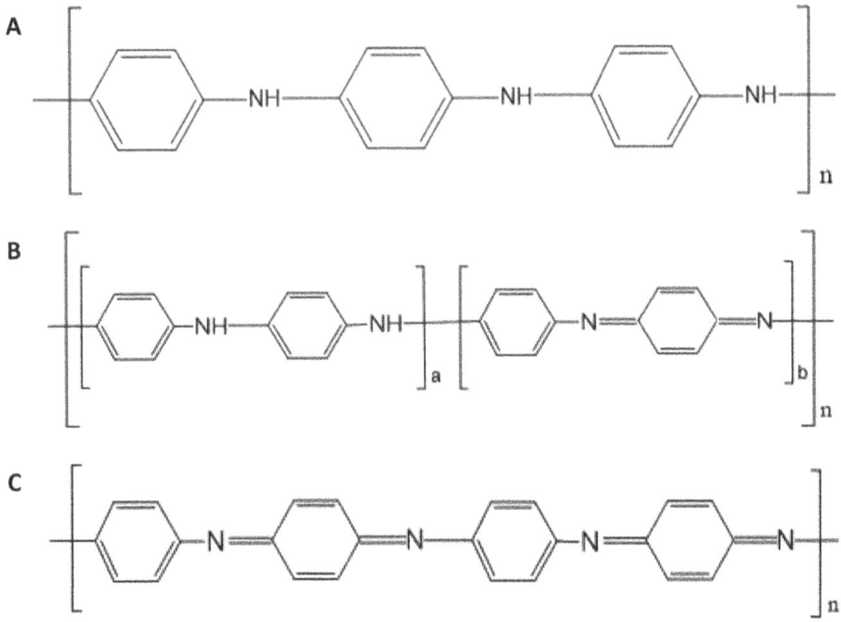

Fig. 2 Chemical structures of polyaniline: **a** Leucoemeraldine (fully reduced); **b** Emeraldine (half-oxidised half-reduced); **c** Pernigraniline (fully oxidised)

exist as imine and in an sp^3 hybridised state, as amines. Due to this, polyaniline can exist in a variety of forms based on the level of oxidation. These forms include: pernigraniline (fully oxidised) Fig. 2c, emeraldine (half-oxidised) Fig. 2b and leucoemeraldine (fully reduced) Fig. 2c [81]. Of these, the green protonated PANI emeraldine has gained the most attention due to its high stability and conductivity. It is half-oxidised i.e. it possesses both the oxidised iminium and the reduced amine nitrogens in equal quantities. In addition to its environmental stability and high conductivity, PANI emeraldine is also cost-effective and relatively easy to synthesise [4]. For these reasons, this conducting polymer has been utilised in numerous biological applications including as protection from corrosion, electrodes in batteries and in microelectronics [14]. However, in comparison to polypyrrole, PANI emeraldine is relatively non-biodegradable, less flexible and less processible [5, 36]. Amongst other conducting polymers, Ppy and PANI have garnered comparatively greater interest in antimicrobial applications and otherwise for their unique properties and facile synthesis in chemical as well as electrochemical techniques. Since their introduction as potential antimicrobial agents, PANI's antibacterial efficacy has been described against gram-positive and gram-negative bacteria as will be discussed further in the chapter [54].

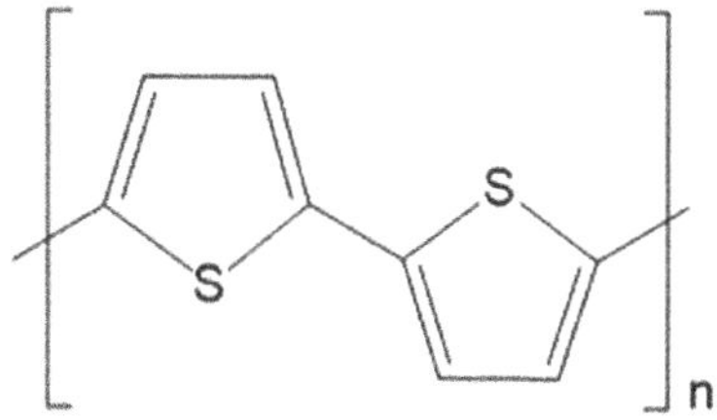

Fig. 3 Chemical structure of polythiophene

1.3 Polythiophene (PTh)

Polythiophene is similar to polypyrrole in structure with the substitution of the NH group in polypyrrole with S in polythiophene Fig. 3. When electrochemically polymerised, PTh has a high level of electrical conductivity, comparable to polypyrrole—greater than 100 S/cm Table 1. However, when chemically polymerised, polythiophene has been found to have a markedly lower conductivity [78]. Further, good environmental stability as well as thermal stability has been observed with polythiophene making it suitable for a range of applications. Among these is the use of PTh in the production of electrochromic devices as well as in organic thin-film transistors [83]. Higher solubilities and capacitor behaviour have been achieved in a number of polythiophene derivatives, including PEDOT, synthesised by chemical or electrochemical techniques [4]. Research on the application of PTh, in its pristine form, for antimicrobial therapy is comparatively limited. Nevertheless, the conducting polymer has been incorporated in formulation with metals including copper and gold in nanocomposites to achieve greater antibacterial potential [1, 53].

1.4 Poly(3,4-Ethyldioxythiophene) (PEDOT)

Derived from polythiophene, PEDOT is synthesised from the polymerisation of the bicyclic monomer 3,4-ethylenedioxythiophene [5, 74]. PEDOT has improved properties relative to PTH as a result of a dioxy alkylene bridging group across the 3- and 4-positions of its heterocyclic ring [52] Fig. 4. The presence of this bridging group has been observed to lower its band gap as well as its reduction and oxidation potential and thus bring about better conductivity as well as stability [21]. PEDOT has been utilised in biosensing as well as for bioengineering applications including in heart muscle patches and nerve grafts for its in vivo biocompatibility [21, 30]. Akin to polythiophene (PTh), studies encompassing the antibacterial capacity of PEDOT have been narrow in comparison the evaluation of PANI and Ppy in the field. Howbeit, more recently, PEDOT-based composites such as PEDOT: poly(styrene-sulfonate) [40] are being appraised for antibacterial proficiency.

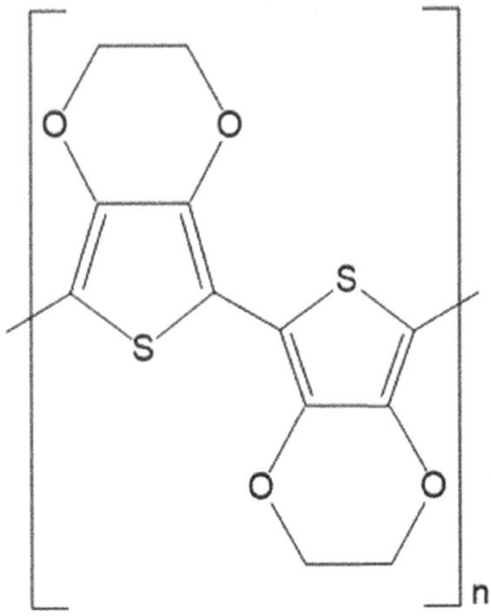

Fig. 4 Chemical structure of PEDOT

1.5 Polycarbazole (PC)

Polycarbazole is an electroactive conducting polymer of carbazole monomers consisting of a five-membered nitrogen-containing ring with two six-membered rings on its sides [67] Fig. 5. It has been polymerised chemically as well as electrochemically to exhibit a conductivity of about 10^{-3} S/cm Table 1. Polycarbazole has been utilised in the production of batteries and, similar to PTh, electrochromic devices [4, 32]. Amid the other conducting polymers discussed, Pristine PCz has been shown to elicit minimal antibacterial activity that was reported to be increased in all cases by its integration in nanocomposites including PCz/TiO_2 which displayed antimicrobial action comparable to that of the well-known antibiotic Ciprofloxacin [25].

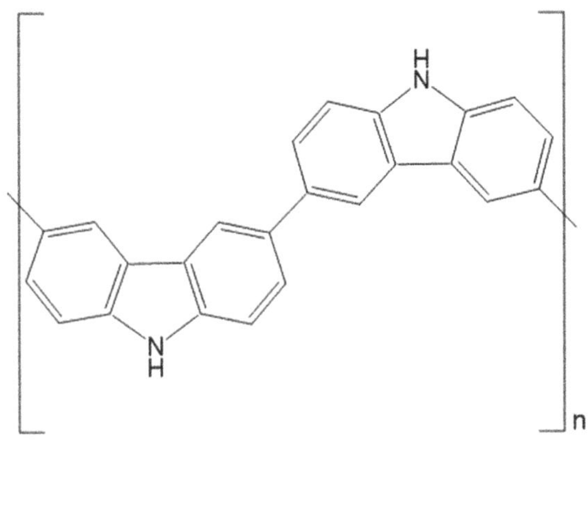

Fig. 5 Chemical structure of Polycarbazole

2 Pathogenic Bacteria

Yearly, approximately 17 million deaths worldwide are attributed to infectious diseases [58]. An era of drug resistance and multi-drug resistance has emerged impeding the treatment of these infections markedly and posing a dire demand for the development of novel therapeutics [55].

ESKAPE is a group comprising of primary bacterial pathogens known to be highly virulent and multi-drug resistant. Namely, these are *Enterococcus faecium*, *Staphylococcus aureus*, *Klebsiella pneumonia*, *Acinetobacter baumannii*, *Pseudomonas aeruginosa* and *Enterobacter* spp. [66].

Staphylococcus aureus is a virulent bacterial species that is gram positive and infects both immunocompetent as well as immunocompromised individuals [23]. Despite being a part of the normal flora of the human body, *S. aureus* is also an opportunistic pathogen and the causative agent of a number of diseases such as infections of the skin and soft tissue as well as infections of the bloodstream which often require immediate attention [76]. Resistance to penicillin is observed in most strains of *S. aureus* through the production of plasmid encoded penicillinase—an enzyme that interrupts the interior structure of penicillin hindering the antimicrobial action of the drug [6]. Methicillin-resistant *S. aureus* (MRSA) is becoming increasingly and especially problematic exhibiting resistance to β-lactam antimicrobials including penicillin, carbapenems and cephalosporins. Resistance is acquired through its expression of a mecA gene that codes for penicillin-binding protein 2A (PBP2A) which has very low affinity for β-lactam antibiotics inhibiting these antimicrobials from affecting bacterial cell wall synthesis [17].

Similar to *S. aureus*, *Enterococcus faecium* is a gram-positive bacterium and exists as a commensal in the human body [27]. *E. faecium* has developed extensive resistance to antimicrobials with the rise of antibiotic resistance in recent decades becoming significant threats in communities. This intrinsic antibiotic resistance in addition to its ability to rapidly acquire additional resistance makes *E. faecium* infections difficult to treat. For instance, *E. faecium* acquires antimicrobial resistance by the acetylation of antibiotics and efflux pumping amongst other resistance mechanisms [46].

Recently, multi-drug resistant strains of *Klebsiella pneumonia*—termed hypervirulent—are being increasingly isolated limiting options for therapeutics. This gram-negative bacterium is found in natural environments such as water and soil ubiquitously and besides pneumonia, also causes infections of the urinary tract and bloodstream [7]. *K. pneumonia* infections occur after the bacteria exist on mucosal surfaces non-pathogenically then disperse to other tissues where infections with consequential mortality and morbidity are elicited [59]. *Klebsiella* species are able to spread antimicrobial resistant genes to other gram-negative bacteria as they are known to be a reservoir of these genes [45]. Additionally, over one third of isolated *K. pneumoniae* were found to be resistant to one or more antibiotic group. Amongst numerous resistance mechanisms, *K. pneumoniae* is able to produce β-lactamases—enzymes that are able to break open the β-lactam ring of β-lactam antibiotics, deactivating the molecule's antibacterial properties [35].

Multi-drug resistant isolates of *Acinetobacter baumannii* have been obtained and seen to be capable of rapidly acquiring supplementary resistance to antibiotics [47]. This opportunistic gram-negative bacterium is the causative factor of skin, bloodstream, and urinary tract infections [62]. They are able to resist drugs through multidrug efflux pumps, enzymatic break down of drugs, and other acquired mechanisms [39, 49].

Infections caused by *Pseudomonas aeruginosa* have notable morbidity and mortality and are observed in mainly in immunocompromised patients [70]. *P. aeruginosa* is inherently resistant to some antimicrobials due to the relative impermeability of its outer membrane which is primarily comprised with small porins (OprB and OprD) [15]. Other resistance mechanisms that *P. aeruginosa* manifest include biofilm production, efflux pumps and the expression of β-lactamases [50].

Similarly, *Enterobacter* spp. are also able to resist antimicrobials through a number of mechanisms which include the production of β-lactamases and efflux pumps [22]. *Enterobacter* spp. are opportunistic pathogens responsible for threatening infections of the urinary and respiratory tracts [34, 82].

Overall, there is a dire need for the development of new antimicrobial agents that are able to overcome these acquired resistance mechanisms. Thus, new avenues have been increasingly investigated for this purpose, including the use of conducting polymers and their nanocomposites.

3 Antibacterial Effects of Conducting Polymers

Amongst many conducting polymers, polypyrrole and polyaniline have been mainly and increasingly used for numerous applications such as chemical sensors, photocatalysts and also as antibacterial agents. Numerous studies have evaluated the potential of conducting polymers as antimicrobial agents.

In a study by da Silva et al. [18], branched polypyrrole as well as conventional polypyrrole were assessed for antibacterial capacity against *E. coli*, *S. aureus* and *K. pneumonia* in disk diffusion assays. This study also investigated the effect of the incorporation of silver nanoparticles on the polymer's antimicrobial abilities. After treatment with branched Ppy for 24 h, inhibition zones on *S. aureus*, *E. coli* and *K. pneumonia* were measured to be 26 mm, 19.1 mm and 18.5 mm respectively. It was noted that *S. aureus* was most susceptible to branched Ppy despite possessing a thicker peptidoglycan cell wall. Conventional Ppy was studied to be more efficacious than its branched variant in inhibiting bacterial growth with zones of inhibition measuring 35 mm, 36 mm and 30 mm in diameter against *S. aureus*, *E. coli* and *K. pneumonia* respectively. Further, the incorporation of silver nanoparticles was found to improve antibacterial action by a maximum of 28% against all species tested [18].

Polypyrrole's antibacterial capacity was also assessed when used in combination with PANI in another study which applied a one pot method of oxidative polymerisation to synthesise a co-polymer of chitosan functionalised polyaniline-polypyrrole (CS/PANI/Ppy) [43]. The polymer was analysed with SEM, X-ray spectroscopy and FT-IR as spherical particles about 0.2 μm in size. Kumar et al. [43] assessed its antimicrobial activity against *E. coli* and *Enterobacter agglomerans* through disk diffusion assays and liquid broth dilution to measure percentage viability. Similar inhibition zones and percentage viability resulted from the treatment of both bacterial species with CS/PANI/Ppy when compared to treatment with PANI/Ppy. For example, against *E. coli*, CS/PANI/Ppy elicited an inhibition zone with diameter 17 0.5 mm whereas PANI/Ppy gave rise to a 17.2 mm zone of inhibition. However, the incorporation of PANI with Ppy allowed for enhanced antibacterial effects. For instance, the viability of *E.* coli after treatment with 50 μg/mL Pristine PANI and pristine Ppy was 20% and 55% respectively. Comparatively, treatment with the same concentration of PANI/Ppy reduced viability of *E. coli* to 5%. Increasing the concentration of PANI/Ppy to 75 μg/mL reduced *E. coli* viability to less than 1% [43]. This study suggests that pristine PANI harbours greater antimicrobial properties against *E. coli* in comparison to pristine Ppy.

PANI, in its pristine form, has also been largely studied for antibacterial efficacy. In one of these studies, Kucekova et al. [41] synthesised and characterised colloidal PANI dispersions through the oxidation of aniline hydrochloride with ammonium persulfate (APS) and using poly(N-vinylpyrrolidone) as a stabilising agent. Broth inoculation techniques and subsequent colony counting to determine MIC were employed in this study to determine the antibacterial activity of PANI against *B. cereus*, *S. aureus*, *E. coli* and *P. aeruginosa*. Amongst the bacterial species, *P. aeruginosa* exhibited most resistance to PANI which caused significant antibacterial effect only at 8500 μg/mL. At this concentration of colloidal PANI, the growth of *S. aureus* was reduced by about 3 log CFU/mL. At a lower concentration of 3500 μg/mL, colloidal PANI reduced the growth of *P. aeruginosa* insignificantly by 2 log CFU/mL. Nevertheless, the same concentration of PANI (3500 μg/mL) was able to fully inhibit the growth of *E. coli* as well as *B. cereus*. This study also investigated the cytotoxicity of colloidal PANI towards a human keratinocyte cell line (HaCat). Through this it was observed that lower concentrations of PANI (70 μg/mL) did not elicit toxicity towards human cells. However, although generally considered biocompatible, higher concentrations (345 μg/mL) affected the viability as well as morphology of HaCat cells [41].

During the oxidation of monomers of PANI and Ppy, cotton fabric was coated with these polymers in-situ [54]. The coating of cotton fabric was confirmed through FT-IR and Raman spectroscopy. The antibacterial potential of the coated and uncoated cotton was assessed against *E. coli* and *S. aureus*. It was observed that uncoated cotton did not exhibit any antibacterial activity. In comparing PANI-coated cotton and Ppy-coated cotton, PANI-coated cotton elicited improved antibacterial effect against both microbes tested. PANI-coated cotton was able to reduce the viability of bacterial pathogens to a minimum of 10%. In comparison, Ppy-coated cotton decreased the viability of cells down to 19%. This study further investigated the effect of the incorporation of silver particles. The inclusion of silver particles was studied to enhance the antimicrobial efficacy of polymer-coated cotton further. This study complements findings by Kumar et al. [43] wherein PANI displayed greater antimicrobial potential than Ppy [54].

Apart from PANI and Ppy, polythiophene (PTh) has also been investigated to exhibit antibacterial properties. A flexible, polymer-based film was synthesised by Liu et al. [51] by depositing polythiophene (PTh) incorporated with porphyrin onto a poly(ethylene terephthalate) sheet (PET) and assessed for antimicrobial potential. The antibacterial potential was assessed under white light irradiation (400–800 nm) against *E. coli*. Subsequent to spin-coating and oxidative polymerisation, a suspension of *E. coli* was dropped onto the films and CFU that arose were enumerated. After irradiation for 20 min, a killing efficiency of 65% was observed. It was also noted that the incorporation of porphyrin into PTh film enhanced its antimicrobial performance [51].

4 Conducting Polymer-Based Nanocomposites (CP-NC)

'Nanocomposites' is the term used to describe materials with two or more constituents of which at least one is nano-sized. In current years, polymer nanocomposites have gained significant interest becoming primary materials in recent nanotechnology. The dispersion of nanoparticles within polymer matrices or their coating by a polymer give rise to progressive functional materials known as CP-NC with combined advantageous properties of its constituents [56]. The potential applications of conducting polymer-based nanocomposites in a wide variety of fields due to their high stability in aqueous solutions, electrical conductivity, as well as electrocatalytic properties have given these materials growing attention. For one, immobilisation of nanoparticles can be achieved on CP matrices due to their porous nature. Moreover, charge can also be transferred between the dispersed nanoparticles and the substrate resulting in an amelioration of the conductivity of the composite system [33]. All in all, as will be further discussed, the incorporation of nanoparticles into CPs diminishes limitations of the individual materials whilst enhancing the overall performance of the hybrid.

Conducting polymers have been studied to be utilised in a number of areas including biomedical applications from drug delivery systems to cell culture technologies. However, these materials present some limitations such as biocompatibility, low processability and degradability. Such drawbacks have been studied to be overcome by the synthesis of their nanocomposites [44]. Drawbacks of CPs involving thermal and mechanical stability have also been seen to be minimised by the combination of CPs and nanoparticles [9]. Ultimately, the resulting composites usually have increased stability and symbioses between its components.

On another note, recently, fields of medicine, electronics and more have geared substantial attention to the use of nanoparticles. They have been seen to—when conjugated with drugs—be effective in treating various diseases including bacterial infections. More specifically, metal and metal oxide nanoparticles namely silver

(Ag), copper (Cu), gold (Au), palladium (Pd), titanium (Ti) and platinum (Pt), have been intensively investigated as therapeutic agents in healthcare due to their attractive catalytic, electrical and antimicrobial abilities [80]. For example, long-lasting biocidal activity, high temperature stability and low volatility is seen with silver-based antimicrobials. However, their limited overall stability as well as propensity for aggregation make them disadvantageous in some aspects [72]. With the incorporation of these nanoparticles into polymer matrices, stabilisation can be attained which in turn can inhibit aggregation and allow the dispersion of the particles in a number of solvents whilst maintaining their advantageous properties [24]. Moreover, the single-step synthesis of nanoparticles can be achieved using suitable functional groups of polymers as targeted reactive sites.

Preparation of CP-NC has been realised in several ways. In situ reduction of metal salts within a polymer matrix [77], mechanical melt mixing of metal nanoparticle with polymer [71] and the in-situ polymerisation of a monomer in the presence of a metal nanoparticle [37] are some of the techniques used to synthesise CP-NC. Precision in selection and combination of the materials allows optimisation of the desired composite. In general, the type of polymer matrix and the size, dispersion and content of the nanoparticles influence the nature of the composite greatly [20].

5 Antibacterial Effects of Conducting Polymer-Based Nanocomposites

5.1 Polypyrrole-Based Nanocomposites

Further studies have utilised conducting polymers in the preparation of their nanocomposites for the well-studied biocidal effects of nanoparticles. These conducting polymer-based nanocomposites were studied to be efficacious antibacterial agents.

Upadhyay et al. [77] synthesised polypyrrole nanotube-silver nanoparticle nanocomposites (PPy-NT: Ag-NP) through the in-situ reduction of $AgNO_3$ in the presence of polypyrrole nanotubes (PPy-NTs). SEM (Scanning electron microscopy) and TEM (Transmission electron microscopy) techniques were utilised in this study and revealed the nanotubular morphology of Ppy and the nano-sized silver nanoparticles. The antibacterial activity of the nanocomposite was determined at varied concentrations of AgNPs using the disk diffusion method against *Escherichia coli* and *Staphylococcus aureus*. The concentrations of AgNPs that were assessed (with respect to PPy-NTs) were 6, 9, 12 and 15 wt% which were attained by varying the concentration of $AgNO_3$ during the in-situ reduction process. An increased bactericidal effect against both bacterial species was observed with increasing concentrations of AgNPs in the composite. The maximum zone of inhibition reported was 23 mm with 15 wt% of AgNPs against both microorganisms. The antibacterial activity of pure PPy-NTs was not determined as no zone of inhibition was produced when using the disk diffusion technique. Therefore, it was concluded that the antimicrobial activity displayed by the composite was attributable to the presence of AgNPs [77].

The same study investigated the minimum inhibitory concentration (MIC) of the nanocomposite using MTT assay. The two bacterial species were incubated with the nanocomposite for 16 h at varying concentrations of AgNPs. The lowest concentration of nanocomposite that was observed to not produce a purple colour (formazan) was regarded as the MIC. Supportive of the disk diffusion results, the MIC values were observed to decrease with increasing concentrations of AgNPs in the composite against both *E. coli* and *S. aureus*. The MIC was observed with 15 wt% of AgNPs at 0.078 mg/mL against *E. coli* and at 0.15625 mg/mL against *S. aureus*. The thicker wall of the gram-positive *S. aureus* is considered to be responsible for its higher MIC in comparison to the gram-negative *E. coli* [77].

In assessing the biocompatibility of the nanocomposite, in vitro haemolysis activity was measured using different concentrations of the composite against mammalian red blood cells. Haemolysis is the situation wherein red blood cells are broken down and as a result release intracellular haemoglobin into their milieu. From these, it was observed that haemolysis increases with increasing concentrations of PPy-NT: Ag-NP as well as with increasing concentrations of silver nanoparticle content within the composite. Thus, it was deduced that, apart from concentration, the surface area to volume ratio of AgNPs also markedly affects biocompatibility. Less than 5% haemolysis was observed with up to 2.5 mg/mL of PPy-NT: Ag-NP. When the concentration was increased to 10 mg/mL with 15 wt % of AgNP, percent haemolysis increased to 15%. A minimal haemolytic activity of 5% is considered permissible for biomaterials.

In another study, polypyrrole zinc oxide chitosan bionanocomposites (Ppy/ZnO/CS BNC) were electrochemically synthesised by Ebrahimiasl et al. [26] through the electrooxidation of 0.1 M pyrrole in an aqueous solution of suitable quantities of ZnO nanoparticles uniformly dispersed in chitosan with indium tin oxide glass substrate. Using the disk diffusion method on Mueller–Hinton agar, the in vitro antibacterial properties of the bionanocomposite was assessed with varying contents of ZnO nanoparticles at 5, 10, 15, 20 wt% and fixed concentrations of Ppy (0.3 M) and CS (0.7% w/v). Streptomycin antibiotic and dimethyl sulfoxide (DMSO) were used as the positive and negative controls in the study. *Staphylococcus aureus* and *Bacillus cereus* were the gram-positive bacteria used whilst *Pseudomonas aeruginosa* and *Escherichia coli* were the gram-negative bacteria used in the study [26].

High antibacterial activities against both gram positive and gram-negative bacteria were demonstrated by Ppy/ZnO/CS BNC. It was also observed that antibacterial effects (zone of inhibition) increased with an increasing concentration of ZnO nanoparticles with the maximum inhibition at 29 mm for 20 wt% against *P. aeruginosa*. This value was comparable to the inhibition obtained using Streptomycin antibiotic against the same species. The zone of inhibition obtained using Ppy/ZnO/CS BNC with 20 wt% of ZnO against *S. aureus* was observed to be larger than that of Streptomycin [26].

Another Ppy-based nanocomposite was prepared by employing the one-pot synthesis of silver-nanoparticle-decorated polypyrrole nanotubes (AgNPs/Ppy NTs). This was achieved through the vapour-deposition polymerisation (VDP)-mediated hard template method [60]. Here, pyrrole monomers were polymerised by chemical oxidation polymerisation via the reduction of Fe/Ag cations. In this process, silver nanoparticles are formed via the reduction of silver ions deposited along the wall of an anodic aluminium oxide (AAO). To assess antimicrobial potential, disk diffusion was utilised with *Escherichia coli* as well as *Staphylococcus aureus*. Clear zones of inhibition were reported to be produced by AgNP/Ppy NTs against both bacterial species. However, reinforcing findings by Upadhyay et al. [77], pristine Ppy NTs did not produce a zone of inhibition against neither gram positive nor gram positive bacteria. Park et al. [60] further assessed antibacterial activity of the nanocomposite in a kinetic test wherein nanocomposite samples were inoculated with bacterial suspensions and aliquots were taken as a function of time from 0 to 30 min with 5 min intervals. The aliquots were then cultured in LB agar plates and colonies observed after overnight incubation were enumerated. In contrast to their first results, Ppy NTs were seen to inhibit bacterial growth as a function of time. Growth inhibition was observed to be enhanced with the incorporation of AgNPs and further increased with increasing concentrations of the nanoparticles [60]. Finally, microbial growth after treatment with the nanocomposite was visually inspected to determine MIC. 30% (wt/wt) AgNP/Ppy NTs had an MIC of 10 μg/mL against *E. coli* and 200 μg/mL against *S. aureus*.

CP-NC have also been assessed for properties involving the evasion of bacterial resistance mechanisms. Similar to numerous other bacterial pathogens, *P. aeruginosa* employs the formation of biofilm as a resistance mechanism against antibiotics. Resistance to antibiotics increases 1000 times when cells are fixed within a biofilm matrix. Khan et al. [38] assessed chitosan-polypyrrole nanocomposites (CS-Ppy) for biofilm inhibition properties. Inhibition of biofilm formation was observed to be the result of treatment with pristine Ppy as well as CS-Ppy in a concentration-dependent manner. At 512 μg/mL, CS-Ppy showed a significant 91% inhibition of biofilm formation relative to the control. Comparatively, CS and Ppy inhibited the formation of biofilm by 70% and 84% respectively. Thus, it was deduced that the inclusion of Ppy enhanced the inhibitory effect of CS-Ppy. The ability of the nanocomposite to eliminate existing biofilm of *P. aeruginosa* was also evaluated in the study. It was noted that lower concentrations of CS could not eradicate existing biofilm. Nevertheless, CS-Ppy and pristine Ppy manifested a dose-dependent disorder of mature biofilm. Different concentrations of CS, Ppy, and CS-Ppy were tested for their capacity to inhibit the production of virulence factors such as rhamnolipid, pyocyanin, as well as pyroverdine by *P. aeruginosa*. The production of these virulence factors was evaluated to be inhibited by CS, Ppy as well as CS-Ppy. Pyocyanin production was inhibited by 79% as a result of 512 μg/mL of CS-Ppy. Likewise, up to 91% inhibition of the production of pyroverdine was attributable to the same concentration of CS, Ppy and CS-Ppy. Rhamnolipid production by *P. aeruginosa* was also inhibited by 85% due to treatment with 512 μg/mL of the nanocomposite [38].

Chitosan was also utilised by another study in the preparation of Chitosan/polyacrylic acid/polypyrrole (CS/PAA/Ppy) which was used with magnetic Fe_3O_4 nanoparticles in the synthesis of a bionanocomposite hydrogel [79]. The magnetic nanoparticles were prepared through a co-precipitation method which were then added to the nanocomposite in different ratios. CS-PAA hydrogel was prepared by polymerisation of acrylic acid in a solution of chitosan whereas polypyrrole was synthesised and dispersed by oxidative phosphorylation. Characterisation of CS/PAA/Ppy based on Fe_3O_4 through FT-IR, XRD revealed spherical nanoparticles fixed within a polymer matrix. Disk diffusion assays were deployed in evaluating the antibacterial potential of the composite at 1, 3 wt% as well as 5 wt% of Fe_3O_4 nanoparticles against *S. aureus* and *P. aeruginosa.* It was noticed that an increasing content of the nanoparticles caused enhanced antimicrobial effects. Moreover, CS/PAA/Ppy was observed to be more effective in inhibiting bacterial growth relative to CS/PAA. Inhibition zones produced by CS/PAA were measured to be 20 mm and 18 mm on *S. aureus* and *P. aeruginosa* respectively. In contrast, CS/PAA/Ppy with 5 wt% Fe_3O_4 increased inhibition zone diameters to 30 mm on *S. aureus* and 29 mm on *P. aeruginosa* [79].

Another instance presenting the antimicrobial action of a Ppy-based nanocomposite is the study by Khan et al. [37]. Here, a gel combustion technique was utilised in the synthesis of copper-doped zinc oxide nanopowders (CZO). A nanocomposite of polypyrrole and CZO (Ppy/CZO) was then fabricated via in-situ polymerisation in the presence of a surfactant Cetrimonium bromide (CTAB) Fig. 6. The nanocomposite was analysed by XRD, FT-IR as well as SEM to be of nanocrystalline nature. In antibacterial activity assessments, Ppy/CZO was used in disk diffusion and colony

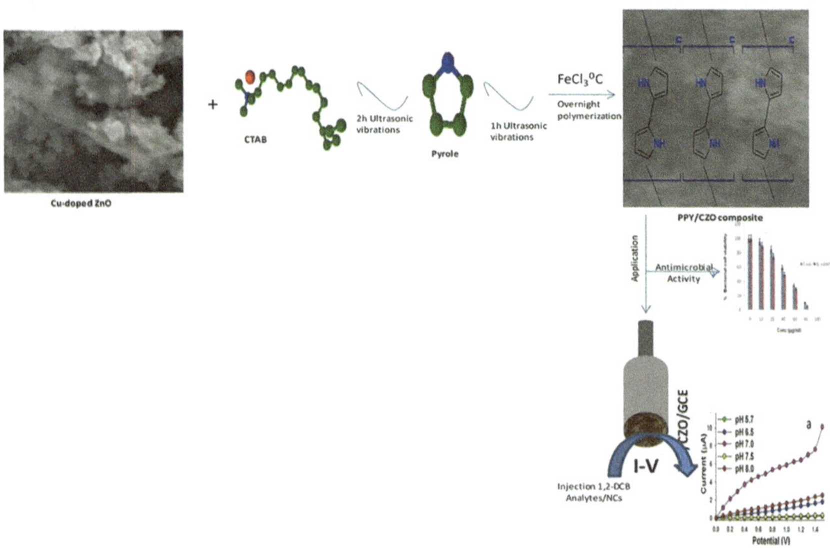

Fig. 6 Schematic representation of the preparation and application of Ppy/CZO [37]

counting assays against *E. coli* and *B. subtilis*. The assays were done in comparison to the commonly used antibiotic—amoxicillin, and percentage viability was calculated. Inhibition zones with diameters 13 mm and 19 mm resulted from the treatment of *E. coli* and *B. subtilis* with Ppy/CZO respectively. Antibacterial efficacy of Ppy/CZO observed was concentration dependent. After treatment with 60 μg/mL of the nanocomposite, *E. coli* had 30% viability and *B. subtilis* exhibited 35% viability. Increasing the concentration of Ppy/CZO resulted in less than 1% viability of both bacterial species [37].

Polypyrrole was also synthesised as a coat of a core–shell structure of carbon nanotubes (CNT/Ppy) by Tondro et al. [75] and its antibacterial and phototherapy effects against *P. aeruginosa* were evaluated. Characterisation by FESEM divulged the deposition of tube-shaped polypyrrole on a smooth surface of carbon nanotubes. CNT/Ppy were used to treat *P. aeruginosa* either in the dark or exposed to laser light irradiation (LIR) with a wavelength of 808 nm for 20 min. Four different groups were established—L^-N^- (Untreated *P. aeruginosa*,in the dark), L^+N^- (Untreated *P. aeruginosa*; exposed to LIR), L^-N^+ (CNT/Ppy treated *P. aeruginosa*; in the dark) and L^+N^+ (CNT/Ppy treated *P. aeruginosa*; exposed to LIR). *P. aeruginosa* from group L^+N^- exhibited a 91% survival rate similar to the control group L^-N^-. From this, it was deduced that the bacterial species was unaffected by laser irradiation only. Treatment of group L^-N^+ with 500 μg/mL of CNT/Ppy, resulted in a decrement in survival rate of the bacterial pathogen—causing a 30–50% killing rate. However, the more marked finding was observed in group L^+N^+ wherein treatment with 50 μg/mL caused a statistically significant decrease in survival of *P. aeruginosa* compared to all other groups. At 250 and 500 μg/mL, a 70% killing rate was displayed [75].

5.2 *Polyaniline-Based Nanocomposites*

Polyaniline (PANI)-based nanocomposites have been increasingly studied for antibacterial potential. These were prepared by Boomi et al. [12] through chemical reduction in the presence of reducing and stabilising agents. Pristine PANI, PANI-platinum (PANI/Pt) and PANI-platinum-palladium (PANI/Pt–Pd) were synthesised and characterised by UV–Vis (Ultraviolet–visible spectroscopy), FTIR (Fourier-transform infrared spectroscopy), XRD (X-ray diffraction), and electron microscopy techniques. The antibacterial potential of PANI and its nanocomposites was assessed in disk diffusion assays against gram-positive *Streptococcus* spp. and *Staphylococcus* spp. as well as gram-negative *E. coli* and *Klebsiella* spp. Fig. 7. Broth micro-dilution was used to evaluate minimum inhibitory concentrations (MIC) against the named bacterial species [12].

No zones of inhibition were observed in disk diffusion assays using colloidal solutions of Pt and Pt–Pd against any of the pathogens studied. Pristine PANI showed zones of inhibition with diameters 14–19 mm. Greater inhibition zones were seen for PANI/Pt and inhibition was enhanced in PANI/Pt–Pd with diameters 18–23 mm and 21–28 mm respectively. The minimum concentration required to inhibit the growth

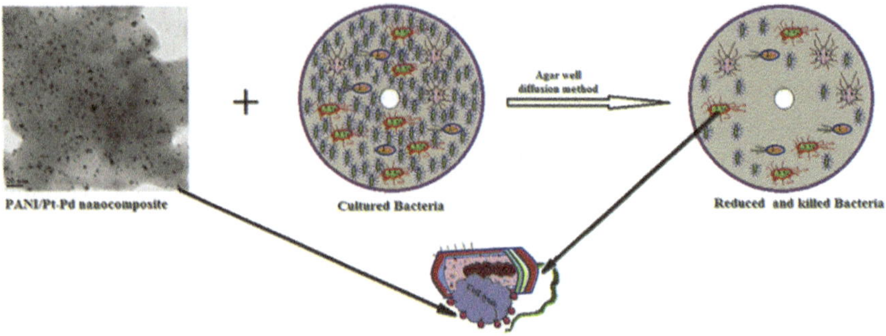

Fig. 7 Illustration of study of antibacterial capacity of PANI/Pt–Pd and its interaction with bacterial cell to cause lysis [12]

of *Staphylococcus* spp. was observed to be 75 μg/mL of pristine PANI, 50 μg/mL of PANI/Pt and 25 μg/mL of PANI/Pt–Pd. From these results it was concluded that although antibacterial capacity is exhibited by pristine PANI, its nanocomposites show relatively improved activities against pathogenic bacteria [12].

In an earlier study by Boomi et al. [11]), polyaniline, polyaniline/silver-platinum (PANI/Ag-Pt) and Ag-Pt colloidal solution prepared in the same technique as [12] were assessed for antibacterial potential. UV–Vis, XRD FTIR and electron microscopy techniques [High-resolution SEM/TEM (HRSEM/HRTEM)] were used to characterise the composites through which a strong interaction between PANI chains and Ag-Pt nanocomposites was observed Fig. 8. Bacterial species used in disk diffusion assays included *S. aureus*, *Streptococcus* spp., *Klebsiella* spp., and *E. coli*. PANI displayed some inhibition of growth in all species tested presenting larger inhibition zones with gram-positive *S. aureus* and *Streptococcus* spp. Ameliorated antimicrobial activity was seen upon treatment of pathogens with PANI/Ag-Pt. Comparatively, against *S. aureus*, *Streptococcus* spp., *Klebsiella* spp., and *E. coli*,

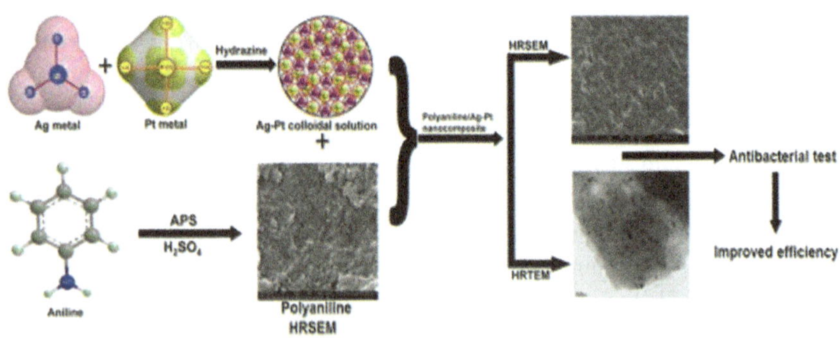

Fig. 8 Diagram illustrating preparation and application of PANI/Ag-Pt [11]

PANI displayed zones of inhibition with diameters 27, 25, 22 and 20 mm respectively, whereas PANI/Ag-Pt presented inhibition zones with diameters 30, 29, 25 and 21 respectively [11].

Another technique to synthesise PANI-based nanocomposite was utilised by Tamboli et al. [73]. Here, the in-situ polymerisation was carried out using APS as an oxidising agent to prepare silver-polyaniline nanocomposites (Ag-PANI) in the presence of silver nitrate ($AgNO_3$). The nanocomposite was characterised by various analytical techniques namely UV–Vis, FT-IR and electron microscopy techniques. The antibacterial activity of the nanocomposite was investigated in disk diffusion assays against *B. subtilis*. Zones of inhibition of about 10–18 mm in diameter were observed upon treating the pathogen with Ag-PANI. MIC and MBC evaluations were done through double-dilution technique wherein the minimum concentration of nanocomposite observed to visually allow cell proliferation is considered the MIC and the minimum concentration at which there is a 3-log reduction in the viability of cells is considered the MBC. The MIC and MBC of Ag-PANI against *B.* subtilis were both found to be 25 μg/mL [73].

In another study evaluating PANI-based nanocomposites for antimicrobial capacity, polyaniline/copper-doped zinc oxide nanocomposites (PANI/$Cu_{0.05}Zn_{0.95}O$) were produced through the in-situ inverse microemulsion method with different contents of copper-doped zinc oxide nanoparticles (CZO)—20, 40, 60 and 80 wt% [48]. The nanocomposites were characterised using XRD, UV–Vis, FT-IR, SEM as well as TEM to affirm the inclusion of CZO into the polymer matrix. *S. aureus* and *E. coli* were used to evaluate the antimicrobial capacity of the nanocomposite in disk diffusion studies. Potent antibacterial activity was observed upon treating the pathogens with PANI/CZO. Copper-doped ZnO exhibited greater antibacterial activity in comparison to pure ZnO against both bacterial species. Further, the antimicrobial action (inhibition zones) of PANI/CZO was found to increase with increasing contents of CZO. Studies investigating MIC and MBC revealed that PANI/CZO (80 wt%) manifested the relatively lowest MIC and MBC at 10 μg/mL and 30 μg/mL respectively. In comparison, PANI displayed inhibitory effects at 1000 μg/mL (MIC) and bactericidal effects at 3000 μg/mL (MBC). CZO was inhibitory at a concentration of 200 μg/mL and bactericidal at 500 μg/mL. Therefore, the involvement of PANI was considered to improve antibacterial performance of the nanocomposite [48].

Similar to [48, 57] also included zinc oxide nanoparticles in the synthesis and analysis of zinc oxide: cerium (IV) oxide: nanocellulose: polyaniline bionanocomposite (ZnO: CeO_2: nanocellulose: PANI). Nanocellulose was isolated from rice husk, ZnO as well as CeO_2 nanoparticles were prepared through a modified sol–gel method and ZnO: CeO_2: nanocellulose: PANI was synthesised using the modified polymerisation of aniline Fig. 9. The bionanocomposite was analysed and characterised to manifest the inclusion of ZnO and CeO_2 nanoparticles in polymeric nanocellulose: PANI matrix by electron microscopy techniques, FT-IR and XRD. TEM analyses confirmed that the in-situ polymerisation of PANI in the presence of nanocellulose, ZnO and CeO_2 allowed for further reduction of ZnO as well as the incorporation of the nanoparticles onto the PANI: nanocellulose matrix [57]. In studying the percentage

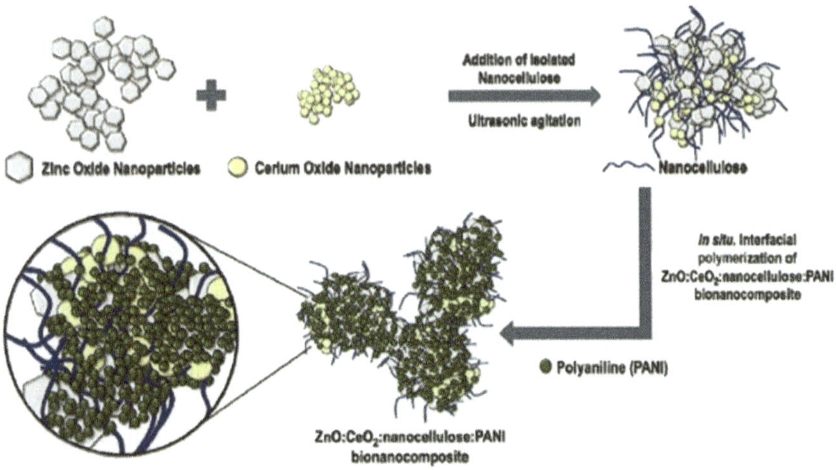

Fig. 9 Schematic diagram of preparation of ZnO: CeO_2: nanocellulose: PANI bionanocomposite [57]

of constituents present, it was observed that 95.75% of ZnO and CeO_2 nanoparticles was successfully incorporated during the polymerisation process onto the PANI matrix. MTT (3-(4,5-dimethylthiazol-2-yl)-2,5-diphenyltetrazolium bromide) assay was conducted to study the antibacterial potential of the composite against *E. coli* and *B. subtilis* and the MIC_{50} was determined from the results. In this assay, the water soluble MTT (yellow) is reduced by mitochondrial dehydrogenase enzyme from live cells into an insoluble formazan (purple). MTT assays revealed that the bionanocomposite (incorporation of PANI) showed improved antibacterial activity relative to the nanoparticles alone. However, pristine PANI did not exhibit detectable antimicrobial action against the bacterial pathogens tested. The MIC_{50} of ZnO: CeO_2: nanocellulose: PANI, ZnO alone and CeO_2 alone against *B. subtilis* were found to be 10.6 μg/mL, 23.7 μg/mL and 22.8 μg/mL respectively. Similarly, the MIC_{50} of the nanocomposite, ZnO alone and CeO_2 alone against *E. coli* were found to be 10.3 μg/mL, 13.47 μg/mL and 14.65 μg/mL respectively. In support of MTT results, disk diffusion assays also displayed larger zones of inhibition upon treating both species with the bionanocomposite in contrast to with nanoparticles alone. The positive control used in these experiments was Ampicillin which presented significant antimicrobial activity at a concentration of 10 mg/mL. This is markedly higher than the MBC found for ZnO: CeO_2: nanocellulose: PANI—40 μg/mL [57].

A copper-PANI nanocomposite (Cu-PANI) was also investigated for antibacterial activity. It was synthesised by oxidative polymerisation in the presence of a Cu salt wherein the polymer and the metal nanoparticle were produced simultaneously [10]. The composite was structurally and optically analysed by UV–Vis, FT-IR and electron microscopy methods to reveal spherical nanoparticles homogenously dispersed throughout a branched and fibrous PANI matrix. The antibacterial effects of Cu-PANI were evaluated in a colony reduction assay against *S. aureus* and *E. coli* in which the percentage of cell growth reduction was calculated (R, %). It was observed from these assays that a reduced number of CFU (colony-forming units) resulted from an increasing concentration of Cu-PANI. For example, 20 ppm of Cu-PANI was calculated to cause a 99.9% reduction of microbial growth when used to treat *E. coli* and a 97.9% reduction against *S. aureus*. After exposure of the bacterial species to pristine PANI for 1 h, a microbial reduction of about 81.5% was observed. No microbial growth was seen to emerge after exposure to 1 ppm of Cu-PANI for 2 h. From these, it was concluded that swift antimicrobial activity arises from Cu-PANI which is ameliorated by the inclusion of PANI. AFM analysis of bacterial cells treated for 1 h showed loss of membrane integrity, indentations on cell wall, ruptured inner membranes which would result in the leakage of cell constituents. After 2 h, *E. coli* cells were severely disrupted with a collapsed membrane and large pits whereas *S. aureus* cells manifested shallow grooves and roughened surface texture [10] Fig. 10.

In a separate study by Poyraz et al. [64], $AgNO_3$ was utilised as both the oxidative as well as the doping agent in the polymerisation of aniline through a one-step

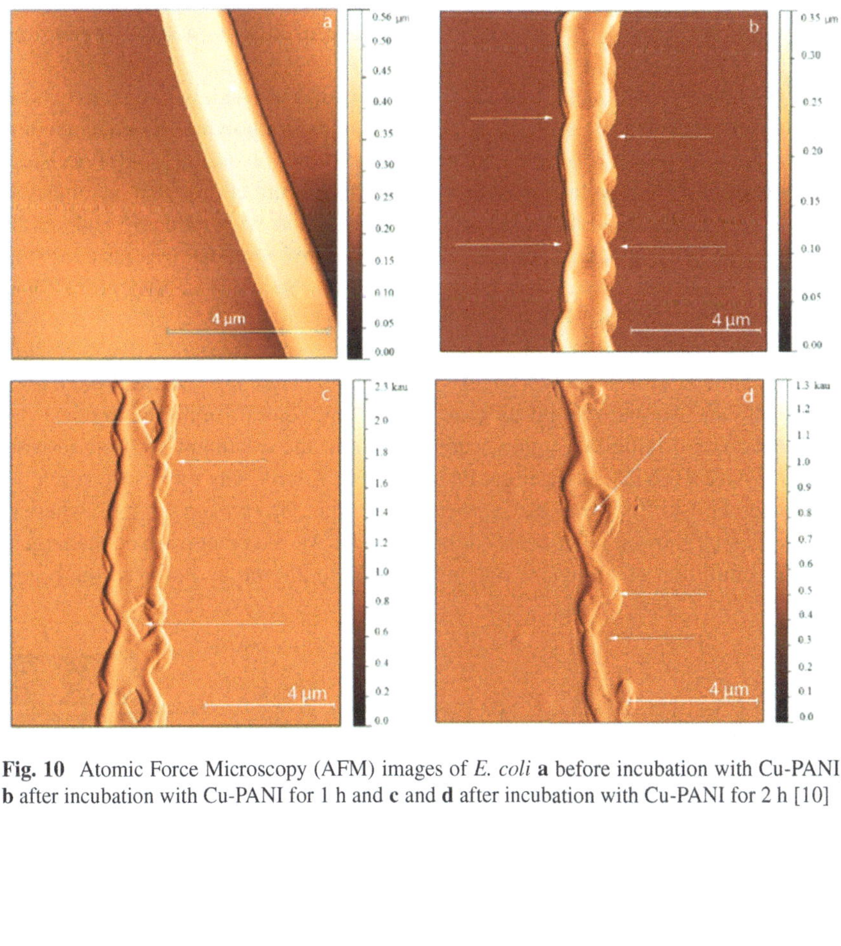

Fig. 10 Atomic Force Microscopy (AFM) images of *E. coli* **a** before incubation with Cu-PANI; **b** after incubation with Cu-PANI for 1 h and **c** and **d** after incubation with Cu-PANI for 2 h [10]

redox process. Polyaniline nanofibers were synthesised in the presence of Vanadium pentoxide (V_2O_5) seeding agents which also acted as a supplementary oxidative agent. The polyaniline: silver nanocomposite (PANI:Ag) was characterised by electron microscopy, FT-IR and UV–Vis to exhibit homogenously distributed silver nanoparticle clusters fixed in a highly dense PANI nanofiber matrix [64]. To assess the antibacterial potential of PANI:Ag against *S.* aureus and *E. coli*, a dynamic flask shaking test and a sandwich test were performed after which emerged colonies were enumerated. Reduction in bacterial growth was calculated after 24 h in log/CFU. It was deduced that PANI nanofibers presented some, albeit limited, reduction in bacterial growth. Antibacterial efficacy was enhanced against both microbes with the incorporation of silver nanoparticles in PANI:Ag. It was also seen that an increasing concentration of silver nanoparticles within the composite elicited improved antibacterial effects. Bacterial reduction also increased with increasing contact time with the nanocomposite. PANI:Ag (0.05 M) caused the complete inactivation of *S. aureus* as well as *E. coli* within 60 min of contact time. Exposure of bacterial species to pristine PANI for 120 min resulted in 1.85 log reduction of *S. aureus* and 2.4 log reduction of *E. coli*. In comparison, PANI:Ag (0.02 M) was able to eliminate *E. coli* completely and cause 1.54 log reduction of *S.* aureus after a 15 min contact time [64].

As previously mentioned, Ag and Au nanoparticles are widely known to exhibit biocidal actions and thus have been employed in studies of polymer-based nanocomposites against pathogenic bacteria. In the preparation of polyaniline/gold (PANI/Au) and polyaniline/gold–palladium (PANI/Au–Pd), Boomi et al. [13] synthesised colloids of Au and Au–Pd through chemical reduction with reducing and stabilising agents. The nanocomposites were then prepared with APS as an oxidising agent Fig. 11. The composite was characterised through UV–Vis, XRD, FT-IR and electron microscopy (HRTEM) to confirm the synthesis of PANI and its composites. Antibacterial potential of the nanocomposite was determined in disk diffusion assays against *E. coli*, *Staphylococcus* spp., *Streptococcus* spp. and *Klebsiella* spp. [13]. Colloidal Au and Au–Pd did not present any zone of inhibition against the pathogens. Good inhibition zones were seen upon the treatment of bacteria with pristine PANI, PANI/Au as well as PANI/Au–Pd. *E. coli* was most susceptible to the nanocomposite followed by *Staphylococcus* spp., *Streptococcus* spp., whereas *Klebsiella* spp. showed lowest susceptibility. PANI/Au–Pd demonstrated the largest zones of inhibition of pathogens ranging from 21 to 25 mm in diameter. Inhibition

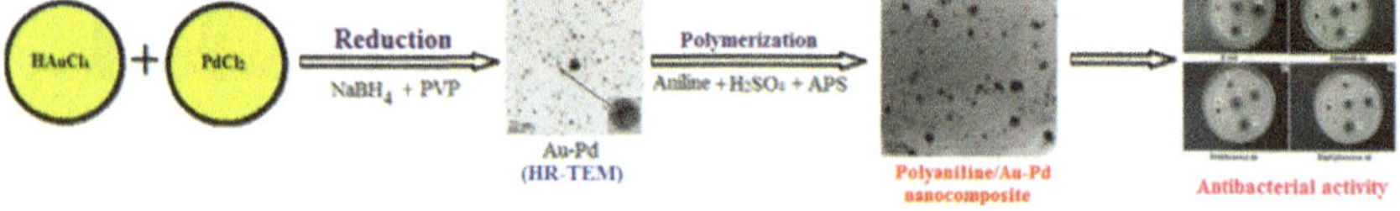

Fig. 11 Graphical illustration summarising the study by Boomi et al. [13] on the synthesis and application of PANI/Au–Pd

zones resulting from treatment with PANI/Au and pristine PANI were smaller with diameters ranging from 13 to 20 mm. *E. coli* was further tested in broth-microdilution techniques to determine the MIC and MBC of PANI, PANI/Au as well as PANI/Au–Pd. Against *E. coli*, 25 μg/mL of PANI/Au–Pd, 75 μg/mL of PANI/Au and 75 μg/mL of PANI elicited inhibitory as well as bactericidal effects (MIC/MBC). Overall, it was reported that Au and Au–Pd did not display antibacterial activity whereas pristine PANI exhibited some activity that was enhanced with the incorporation of Au and Au–Pd in the composite [13].

5.3 Polythiophene-Based Nanocomposites

Although not as largely investigated as PANI/Ppy-based nanocomposites, PTh-based nanocomposites have also been seen to be antibacterial in a number of studies. In a study by Ma et al., Cu-doped zinc oxide polythiophene nanocomposites (CZ/PTh) were tested for antibacterial activity. Cu-doped zinc oxide nanoparticles (CZ) were synthesised by the Sol–Gel method followed by the in-situ polymerisation of thiophene to prepare CZ/PTh with varying concentrations of CZ (20, 40, 60, 80 wt% CZ). The antimicrobial properties of CZ, PTh and CZ/PTh were evaluated in disk diffusion assays as well as measures of minimum inhibitory concentrations (MIC) and minimum bactericidal concentrations (MBC) against *E. coli* and *S. aureus* [53].

Minimal antibacterial activity was observed to be produced by PTh alone and CZ alone, but the combined nanocomposite spawned a notably higher antimicrobial effect. Larger zones of inhibition were noticed with increasing content of CZ up to 80 wt%. Through MIC and MBC studies, it was reinforced that greater bactericidal and inhibitory potential arises with the incorporation of PTh with CZ. In the absence of PTh, minimum inhibitory concentrations are 10 times greater and minimum bactericidal concentrations are also 10 times greater against both *S. aureus* as well as *E. coli*. The MIC and MBC observed of CZ/PTh against both bacterial pathogens tested were 10 μg/mL and 20 μg/mL respectively. Relatively, CZ alone displayed an MIC and MBC of 100 μg/mL and 200 μg/mL respectively against both *S. aureus* and *E. coli* [53].

Another study by Adhikari et al. [1] assessed the bactericidal potential of water-soluble gold nanoparticle-polythiophene (AuNP-PTh) against a range of pathogenic bacteria namely *E. coli*, *B. cereus*, *Yersinia enterocolitica*, *Enterobacter aerogenes*, *S. aureus*, *Listeria monocytogenes*. 45, 75 and 150 μM of AuNP-PTh were noted to cause significant growth inhibition to all bacterial pathogens tested in a dose-dependent manner. To eliminate influences by cell proliferation, the same tests were conducted in phosphate buffered saline (PBS) and results of these were complementary to the former. That is, an increase in inhibitory effects with increasing concentrations of nanocomposites was observed. Moreover, in these, higher concentrations of the nanocomposite (150 μM) were observed to eradicate the bacterial population tested—leaving less than 1.0 $\log_{10}$ CFU viable. The MBC of AuNP-PTh was determined to be 112 μM against *E. coli* and *L. monocytogenes* [1].

To further investigate the mechanism of action of AuNP-PTh, Adhikari et al. illustrated the membrane-targeting activity of the nanocomposite in fluorescence and electron-microscopy-based studies. In these studies, *E. coli* and *L. monocytogenes* were treated with increasing concentrations of the nanocomposite before dye uptake was assessed. The membrane impermeant dye used, propidium iodide (PI) is consumed by cells with damaged membranes and subsequently binds to nucleic acids resulting in enhanced fluorescence. Increasing uptake of PI was observed to be the result of an increase in AuNP-PTh concentration indicating concentration-dependent membrane damage. Through transmission electron microscopy (TEM), loss of membrane integrity was observed to escalate with time and a reduction of intracellular electron density was also noted and discussed to be the result of leakage of intracellular components [1].

Lastly, various concentrations of AuNP-PTh were assessed for cytotoxicity in XTT assays against HT-29 cells (adenocarcinoma cells) as a model human cell line. It was observed that the viability of HT-29 cells were unaffected after exposure to AuNP-PTh for up to 48 h, even at a concentration of 112 μM (MBC of AuNP-PT) [1].

5.4 PEDOT-Based Nanocomposites

The polythiophene derivative PEDOT has also been investigated in some studies for antimicrobial potency. For instance, Kumar et al. [42] electrochemically synthesised a novel PEDOT based nanocomposite with varying concentrations of fluoro hydroxyapatite (FHA) nanoparticles for coating a Ti-Nb-Zr (TNZ) alloy in orthopaedic implant applications. Hydroxy apatite (HA) is commonly used in this field as a coating on metallic implants due to its resemblance to the inorganic composition of natural bone [42]. This study also used fluorine for its natural presence in human bones and teeth. However, limitations arise with the use of HA in part due to its brittle nature and weak adhesion to metallic implants. To counter this, conducting polymers have recently been utilised to make HA composites with greater strength and adhesion properties.

The formation of PEDOT/FHA nanocomposite was characterised in this study by FTIR, XRD and Raman analyses and the antibacterial potential of the composite was assessed using surface spread plate techniques against *E. coli* and *B. subtilis*. Uncoated TNZ was observed to allow the greatest amount of bacterial growth having the least antibacterial activity. TNZ substrates coated with PEDOT showed notable inhibition of bacterial growth which was further enhanced with the incorporation of FHA in PEDOT/FHA nanocomposites which elicited an inhibition of 84.1 and 85.41% on *B. subtilis* and *E. coli* respectively. To assess the biocompatibility of the nanocomposites, in vitro assays with human osteosarcoma cells (MG63) were conducted in which the cells were seeded on coated as well as uncoated TNZ substrates before cells were enumerated. In contrast to uncoated TNZ wherein proliferation rate was relatively low, coating with PEDOT allowed MG63 cells to proliferate to a higher cell density uniformly around the coated substrate. The proliferation rate of cells seeded on PEDOT/FHA coating was observed to be greater than that of cells growing on PEDOT-coated and uncoated TNZ after 5 days. Even after 7 days, cells seeded on PEDOT/FHA coating continued to proliferate to a large extent [42].

On another note, recently, significant attention has geared towards the use of near-infrared (NIR) induced photothermal therapy against pathogenic bacteria. This is in part due to the harmless and penetrable nature of NIR irradiation in living tissues at a range of 700–1100 nm [40]. In order to effectively, selectively and rapidly achieve hyperthermal killing of bacteria, a photothermal agent that is able to absorb NIR light and in turn emit photothermal heat is required. Ko et al. synthesised an organic nanocomposite hydrogel comprising of PEDOT: poly(styrene-sulfonate) (PEDOT: PSS) with agarose as the backbone Fig. 12(a–b). This hydrogel nanocomposite exhibited high NIR absorbance, photothermal conversion, photostability and biocompatibility. Gelation of the hydrogel was obtained when cooled to room temperature and reversible switching was achieved via a thermal stimulus Fig. 12c. Additionally, excellent bending and twisting of the hydrogel was observed Fig. 12d. Efficiency in photothermal conversion was observed to heighten with increasing PEDOT: PSS content.

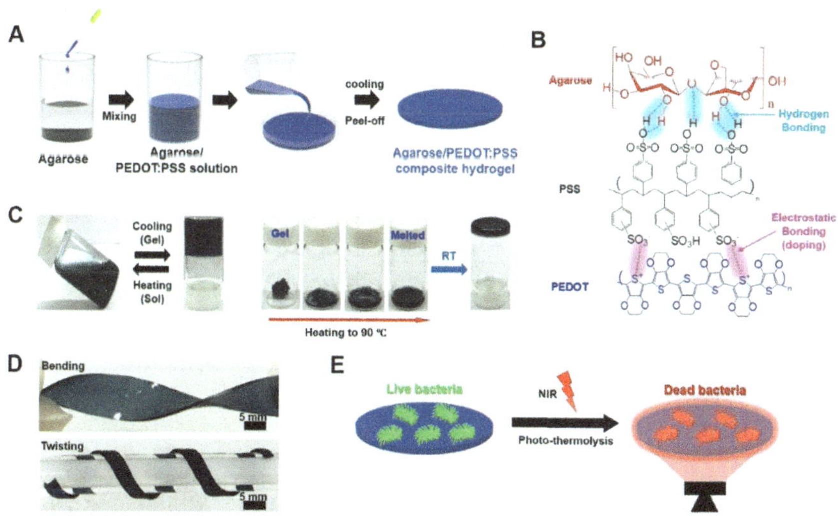

Fig. 12 Illustrative diagram displaying **a** the synthesis of Agarose/PEDOT:PSS composite hydrogel, **b** the interaction between components of the composite, **c** the thermoplastic reversibility of the hydrogel (sol–gel transition), **d** excellent bending and twisting ability of the hydrogel, **e** a schematic representation of the photo-thermolysis ability of Agarose/PEDOT:PSS on surface bacteria when exposed to NIR laser irradiation [40]

E. coli and *S. aureus* were used in the assessment of photothermal antibacterial activity of PEDOT: PSS/agarose (40 v/v %) by Ko et al. [40] Fig. 12e. Bacterial suspensions were prepared and spread onto pure agarose plates and PEDOT: PSS/agarose plates before irradiation with NIR laser (808 nm) for 3 min. NIR irradiation did not affect the viability of either bacterial species on pure agarose. Interestingly, after 1 min of NIR irradiation, the viability of *E. coli* and *S. aureus* on the nanocomposite hydrogel reduced to 44% and 37% viability respectively. Complete bacterial death was observed after two minutes of NIR irradiation [40].

5.5 Polycarbazole-Based Nanocomposites

Polycarbazole-based nanocomposites were evaluated for antimicrobial action in a study by Iram et al. in-situ oxidative polymerisation was used to synthesise polycarbazole titanium dioxide nanocomposites (PCz/TiO_2-8) in the presence of TiO_2 nanoparticles using ammonium persulfate (APS) as an oxidising agent Fig. 13. In disk diffusion assays, the antibacterial capacity of the nanocomposite was determined against the gram-negative bacteria *Proteus vulgaris* and *P. aeruginosa* and the gram-positive bacteria *Bacillus megaterium* and *Bacillus subtilis* [25]. Pristine PCz elicited minimal antibacterial activity on the species assessed whereas PCz/TiO_2 displayed antimicrobial action comparable to that of Ciprofloxacin,an antibiotic used as a positive control in this study. PCZ displayed a percentage area of inhibition of 10–30%

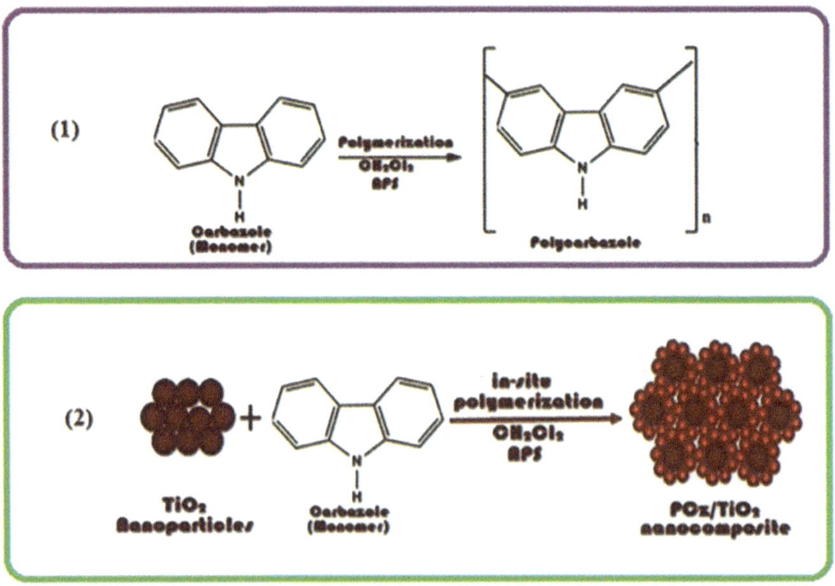

Fig. 13 Schematic illustration of synthesis of (1) Polycarbazole and (2) Polycarbazole/TiO_2 [25]

against the bacterial pathogens assessed. In comparison, PCz/TiO_2 elicited 90–100% area of inhibition against the microbes. Thus, inhibitory effects were regarded to be attributable to the incorporation of TiO_2 in the composite.

Further studies were completed to study DNA binding through biophysical techniques. It was ascertained that PCz/TiO_2 possessed superior DNA binding qualities in contrast to PCz. The mode of interaction between PCz/TiO_2 and DNA helices was observed through molecular docking simulation in this study and described as 'groove binding'—a mode of non-covalent interaction [25].

6 Mechanism of Action

In most cases, the presence of nanomaterials is studied to be the primary cause of antibacterial efficacy in conducting-polymer nanocomposites (CP-NC). It is well known that metal and metal oxide nanoparticles are active biocidal agents and, as will be discussed, act upon target cells in a number of mechanisms.

To begin, conducting polymers act as effective capping agents, preventing the agglomeration and increasing the steric stabilisation of nanoparticles in CP-NC [77]. This brings about well-dispersed nanoparticles with larger exposed surface areas. Nanoparticles, when effectively dispersed, possess the ability to adhere to the cell walls of bacterial pathogens and give rise to structural alterations such as the creation of depressions which likely increase the permeability of the membrane resulting in possible cell death. Nanoparticles have also been seen to induce oxidative stress upon contact. Reactive oxygen species target proteins, lipids, DNA/RNA and are also causative of the oxidation of DNA and membrane damage [42].

Alternatively, an incline or disruption in the permeability and internal osmotic balance of bacterial cells may be attributable to an electrostatic adherence between the negatively charged cell membrane and positively charged polymer molecule. The polymer molecule attains a positive charge during the oxidative polymerisation of its monomer through which positive charges are formed along the backbone of the polymer chain. In contrast, bacterial cells are negatively charged due to the presence of lipopolysaccharides in gram-negative bacteria and the presence of teichoic acid in gram-positive bacteria. Comparatively, zwitterionic phospholipids in the cytoplasmic membrane of mammalian cells make them neutral at physiological pH. This dissimilarity in charge possibly account for selective anti-bacterial activity of CP-NC.

Antibacterial actions may also be the result of the DNA and cellular enzymes of bacterial cells being deactivated or damaged as a result of the interaction of Ag^+ ions from the nanoparticles with electron donors including thiols, amides, carboxylates, etc. [77].

Further, it has been proposed that the nano-structure and low molecular weight of polymers within nanocomposites allow them to penetrate the cell wall of bacterial cells and elicit disruption by interacting with phosphorous and sulphur containing compounds including proteins, enzymes and DNA molecules Fig. 14. Specifically,

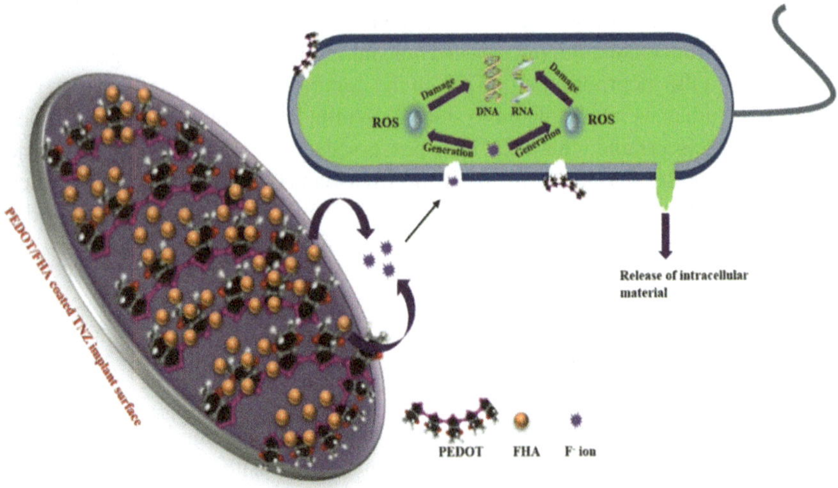

Fig. 14 Illustrative diagram presenting proposed mechanism of antibacterial activity of PEDOT/FHA coatings [42]

nitrogen-containing groups within conducting polymers increase their antibacterial efficacy through direct physical interaction with the polymer [42].

Additionally, depending of the structure of conducting polymer synthesised, their large surface area may cover bacterial cells, hinder cell proliferation, prevent their uptake of nutrients and in turn increase their exposure to nanomaterials within the composite.

Poyraz et al. [64] suggested that the surface hydrophobicity, length, low molecular weight and the presence of amino groups of the polymer chain are factors that reduce potential development of resistance by eliciting inhibitory effects on stress-response genes which function mainly in repair, energy metabolism, transport and the formation of biofilm.

Overall, antibacterial activity of CP-NC arises through oxidative stress, protein/DNA inactivation, membrane damage and/or the inhibition of stress-response genes (Table 2 and Fig. 15).

7 Conclusion

Electrically conducting polymers have recently gained significant attention in the field of nanotechnology for their ability to conduct electricity while retaining properties of conventional polymers and for their suitability as matrices for the immobilisation of nanoparticles. Thus, CP-NC have been increasingly synthesised, and their potency as antibacterial agents assessed.

Table 2 Reviewed conducting polymer-based nanocomposites and their studied antibacterial and biocompatibility properties

	Conducting polymer—nanocomposite	Bacterial species tested	Antibacterial effects	Biocompatibility/cytotoxicity	References
Polypyrrole (Ppy)	Polypyrrole nanotube-silver nanoparticles (Ppy-NT: AgNP)	*Escherichia coli* *Staphylococcus aureus*	MIC against *E. coli*: 0.078 mg/mL MIC against *S. aureus*: 0.15625 mg/mL	Haemolysis: less than 5% haemolysis was elicited up to 2.5 mg/mL of nanocomposite	[77]
	Polypyrrole zinc oxide chitosan (Ppy/ZnO/CS)	*Staphylococcus aureus* *Bacillus cereus* *Pseudomonas aeruginosa* *Escherichia coli*	Maximum ZOI: 29 mm against *P. aeruginosa*	N/A	[26]
	Polypyrrole nanotube silver nanoparticles (AgNP/Ppy NT)	*Escherichia coli* *Staphylococcus aureus*	MIC against *E. coli*: 10 μg/mL MIC against *S. aureus*: 200 μg/mL	N/A	[60]
	Chitosan-polypyrrole nanocomposite (CS-Ppy)	*Pseudomonas aeruginosa*	Inhibition of biofilm production: 91% inhibition caused by 512 μg/mL of nanocomposite Inhibition of virulence factor production: 79% inhibition of Pyocyanin production; 91% inhibition of Pyoverdine production; 85% inhibition of Rhamnolipid production	N/A	[38]

(continued)

Table 2 (continued)

	Conducting polymer—nanocomposite	Bacterial species tested	Antibacterial effects	Biocompatibility/cytotoxicity	References
	Chitosan polyacrylic acid polypyrrole- Fe_3O_4 nanoparticles	*Staphylococcus aureus* *Pseudomonas aeruginosa*	Maximum ZOI: 30 mm against *S. aureus*; 29 mm against *P. aeruginosa*	N/A	[79]
	Polypyrrole/copper-doped zinc oxide nanocomposite (Ppy/CZO)	*Escherichia coli* *Bacillus subtilis*	Viability: 60 μg/mL of nanocomposite resulted in 30% and 35% viability of *E. coli* and *B. subtilis* respectively	N/A	[37]
	Carbon nanotubes coated with polypyrrole (CNT/Ppy)	*Pseudomonas aeruginosa*	Killing rate: 250 μg/mL of CNT/Ppy with irradiation caused 70% killing	N/A	[75]
Polyaniline (PANI)	Polyaniline/platinum (PANI/Pt) Polyaniline/platinum-palladium (PANI/Pt–Pd)	*Staphylococcus aureus* *Streptococcus* spp. *Escherichia coli* *Klebsiella* spp.	PANI MIC: 75 μg/mL PANI/Pt MIC: 50 μg/mL PANI/Pt–Pd MIC: 25 μg/mL	N/A	[12]

(continued)

Table 2 (continued)

	Conducting polymer—nanocomposite	Bacterial species tested	Antibacterial effects	Biocompatibility/cytotoxicity	References
	Polyaniline/silver-platinum (PANI/Ag-Pt)	*Staphylococcus aureus* *Streptococcus* spp. *Escherichia coli* *Klebsiella* spp.	PANI ZOI: 27 mm against *Staphylococcus aureus*; 25 mm against *Streptococcus* spp.; 20 mm against *Escherichia coli*; 22 mm against *Klebsiella* spp. PANI/Ag-Pt ZOI: 30 mm against *Staphylococcus aureus*; 29 mm against *Streptococcus* spp.; 21 mm against *Escherichia coli*; 25 mm against *Klebsiella* spp.	N/A	[11]
	Silver-polyaniline nanocomposites (Ag-PANI)	*Bacillus subtilis*	MIC against *B. subtilis*: 25 μg/mL MBC against *B. subtilis*: 25 μg/mL	N/A	[73]
	Polyaniline/copper-doped zinc oxide nanocomposites (PANI/CZO)	*Escherichia coli* *Staphylococcus aureus*	PANI MIC: 1000 μg/mL CZO MIC: 200 μg/mL PANI/CZO MIC: 10 μg/mL PANI MBC: 3000 μg/mL CZO MBC: 500 μg/mL PANI/CZO MBC: 30 μg/mL	N/A	[48]

(continued)

Table 2 (continued)

	Conducting polymer—nanocomposite	Bacterial species tested	Antibacterial effects	Biocompatibility/cytotoxicity	References
	Zinc oxide: Cerium (IV) oxide: nanocellulose: polyaniline (ZnO: CeO_2: nanocellulose: PANI)	*Escherichia coli* *Bacillus subtilis*	Against *B. subtilis*: ZnO: CeO2: nanocellulose: PANI MIC_{50}: 10.6 μg/mL ZnO MIC_{50}: 23.7 μg/mL CeO2 MIC_{50}: 22.8 μg/mL Against *E. coli*: ZnO: CeO2: nanocellulose: PANI MIC_{50}: 10.3 μg/mL ZnO MIC_{50}: 13.47 μg/mL CeO2 MIC_{50}: 14.65 μg/mL	N/A	[57]
	Copper-polyaniline (Cu-PANI)	*Escherichia coli* *Staphylococcus aureus*	20 ppm Cu-PANI Cell growth reduction: 99.9% reduction of *E. coli* 97.9% reduction of *S. aureus*	N/A	[10]
	Polyaniline: silver nanocomposite (PANI: Ag)	*Escherichia coli* *Staphylococcus aureus*	PANI: Ag (0.05 M): Complete inactivation of *S. aureus* and *E. coli* PANI: 1.85 log reduction of *S. aureus*; 2.4 log reduction of *E. coli*	N/A	[64]

(continued)

Table 2 (continued)

	Conducting polymer—nanocomposite	Bacterial species tested	Antibacterial effects	Biocompatibility/cytotoxicity	References
	Polyaniline/gold (PANI/Au) Polyaniline/gold–palladium (PANI/Au–Pd)	*Staphylococcus* spp. *Streptococcus* spp. *Escherichia coli* *Klebsiella* spp.	PANI MIC/MBC: 75 μg/mL PANI/Au MIC/MBC: 75 μg/mL PANI/Au–Pd MIC/MBC: 25 μg/mL	N/A	[13]
Polythiophene (PTh)	Copper-doped zinc oxide polythiophene nanocomposite (CZ/PTh)	*Escherichia coli* *Staphylococcus aureus*	CZ/PTh MIC: 10 μg/mL CZ MIC: 100 μg/mL CZ/PTh MBC: 20 μg/mL CZ MBC: 200 μg/mL	N/A	[53]
	Gold nanoparticle-polythiophene nanocomposite (AuNP-PTh)	*Escherichia coli* *Bacillus cereus* *Yersinia enterocolitica* *Enterobacter aerogenes* *Staphylococcus aureus* *Listeria monocytogenes*	MBC AuNP-PTh: 112 μM 150 μM AuNP-PTh: < 1 log_{10} CFU	Cytotoxicity towards HT-29 (human adenocarcinoma cells) evaluated through XTT assay: No effect of HT-29 viability even at 112 μM (MBC)	[1]

(continued)

Table 2 (continued)

	Conducting polymer—nanocomposite	Bacterial species tested	Antibacterial effects	Biocompatibility/cytotoxicity	References
Poly(3,4-ethylidoxythiophene)	PEDOT:poly(styrene-sulfonate) nanocomposite hydrogel (PEDOT:PSS)	*Escherichia coli* *Staphylococcus aureus*	1 min of near-infrared (NIR) irradiation reduced viability to: *E. coli* on hydrogel—44% *S. aureus* on hydrogel—37% 2 min of NIR irradiation: Complete bacterial death	N/A	[40]
	PEDOT/fluoro hydroxyapatite (PEDOT/FHA)	*Escherichia coli* *Bacillus subtilis*	Inhibition of viability: *B. subtilis*: 84.1% inhibition *E. coli*: 85.41% inhibition	Biocompatibility towards MG63 (human osteosarcoma cells): coating of Ti-Nb-Zr alloy (TNZ) with PEDOT/FHA allowed the proliferation of MG63 in contrast to uncoated TNZ	[42]
Poly-Carbazole (PCz)	Polycarbazole titanium dioxide (PCz/TiO_2)	*Proteus vulgaris* *Pseudomonas aeruginosa* *Bacillus megaterium* *Bacillus subtilis*	% Area of inhibition: PCz/TiO_2: 90—100% inhibition PCz: 30% inhibition	N/A	[25]

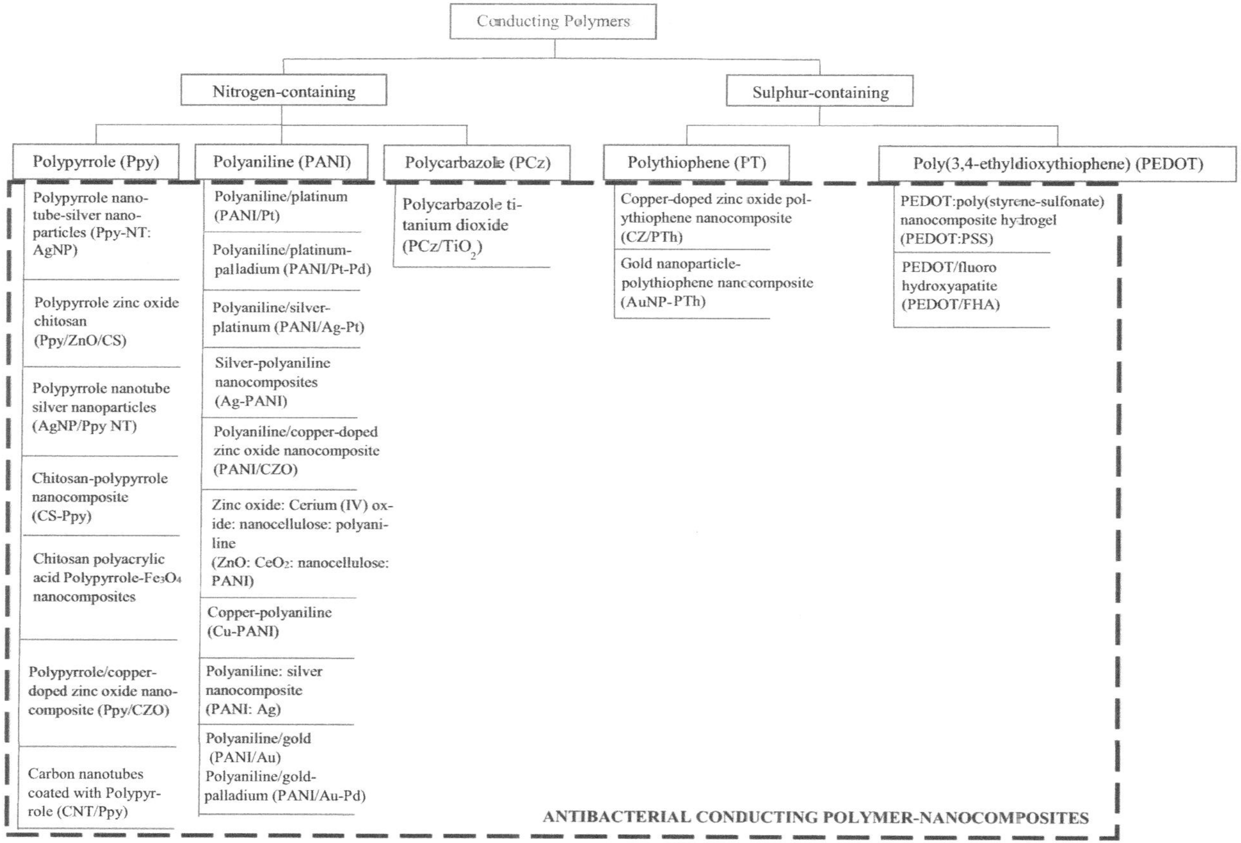

Fig. 15 Schematic flowchart distinguishing nitrogen-containing and sulphur-containing conducting polymers and their respective studied antibacterial nanocomposites

Conducting polymers, mainly PANI and Ppy have been studied to exhibit antibacterial properties in their pristine forms. However, overall, conducting polymer-based nanocomposites have been observed to be more efficacious as antimicrobial agents relative to their pristine variants. In addition, the incorporation of conducting polymers within the CP-NC has also been seen to ameliorate the antibacterial activity of the nanomaterial possibly through its ability to stabilise and immobilise nanoparticles, hindering their agglomeration.

The mechanism through which CP-NC elicit such antimicrobial effects is not fully understood. However, it has been proposed that this is attributable to oxidative stress, membrane damage, protein/DNA inactivation and/or the inhibition of stress-response genes caused by the CP-NC.

The biocompatibility and cytotoxicity of CP-NC has not been studied extensively in these reports. Nevertheless, available studies show minimal effects on human cells. Together, these properties make CP-NC an interesting avenue of research for antibacterial agents. With biocompatibility further studied, these materials would be useful to investigate for effects on other pathogens in the treatment of various infectious diseases that currently pose serious threats to society.

References

1. Adhikari, M.D., Goswami, S., Panda, B.R., Chattopadhyay, A., Ramesh, A.: Membrane-directed high bactericidal activity of (gold nanoparticle)–polythiophene composite for niche applications against pathogenic bacteria. Adv. Healthcare Mater. **2**(4), 599–606 (2013)
2. Akkouch, A., Shi, G., Zhang, Z., Rouabhia, M.: Bioactivating electrically conducting polypyrrole with fibronectin and bovine serum albumin. J. Biomed. Mater. Res. **92**(1), 221–231 (2010)
3. Ateh, D., Navsaria, H., Vadgama, P.: Polypyrrole-based conducting polymers and interactions with biological tissues. J. Royal Soc. Interface **3**(11), 741–752 (2006)
4. Ates, M., Karazehir, T., Sezai Sarac, A.: Conducting polymers and their applications. Current Phys. Chem. **2**(3), 224–240 (2012)
5. Balint, R., Cassidy, N.J., Cartmell, S.H.: Conductive polymers: towards a smart biomaterial for tissue engineering. Acta Biomater. **10**(6), 2341–2353 (2014)
6. Bamberger, D.M., Boyd, S.E.: Management of *Staphylococcus aureus* infections. Am. Family Phys. **72**(12), 2474–2481 (2005)
7. Bengoechea, J.A., Sa Pessoa, J.: *Klebsiella pneumoniae* infection biology: living to counteract host defences. FEMS Microbiol. Rev. **43**(2), 123–144 (2018)
8. Bettinger, C.J., Bruggeman, J.P., Misra, A., Borenstein, J.T., Langer, R.: Biocompatibility of biodegradable semiconducting melanin films for nerve tissue engineering. Biomaterials, **30**(17), 3050–3057 (2009)
9. Bhat, N., Gadre, A., Bambole, V.: Structural, mechanical, and electrical properties of electropolymerized polypyrrole composite films. J. Appl. Polym. Sci. **80**(13), 2511–2517 (2001)
10. Bogdanovic, U., Vodnik, V., Mitric, M., Dimitrijevic, S., Skapin, S. D., Zunic, V., ... Stoiljkovic, M.: Nanomaterial with high antimicrobial efficacy copper/polyaniline nanocomposite. ACS Appl. Mater. Interfaces, **7**(3), 1955–1966 (2015)
11. Boomi, P., Prabu, H.G., Mathiyarasu, J.: Synthesis and characterization of polyaniline/Ag–Pt nanocomposite for improved antibacterial activity. Colloids Surfaces B: B **103**, 9–14 (2013)

12. Boomi, P., Prabu, H.G., Mathiyarasu, J.: Synthesis, characterization and antibacterial activity of polyaniline/Pt–Pd nanocomposite. Eur. J. Med. Chem. **72**, 18–25 (2014)
13. Boomi, P., Prabu, H.G.: Synthesis, characterization and antibacterial analysis of polyaniline/Au–Pd nanocomposite. Colloids Surf. A: Physicochemical Eng. A **429**, 51–59 (2013)
14. Borriello, A., Guarino, V., Schiavo, L., Alvarez-Perez, M., Ambrosio, L.: Optimizing PANi doped electroactive substrates as patches for the regeneration of cardiac muscle. J. Mater. Sci. Mater. Med. **22**(4), 1053–1062 (2011)
15. Breidenstein, E.B., de la Fuente-Núñez, C., Hancock, R.E.: *Pseudomonas aeruginosa*: all roads lead to resistance. Trends in Microbiol. **19**(8), 419–426 (2011)
16. Cetiner, S., Kalaoglu, F., Karakas, H., Sarac, A.S.: Electrospun nanofibers of polypyrrole-poly (acrylonitrile-co-vinyl acetate). Text. Res. J. **80**(17), 1784–1792 (2010)
17. Chen, F. J., Wang, C. H., Chen, C.-Y., Hsu, Y.-C., Wang, K.-T.: Role of the mecA gene in oxacillin resistance in a *Staphylococcus aureus* clinical strain with a pvl-Positive ST59 genetic background. Antimicrob. Agents Chemother. **58**(2), 1047–1054 (2014)
18. da Silva Jr, F.A., Queiroz, J. C., Macedo, E.R., Fernandes, A.W., Freire, N.B., da Costa, M.M., de Oliveira, H.P.: Antibacterial behavior of polypyrrole: the influence of morphology and additives incorporation. Mater. Sci. Eng. C, **62**, 317–322 (2016)
19. Dai, L.: Intelligent Macromolecules for Smart Devices: From Materials Synthesis to Device Applications. Springer Science & Business Media (2004)
20. Dallas, P., Sharma, V.K., Zboril, R.: Silver polymeric nanocomposites as advanced antimicrobial agents: classification, synthetic paths, applications, and perspectives. Adv. Colloid Interface Sci. **166**(1–2), 119–135 (2011)
21. Das, T.K., Prusty, S.: Review on conducting polymers and their applications. Polymer-plastics Technol. Eng. **51**(14), 1487–1500 (2012)
22. Davin-Regli, A., Lavigne, J.-P., Pagès, J.-M.: Enterobacter spp.: update on taxonomy, clinical aspects, and emerging antimicrobial resistance. Clin. Microbiol. Rev. **32**(4), e00002–00019 (2019)
23. Dayan, G.H., Mohamed, N., Scully, I.L., Cooper, D., Begier, E., Eiden, J., Jansen, K.U., Gurtman, A., Anderson, A.S.: *Staphylococcus aureus*: the current state of disease, pathophysiology and strategies for prevention. Expert Rev. Vaccines **15**(11), 1373–1392 (2016)
24. Domènech, B., Muñoz, M., Muraviev, D., Macanás, J.: Polymer-silver nanocomposites as antibacterial materials. In: Microbial Pathogens Strategies for Combating Them: Science, Technology and Education (2013)
25. e Iram, N., Khan, M.S., Jolly, R., Arshad, M., Alam, M., Alam, P., Khan, R.H., Firdaus, F. Firdaus, F.: Interaction mode of polycarbazole–titanium dioxide nanocomposite with DNA: molecular docking simulation and in-vitro antimicrobial study. J. Photochem. Photobiol. B: Biol. **153**, 20–32 (2015)
26. Ebrahimiasl, S., Zakaria, A., Kassim, A., Basri, S.N.: Novel conductive polypyrrole/zinc oxide/chitosan bionanocomposite: synthesis, characterization, antioxidant, and antibacterial activities. Int. J. Nanomedicine **10**, 217 (2015)
27. Gao, W., Howden, B.P., Stinear, T.P.: Evolution of virulence in *Enterococcus faecium*, a hospital-adapted opportunistic pathogen. Current Opin. Microbiol. **41**, 76–82 (2018)
28. Ghasemi-Mobarakeh, L., Prabhakaran, M.P., Morshed, M., Nasr-Esfahani, M.H., Baharvand, H., Kiani, S., Al-Deyab, S.S., Ramakrishna, S.: Application of conductive polymers, scaffolds and electrical stimulation for nerve tissue engineering. J. Tissue Eng. Regenerative Med. **5**(4), e17–e35 (2011)
29. Gomez, N., Lee, J.Y., Nickels, J.D., Schmidt, C.E.: Micropatterned polypyrrole: a combination of electrical and topographical characteristics for the stimulation of cells. Adv. Funct. Mater. **17**(10), 1645–1653 (2007)
30. Groenendaal, L., Zotti, G., Aubert, P.H., Waybright, S.M., Reynolds, J.R.: Electrochemistry of poly (3, 4-alkylenedioxythiophene) derivatives. Adv. Mater. **15**(11), 855–879 (2003)
31. Guimard, N.K., Gomez, N., Schmidt, C.E.: Conducting polymers in biomedical engineering. Prog. Polym. Sci. **32**(8–9), 876–921 (2007)

32. Gupta, B., Prakash, R.: Interfacial polymerization of carbazole: morphology controlled synthesis. Synth. Met. **160**(5–6), 523–528 (2010)
33. Hatchett, D.W., Josowicz, M.: Composites of intrinsically conducting polymers as sensing nanomaterials. Chem. Rev. **108**(2), 746–769 (2008)
34. Hoffmann, H., Schmoldt, S., Trülzsch, K., Stumpf, A., Bengsch, S., Blankenstein, T., Heesemann, J., Roggenkamp, A.: Nosocomial urosepsis caused by *Enterobacter kobei* with aberrant phenotype. Diagn. Microbiol. Infect. Dis. **53**(2), 143–147 (2005)
35. Holt, K.E., Wertheim, H., Zadoks, R.N., Baker, S., Whitehouse, C.A., Dance, D., Jenney, A., Connor, T.R., Hsu, L.Y., Severin, J.: Genomic analysis of diversity, population structure, virulence, and antimicrobial resistance in *Klebsiella pneumoniae*, an urgent threat to public health. Proc. Na. Acad. Sci. **112**(27), E3574–E3581 (2015)
36. Humpolicek, P., Kasparkova, V., Saha, P., Stejskal, J.: Biocompatibility of polyaniline. Synthetic Met. **162**(7–8), 722–727 (2012)
37. Khan, A.A.P., Khan, A., Rahman, M.M., Asiri, A.M., Oves, M.: Sensor development of 1, 2 dichlorobenzene based on polypyrole/Cu-doped ZnO (PPY/CZO) nanocomposite embedded silver electrode and their antimicrobial studies. Int. J. Biolog. Macromol. **98**, 256–267 (2017)
38. Khan, F., Manivasagan, P., Pham, D.T.N., Oh, J., Kim, S.-K., Kim, Y.-M.: Antibiofilm and antivirulence properties of chitosan-polypyrrole nanocomposites to *Pseudomonas aeruginosa*. Microbial Pathog. **128**, 363–373 (2019)
39. Kim, Y., Kim, S., Kim, Y., Hong, K., Wie, S., Park, Y., Jeong, H., Kang, M.: Carbapenem-resistant Acinetobacter baumannii: diversity of resistant mechanisms and risk factors for infection. Epidemiol. Infect. **140**(1), 137–145 (2012)
40. Ko, Y., Kim, J., Jeong, H.Y., Kwon, G., Kim, D., Ku, M., Yang, J., Yamauchi, Y., Kim, H.Y., Lee, C.: Antibacterial poly (3, 4-ethylenedioxythiophene): poly (styrene-sulfonate)/agarose nanocomposite hydrogels with thermo-processability and self-healing. Carbohydr. Polym. **203**, 26–34 (2019)
41. Kucekova, Z., Humpolicek, P., Kasparkova, V., Perecko, T., Lehocký, M., Hauerlandová, I., Saha, P., Stejskal, J.: Colloidal polyaniline dispersions: antibacterial activity, cytotoxicity and neutrophil oxidative burst. Colloids Surf. B: B. **116**, 411–417 (2014)
42. Kumar, A.M., Adesina, A.Y., Hussein, M., Ramakrishna, S., Al-Aqeeli, N., Akhtar, S., Saravanan, S.: PEDOT/FHA nanocomposite coatings on newly developed Ti-Nb-Zr implants: Biocompatibility and surface protection against corrosion and bacterial infections. Mater. Sci. Eng. C, **98**, 482–495 (2019)
43. Kumar, R., Oves, M., Almeelbi, T., Al-Makishah, N.H., Barakat, M.: Hybrid chitosan/polyaniline-polypyrrole biomaterial for enhanced adsorption and antimicrobial activity. J. Colloid Interface Sci. **490**, 488–496 (2017)
44. Lalegül-Ülker, Ö., Elçin, A.E., Elçin, Y.M.: Intrinsically conductive polymer nanocomposites for cellular applications. In: Cutting-Edge Enabling Technologies for Regenerative Medicine, pp. 135–153. Springer (2018)
45. Lam, M., Wick, R., Wyres, K., Gorrie, C., Judd, L., Jenney, A.: Genetic diversity, mobilisation and spread of the yersiniabactin-encoding mobile element ICEKp in *Klebsiella pneumoniae* populations. Microb. Genom. **4**(9) (2018)
46. Lebreton, F., van Schaik, W., McGuire, A.M., Godfrey, P., Griggs, A., Mazumdar, V., Corander, J., Cheng, L., Saif, S., Young, S.: Emergence of epidemic multidrug-resistant *Enterococcus faecium* from animal and commensal strains. MBio **4**(4), e00534–00513 (2013)
47. Lee, C.-R., Lee, J.H., Park, M., Park, K.S., Bae, I.K., Kim, Y.B., Cha, C.J., Jeong, B.C., Lee, S.H.: Biology of *Acinetobacter baumannii*: pathogenesis, antibiotic resistance mechanisms, and prospective treatment options. Front. Cell. Infect. Microbiol. **7**, 55 (2017)
48. Liang, X., Sun, M., Li, L., Qiao, R., Chen, K., Xiao, Q., Xu, F.: Preparation and antibacterial activities of polyaniline/Cu 0.05 Zn 0.95 O nanocomposites. Dalton Trans. **41**(9), 2804–2811 (2012)
49. Lin, M.-F., Lan, C.-Y.: Antimicrobial resistance in *Acinetobacter baumannii*: from bench to bedside. World J. Clin. Cases: WJCC, **2**(12), 787 (2014)

50. Lister, P.D., Wolter, D.J., Hanson, N.D.: Antibacterial-resistant *Pseudomonas aeruginosa*: clinical impact and complex regulation of chromosomally encoded resistance mechanisms. Clin. Microbiol. Rev. **22**(4), 582–610 (2009)
51. Liu, L., Chen, J., Wang, S.: Flexible antibacterial film deposited with polythiophene–porphyrin composite. Adv. Healthcare Mater. **2**(12), 1582–1585 (2013)
52. Lota, K., Khomenko, V., Frackowiak, E.: Capacitance properties of poly (3, 4-ethylenedioxythiophene)/carbon nanotubes composites. J. Phys. Chem. Solids **65**(2–3), 295–301 (2004)
53. Ma, G., Liang, X., Li, L., Qiao, R., Jiang, D., Ding, Y., Chen, H.: Cu-doped zinc oxide and its polythiophene composites: preparation and antibacterial properties. Chemosphere, **100**, 146–151 (2014)
54. Maráková, N., Humpolíček, P., Kašpárková, V., Capáková, Z., Martinková, L., Bober, P., Trchová, M., Stejskal, J.: Antimicrobial activity and cytotoxicity of cotton fabric coated with conducting polymers, polyaniline or polypyrrole, and with deposited silver nanoparticles. Appl. Surface Sci. **396**, 169–176 (2017)
55. Mulani, M.S., Kamble, E.E., Kumkar, S.N., Tawre, M.S., Pardesi, K.R.: Emerging strategies to combat ESKAPE pathogens in the era of antimicrobial resistance: a review. Front. Microbiol. **10** (2019)
56. Naseri, M., Fotouhi, L., Ehsani, A.: Recent Progress in the development of conducting polymer-based nanocomposites for electrochemical biosensors applications: a mini-review. Chem. Rec. **18**(6), 599–618 (2018)
57. Nath, B., Chaliha, C., Kalita, E., Kalita, M.: Synthesis and characterization of ZnO: CeO_2: nanocellulose: PANI bionanocomposite. A bimodal agent for arsenic adsorption and antibacterial action. Carbohydr. Polym. **148**, 397–405 (2016)
58. Navidinia, M.: The clinical importance of emerging ESKAPE pathogens in nosocomial infections. Arch. Adv. Biosci. **7**(3), 43–57 (2016)
59. Paczosa, M.K., Mecsas, J.: *Klebsiella pneumoniae*: going on the offense with a strong defense. Microbiol. Mol. Biology Rev. **80**(3), 629–661 (2016)
60. Park, E., Kim, H., Song, J., Oh, H., Song, H., Jang, J.: Synthesis of silver nanoparticles decorated polypyrrole nanotubes for antimicrobial application. Macromol. Res. **20**(10), 1096–1101 (2012)
61. Patil, A.O.: Electrically conducting polymers. Appl. Polym. Sci. 21st Century, 325 (2000)
62. Peleg, A.Y., Seifert, H., Paterson, D.L.: Acinetobacter baumannii: emergence of a successful pathogen. Clin. Microbiol. Rev. **21**(3), 538–582 (2008)
63. Porri, L.: Giulio natta–his life and scientific achievements. Paper Presented at the Macromolecular Symposia (2004)
64. Poyraz, S., Cerkez, I., Huang, T.S., Liu, Z., Kang, L., Luo, J., Zhang, X.: One-step synthesis and characterization of polyaniline nanofiber/silver nanoparticle composite networks as antibacterial agents. ACS Appl. Mater. Interfaces **6**(22), 20025–20034 (2014)
65. Ramakrishnan, S.: From a laboratory curiosity to the market place. Resonance, **16**(12), 1254–1265 (2011)
66. Rice, L.B.: Federal Funding for the Study of Antimicrobial Resistance in Nosocomial Pathogens: No ESKAPE. The University of Chicago Press (2008)
67. Sangwan, W., Paradee, N., Sirivat, A.: Polycarbazole by chemical oxidative interfacial polymerization: morphology and electrical conductivity based on synthesis conditions. Polym. Int. **65**(10), 1232–1237 (2016)
68. Seshadri, D.T., Bhat, N.V.: Use of polyaniline as an antimicrobial agent in textiles. Ind. J. Fib. Text. Res. **30**(2), 204–206 (2005)
69. Shirakawa, H., Louis, E.J., MacDiarmid, A.G., Chiang, C.K., Heeger, A.J.: Synthesis of electrically conducting organic polymers: halogen derivatives of polyacetylene, (CH) x. J. Chem. Soc. Chem. Commun. **16**, 578–580 (1977)
70. Smith, W.D., Bardin, E., Cameron, L., Edmondson, C.L., Farrant, K.V., Martin, I., Murphy, R.A., Soren, O., Turnbull, A.R., Wierre-Gore, N.: Current and future therapies for *Pseudomonas aeruginosa* infection in patients with cystic fibrosis. FEMS Microbiol. Lett. **364**(14) (2017)

71. Son, W.K., Youk, J.H., Park, W.H. (2006). Antimicrobial cellulose acetate nanofibers containing silver nanoparticles. Carbohydr. Polym. **65**(4), 430–434
72. Tamayo, L., Zapata, P., Vejar, N., Azócar, M., Gulppi, M., Zhou, X., Thompson, G.E., Rabagliati, F.M., Páez, M.: Release of silver and copper nanoparticles from polyethylene nanocomposites and their penetration into *Listeria monocytogenes*. Mater. Sci. Eng. C, **40**, 24–31 (2014)
73. Tamboli, M.S., Kulkarni, M.V., Patil, R.H., Gade, W.N., Navale, S.C., Kale, B.B.: Nanowires of silver–polyaniline nanocomposite synthesized via in situ polymerization and its novel functionality as an antibacterial agent. Colloids Surfaces B: Biointerfaces, **92**, 35–41 (2012)
74. Thomas, C., Zong, K., Schottland, P., Reynolds, J.: Poly (3, 4-alkylenedioxypyrrole) s as highly stable aqueous-compatible conducting polymers with biomedical implications. Adv. Mater. **12**(3), 222–225
75. Tondro, G., Behzadpour, N., Keykhaee, Z., Akbari, N., Sattarahmady, N.: Carbon@ polypyrrole nanotubes as a photosensitizer in laser phototherapy of *Pseudomonas aeruginosa*. Colloids Surfaces B: Biointerfaces, **180**, 481–486 (2019)
76. Tong, S.Y., Davis, J.S., Eichenberger, E., Holland, T.L., Fowler, V.G.: *Staphylococcus aureus* infections: epidemiology, pathophysiology, clinical manifestations, and management. Clinic. Microbiol. Rev. **28**(3), 603–661 (2015)
77. Upadhyay, J., Kumar, A., Gogoi, B., Buragohain, A.: Antibacterial and hemolysis activity of polypyrrole nanotubes decorated with silver nanoparticles by an in-situ reduction process. Mater. Sci. Eng. C, **54**, 8–13 (2015)
78. Yi, N., Abidian, M.: Conducting polymers and their biomedical applications. In: Biosynthetic Polymers for Medical Applications, pp. 243–276. Elsevier (2016)
79. Youssef, A., Abdel-Aziz, M., El-Sayed, E., Abdel-Aziz, M., El-Hakim, A.A., Kamel, S., Turky, G.: Morphological, electrical & antibacterial properties of trilayered Cs/PAA/PPy bionanocomposites hydrogel based on Fe3O4-NPs. Carbohydr. Polym. **196**, 483–493 (2018)
80. Zare, Y., Shabani, I.: Polymer/metal nanocomposites for biomedical applications. Mater. Sci. Eng. C, **60**, 195–203 (2016)
81. Zhou, D.D., Cui, X.T., Hines, A., Greenberg, R.J.: Conducting polymers in neural stimulation applications. In: Implantable Neural Prostheses, vol. 2, pp. 217–252. Springer (2009))
82. Zhu, B., Wang, S., Li, O., Hussain, A., Hussain, A., Shen, J., Ibrahim, M.: High-quality genome sequence of human pathogen *Enterobacter asburiae* type strain 1497-78T. J. Glob. Antimicrob. Resist. **8**, 104–105 (2017)
83. Zou, Y., Sang, G., Wu, W., Liu, Y., Li, Y.: A polythiophene derivative with octyloxyl triphenylamine-vinylene conjugated side chain: synthesis and its applications in field-effect transistor and polymer solar cell. Synthetic Met. **159**(3-4), 182–187 (2009)

Towards the Conducting Polymer Based Catalysts to Eliminate Pt for Dye Sensitized Solar Cell Applications

Muhammad Shakeel Ahmad, Khairus Syifa Hamdan, and Nasrudin Abd Rahim

Abstract The extended π-conjugated intrinsic conducting polymers (CP) have increasingly been employed to replace Pt-based counter electrodes (CEs) in dye sensitized solar cell (DSSC) due to their ease in processing, control over electrical properties, structural porosity and resistance to corrosion. Most conductive polymers are derivatives of polyacetylene, polyaniline, polypyrrole or polythiophenes. The Pt-based counter electrode accounts for 46% of total cost of DSSC. Various materials classes and their combinations have been investigated as potential replacement with increasingly favorable claims. Herein, a brief account of study has been presented with focus on conducting polymers as potential replacement for Pt-based CEs in DSSCs. Various nano-reinforcements have also been presented and their perceived effects will be discussed in detail. The CPs accounts for 11.19% of total published articles related to CEs for DSSCs and 1.23% share in granted patents. In this chapter, all the important conducting polymers (which have been used as CE) have been discussed section-wise with nano-reinforcements being utilized to improve the performance.

Keywords Conducting polymers · Catalytic activity · Electrical property · Nanocomposites

M. S. Ahmad (✉) · K. S. Hamdan · N. A. Rahim
UM Power Energy Dedicated Advanced Center (UMPEDAC), Wisma R&D University of Malaya, Jalan Pantai Baharu, 59990 Kuala Lumpur, Malaysia
e-mail: shakeelalpha@gmail.com

N. A. Rahim
e-mail: shakeelalpha@gmail.com

M. S. Ahmad
U.S.—Pakistan Center for Advanced Studies in Energy (USPCAS-E) University of Engineering and Technology (UET), Peshawar, Pakistan

© Springer Nature Switzerland AG 2021

S. Shahabuddin et al. (eds.), *Advances in Hybrid Conducting Polymer Technology*, Engineering Materials, https://doi.org/10.1007/978-3-030-62090-5_12

1 Introduction

The DSSC device consists of five main constituents i.e., (1) conducting substrate, (2) semi-conductor nanostructure, (3) dye sensitizer, (4) electrolyte and (5) catalyst. The working principle is based on the regeneration process of electron. In brief, the incident light excites the dye sensitizer which in-turn gets oxidized, losing the electron to semi-conductor nanostructure. The dye gets reduced by accepting electron from electrolyte. The electron travels to outer circuit through conducting substrate and reclaimed by the electrolyte with the help of counter electrode. The counter electrode in dye sensitized solar cell is essentially act as cathode which is responsible to accept electrons from the outer circuit and re-generate oxidized electrolyte present in the sandwich construct of the DSSC. In principle, an ideal counter electrode should perform three main functions i.e., (1) as catalyst, it must accelerate reduction process of electrolyte, (2) it must collect electrons from the outer circuit in an efficient way and (3) it must as a mirror and reflects unabsorbed light back to the sandwich construct to improve light harvesting properties. Still another property of counter electrode is its transparency which facilitated the construction of bi-facial DSSCs. These bi-facial construct is consists of two layers of photoanode bi-facially on one substrate, sandwiched in between two transparent counter electrodes. Various organic and inorganic elements and their combinations have been trialed and successfully implemented as suitable candidates for counter electrode [1]. Presently, platinum is thought to be the best catalyst due to its high catalytic activity and corrosion resistance. But Pt is rare earth element and is too expensive to be commercially employed in DSSC technology. Scientists are heavily engaged in the research and development to replace Pt based counter electrodes with other suitable materials and claimed much of the success. In this regards, conducting polymers and their nanocomposites with other inorganic and organic micro/nano particulates have been extensively studied with encouraging results [2].

In 1977, the fourth generation of polymers, the conductive polymers were discovered and have been industrialized from the materials laboratory. Later, in year 2000, Nobel Prize has honored the fundamental discovery of this conductive polymers or more precisely, intrinsic conducting polymers, to Shirakawa, MacDiarmind, and Heeger. Basically, conductive polymers are organic polymers that conduct electricity where such compounds may have metallic conductivity or can be semiconductors. By using the methods of organic synthesis and by advanced dispersion techniques, the electrical properties of the polymers can be fine-tuned. Dispersion technique is mostly used to process conductive polymers make it as one of its important advantageous. Polyacetylene, polyaniline, polypyrrole or polythiophenes are the main derivatives of conductive polymers. The molecular structure of these polymers is basically a conjugated double bond structure for conduction. The fact of having facile synthesis process, porous structure, electrical conductivity, low cost, abundance, and favorable catalytic properties, conductive polymers turn out to be very potential candidates to replace Pt for CE materials in DSSCs [3].

Conducting polymers, though with their inherited advantages of suitable conductivity, excellent chemical and photo corrosion, easy processability and cost effective nature are still not suitable to replace Pt as catalyst. This is because of inferior catalytic activity and semi-conducting nature. Various inorganic and organic elements and their combinations in the form of micro/nano particulates have been tried with encouraging observations. The intense research and development in this field finally made its way to replace Pt as catalyst in DSSCs. The preceding chapter was dedicated to synthesis and characterization of conducting polymers. This chapter is dedicated to discuss the micro/nano composites of conducting polymers in detail with possible mechanisms and impact.

2 Most Common Conducting Polymers

In this section, some of the most common conducting polymer has been discussed in brief to facilitate a bird eye view to the reader.

2.1 Polypyrrole (PPy)

Besides having an easy synthesis process, polypyrrole (PPy) is also have a favorable catalytic activity, high yield, and environmentally stable. Thus, PPy conductive polymer also can be a promising alternative of Pt CE. The most common way to from PPy is by polymerizing the pyrrole monomer, accompany with the color changing from yellow to blue or black. There are two categories of polymerization noted as chemical polymerization and electrochemical polymerization. Various oxidants can be used in chemical polymerization of PPy as has been demonstrate by Peng et al. where Iron (III) chloride hexahydrate ($FeCl_3$ $6H_2O$) is being used to oxidized pyrrole monomer in the synthesizing process of free-standing paper-like polypyrrole nanotube membrane. They manage to achieve efficiency of 5.27% for device using the PPy nanotube membrane CE which is comparable to Pt CE. Whereas, Wu and colleagues have successfully fabricated PPy CE and the device efficiency achieves 7.66% which is higher than 6.90% of Pt CE. They prepared the PPy CE by coating the iodine oxidation polymerized PPy nanoparticles on FTO [4]. A bifacial assemble of DSSC with the transparent and stable PPy CE has successfully fabricated by Bu et al. giving the efficiency of 5.74%. The ammonium persulfate was used as the oxidant, where FTO was immersed into the mixed solution of APS and pyrrole monomer 0 °C for 3 h [5]. In the other hand, Wang has used the Vanadium pentoxide (V_2O_5) nanofiber and FE Cl_3 to oxidize pyrrole monomer step by step and produced hierarchical nanostructure PPy CE giving the efficiency of 6.78%, which is higher than that of PPy nanoparticles (5.41%), presence 92.5% of Pt CE (7.33%) [6]. By using electrochemical polymerization Xu et al. has prepared the PPy-coated cotton fabrics CE, obtaining the efficiency up to 3.83% of the corresponding device [7]. Although abundant of researchers put much efforts on the PPy CE, the device efficiencies based on the PPy CEs are disadvantaged as it is concentrated on the high charge transfer resistance (RCT) and low conductivity of PPy [8].

2.2 Polyaniline (PANI)

Another most studied conductive polymers is polyaniline (PANI) due to its relatively low cost, multiple oxidation states with different colors, and acid/base doping responsiveness. An average diameter size of 100 nm microporous PANI has been successfully developed by LI and Wu et al. using perchloride acid doped APS oxidative polymerization. This PANI CE device achieved 7.15% efficiency which is higher than that of Pt CE (6.90%) [9]. The enhanced device performance of in PANI nanoparticles CE are given by oriented morphologies which is more conducive to transport the electron transfer efficiency compared to random nanoparticles. A polyaniline nanoribbon (PANINR) CE with serrated, flexible and ultrathin nanostructures has successfully synthesized by Hou et al. using the electrospun V_2O_5 as template and oxidant to comprehend the in-situ polymerization of aniline and etching the V_2O_5 template by acid. The PANI NR based DSSC achieved efficiency of 7.23%, which is comparable enough with the Pt-based DSSC (7.42%) was due to lots of active sites, good contact performance with substrate, and the oriented electron transfer along the nanoribbons. The oriented PANI nanowire counter electrode. in-situ fabricated by Wang et al., where FTO glass was immersed into the aniline/APS solution with the fixed mole ratio of 1.5 at 0–5 °C for 24 h. High catalytic activity exhibited by the obtained PANI nanowire CE has increased the device performances of 8.24% efficiency higher than that of random PANI nanofibers (5.97%) and Pt CE (6.78%). In addition, the effectively exposed PANI nanowires and the rapid electron transfer along the oriented nanowire contributes to the enhanced performance of this device. In the other hand, Xiao et al. has used the pulse potentio static electro polymerization method to fabricated the PANI CE with nanofiber structure giving the efficiency of 5.19%, up to 90% of Pt CE [10].

2.3 Poly(3,4-Ethylene-Dioxythiophene) (PEDOT)

The poly(3,4-ethylenedioxythiophene)(PEDOT) is another important conductive polymer where it presents the highest conductivity (300–500 S_cm^{-1}) than those of PANI (0.1–5 Scm^{-1}) and PPy (10–50 Scm^{-1}) and highest catalytic activity for I^{3-} reduction. With poly(styrene sulfonate) (PSS) dopant, the PEDOT manage to increase the existing conductivity to more than 4600 Scm^{-1}. Besides, the other reasons that made PEDOT as an ideal material for DSSC CE are, the excellent thermal and chemical stability and also the electrochemical reversibility. Honeycomb-like PEDOT CE fabricated by using cyclic voltammetry electrodeposition and sacrificial template methods has prepared by Li et al. where the resultant device gives great performance. Having the front efficiency of 9.12% and rear efficiency of 5.75%, outperforming than that of flat PEDOT (8.05% and 3.78%, respectively), this honeycomb-like PEDOT CE has lights up the roads of bifacial and tandem devices. PEDOT nanotube arrays fabricated by Trevisan et al. using combining the electrodeposition and template etching technique has yielded the efficiency of 8.3%. The influence of doping ions on the morphologies of PEDOT and corresponding device performances were studied by Xia et al. and concluded that ClO^{4-} and PSS doped PEDOT CE present higher efficiencies than those of TsO^- doped PEDOT [11].

3 Polypyrrole (PPy) Based Nanocomposites

Carbon in its most abounded form i.e., carbon black and allotropic form has been extensively employed to improve the conductivity and catalytic activity of conducting polymer such as PPy. In a study, graphene oxide has been incorporated in PPy matrix through simple mixing method and converted to reduced graphene oxide (rGO) using in-situ reduction. Figure 1 shows scanning electron microscopic images of composite along with current–voltage (I–V) characteristics of prepared DSSC devices. Authors claimed 8.14% conversion efficiency which was ~14% higher than bare PPy and comparable to Pt based counter electrode (8.34%). This is due to development of conducting channels and high catalytic activity of rGO [12].

Graphene quantum-dot has also been incorporated in PPy as the catalytic layer on the CE in DSSC. The achieved cell efficiency was 5.27% [13]. Graphene doped with nitrogen and polypyrrole composites have been tried and the cell efficiency was claimed to be 6.8% with high reliability and durability and comparable to that of Pt conter electrode (7.0%) [14]. In a recent study, graphene coated alumina nanoparticles have been incorporated in PPy matrix to avoid agglomeration. The effect of graphene concentration has also been observed in the same study. Figure 2 shows the change in

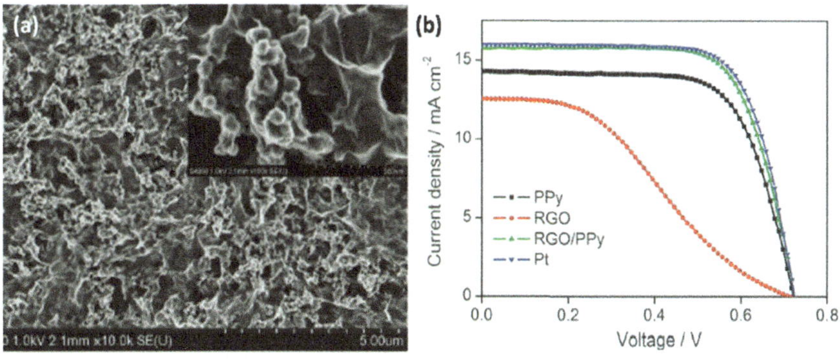

Fig. 1 **a** SEM micrographs of rGO incorporated in PPy and **b** I-V characteristics of DSSCs [12]

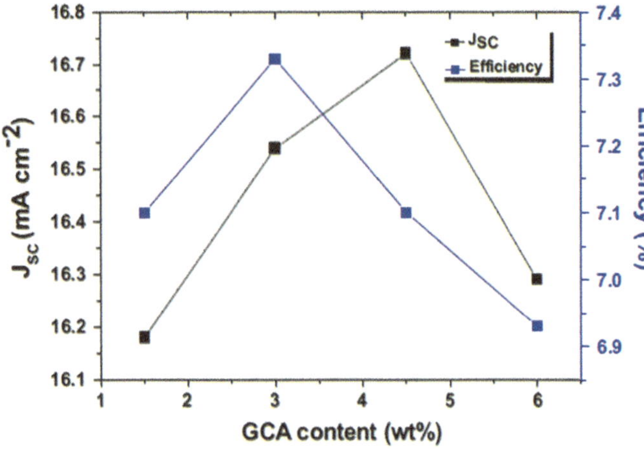

Fig. 2 Current density and conversion efficiency with concentration of graphene coated alumina [15]

conversion efficiency with concentration of graphene coated alumina nanoparticles. The claimed efficiency at optimum concentration is 7.33% which is 13% higher than bare PPy and comparable with Pt based DSSC (7.54) [15].

Multiwalled carbon nanotubes (MWCNTs) have also been investigated to improve the performance of PPy. PPy has been electro-polymarized on the MWCNTs and then deposited on FTO glass. Figure 3 shows SEM and I-V characteristics of prepared counter electrode and compared with bare PPy, MWCNTs and Pt based counter electrode. The reported conversion efficiency of PPy/MWCNTs counter electrode was 7.15% which is comparable to Pt based DSSCs (7.76) [16].

In another study, a remarkable achievement has been reported. Authors claimed conversion efficiency higher than that of Pt based counter electrode by incorporating

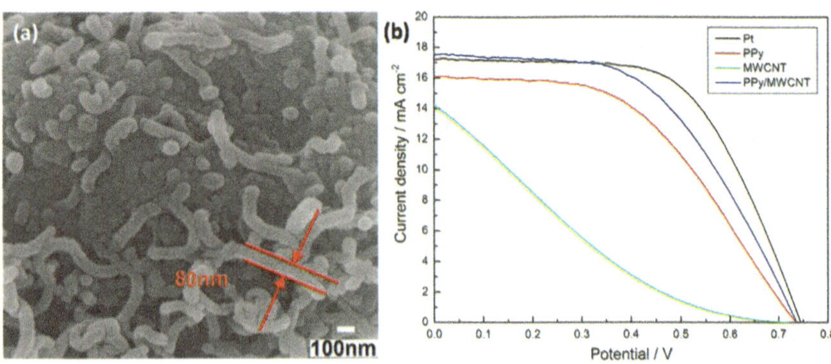

Fig. 3 **a** SEM micrographs of MWCNTs incorporated in PPy and **b** I-V characteristics of DSSCs [16]

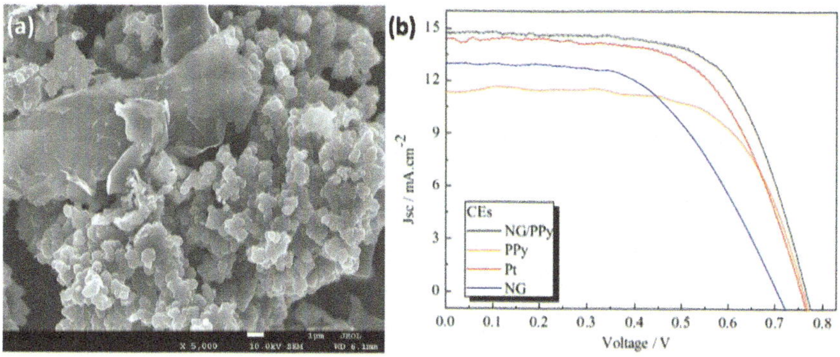

Fig. 4 **a** SEM micrographs of NGs incorporated in PPy and **b** I-V characteristics of DSSCs [17]

nanographite (NG) in PPy. Figure 4 shows SEM and I-V characteristics of prepared counter electrode and compared with bare PPy, NG and Pt based counter electrode. The claimed conversion efficiency of PPy/NG nanocomposite is 7.40 with 5% NG concentration, which is much higher than Pt based counter electrode (6.98) [17]. This novel nanocomposite may prove to be a cost effective alternative to Pt.

Another successful study reported higher conversion efficiency compared to Pt by introducing copper nanostructure in PPy matrix. Two step electrochemical synthesis method has been employed to first deposite PPy layer and then copper nano-rods on the prepared PPy layer using potentiostate. The conversion efficiency at optimum layer thickness was claimed to be 7.42% which is much higher than Pt based counter electrode. Figure 5 shows SEM micrograph and I-V characteristics of prepared DSSCs along with catalytic properties of counter electrodes [18].

4 Polyaniline (PANi) Based Nanocomposites

PANI is one of the most studied conducting polymers in its class. This is because of its facile synthesis and remarkable electro-chemical properties. Recently, rGO aero gel has been incorporated in PANI nanotubes matrix to improve it electrical and catalytic properties. The effect of various concentrations of PANI nanotubes has also been presented. The reported conversion efficiency was 5.47% which is slightly lower than Pt based reference counter electrode i.e., 5.54%. The improved catalytic activity is due to high surface area of rGO aero gel which provides added catalytic cites for redox reaction. Figure 6 shows the TEM micrographs and I-V characteristics of PANI nanotube/rGO aero gel composite counter electrode based quasi solid state DSSC [19].

In another study, free standing PANI based thin film supported by few layers of graphene has been developed and investigated as counter electrode in DSSC. These types of counter electrodes do not require any conducting substrate as the film can

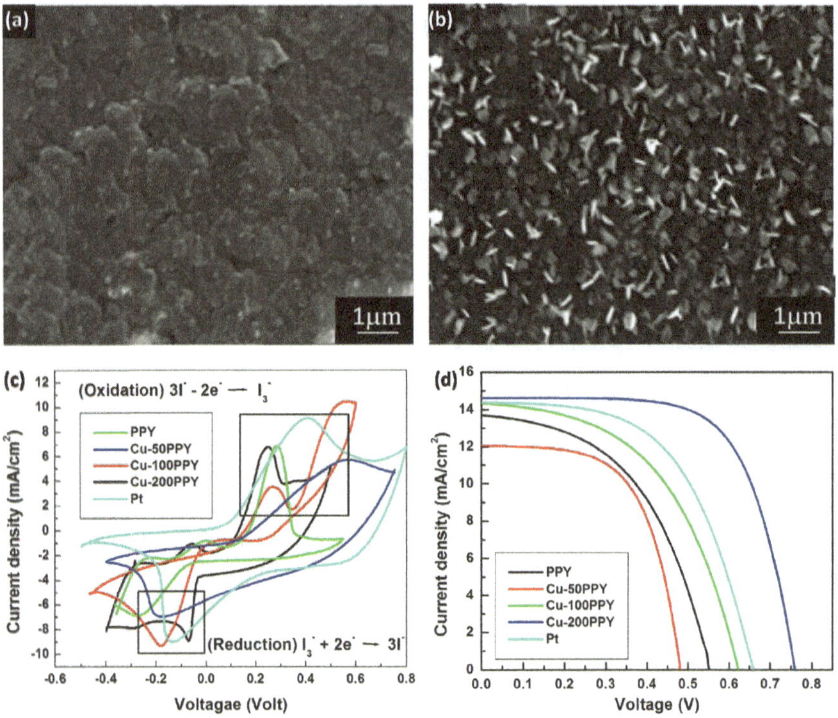

Fig. 5 **a** SEM micrograph of PPy layer (200 nm), **b** SEM micrograph of Cu/PPy layer (200 nm), **c** cyclic voltamograms and **d** I-V characteristics of DSSCs [18]

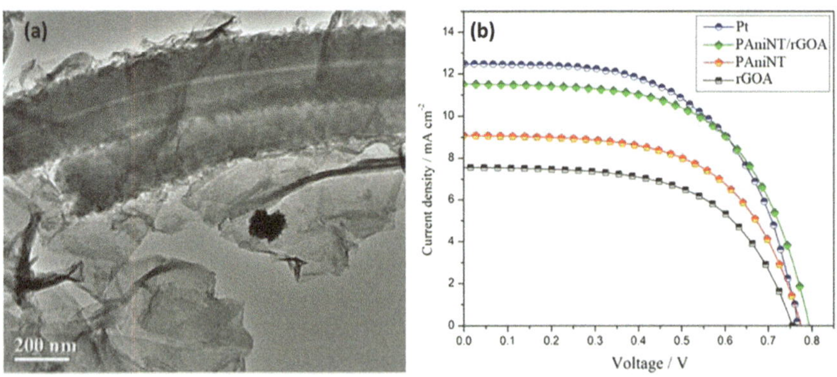

Fig. 6 **a** TEM image of PANI nanotube/rGO aero gel and **b** I-V characteristics of prepared DSSC devices [19]

support its own weight. The reported conversion efficiency was 3.58%, which was 90% equivalent to Pt based reference counter electrode [20]. In still another recent study, the concentration of graphene in PANI has been observed as Pt and TCO free counter electrode. The observed conversion efficiency at 9% graphene concentration was 7.45 which is slightly lower than that of Pt based reference counter electrode (7.63). The advantage of these types of catalyst materials is their ability to replace both the expensive Pt and TCO layer [21]. The combined effect of graphene nano platelets (GNP) and MWCNTs in PANI has also been studied and a significant improvement in conversion efficiency has been observed. The reported conversion efficiency at optimum concentration ratio was 7.67% which was higher than that of Pt based reference counter electrode (7.62) [22].

The effect of MWCNTs and doped MWCNTs with transition metal elements in PANI matrix has been investigated as Pt free counter electrode. Transition metal elements have been ion deposited on MWCNTs. It has been observed that with the addition of ion doping, the catalytic activity improved due to availability of high catalytic cites and high catalytic activity of transition metals [23]. Figure 7 shows FTIR spectra and I-V characteristics of PANI, MWCNTs and ion doped nanocomposites. Table 1 details the I-V parameters of the devices.

Other than ion deposition of transition metals, PANi has also been doped with Pt based transition metal alloys such as Molybdenum, Palladium and cobalt. Two step electro-chemical method has been employed to first develop PANi layer and then

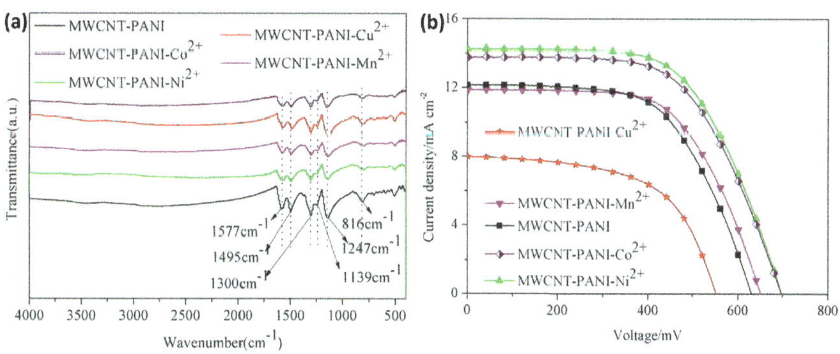

Fig. 7 **a** FTIR spectra of prepared counter electrodes and **b** I-V characteristics of DSSCs [23]

Table 1 Detailed I-V parameters and conversion efficiency [23]

Counter electrodes	Voc	Jsc	FF%	PCE %
MWCNT-PANI	0.631	12.13	60	4.58
MWCNT-PANI-Ni_{2+}	0.700	14.25	60	6.00
MWCNT-PANI-Co_{2+}	0.698	13.75	60	5.75
MWCNT-PANI-Mn_{2+}	0.652	11.88	62	4.77
MWCNT-PANI-Cu_{2+}	0.553	8.00	58	2.57

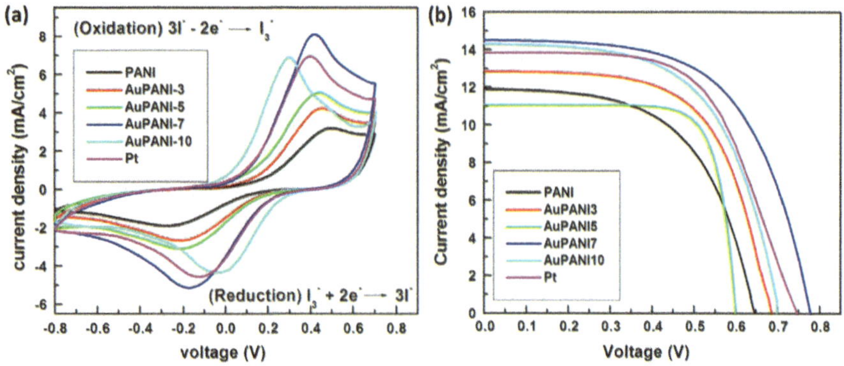

Fig. 8 **a** cyclic voltamograms and **b** I-V curves [24]

deposite alloy particles using metal solutions in chloroplatinic acid. The claimed conversion efficiencies were remarkable with exceptionally competitive efficiency showed by PANi/Platinum/cobalt based with pure Pt based counter electrodes. Gold nanoparticles due to their high catalytic activity and plasmonic effect are the material of choice. Also gold is less expensive than Pt and possess good chemical and photo corrosion resistance. Two step electro-chemical method has been employed to first deposit PANi layer using three point potentiostate configuration and then gold nanoaprticles were deposited using the same apparatus with chloroauric acid as gold precursor. Figure 8 shows catalytic properties and I-V characteristics of devices prepared using Au/PANi counter electrode. An impressive conversion efficiency of 6.71% has been reported with 0.07 gold concentrations in solution which is higher than that of reference Pt based counter electrodes (6.18) [24].

In another study, PANI has been synthesized along with ternary architecture of Pt(Ni and FeNi flower like) alloys using one-step method via galvanice displacement reaction for DSSCs. The counter electrode based on PtNi nanowire/PANI showed impressive efficiency of the order of 8.52 which is due to high catalytic activity of both Pt and Ni. Other than this, the nanowires provide added catalytic cites and improved surface area. PtNi/PANI in flower like configuration also displayed impressive 8.39% conversion efficiency which is higher than reference PANI based counter electrodes [25]. Figure 9 shows schematic synthesis route of ternary configuration along with I-V curves and electro-chemical impedance spectroscopy (EIS) pattern.

The effect of Molybdenum sulfide (MoS2) has also been investigated with impressive conversion efficiency and high transparency. The high electrocatalytic activity toward I3 − reduction of PANi-MoS_2 complex CE is due to the fast charge transfer between PANi (N atoms) and MoS_2 (Mo atoms) by the metal (dπ)-nitrogen (pπ) anti-bonding interaction. An impressive 9.71% conversion efficiency has be exhibited by PANi/MoS_2 (10 wt% MoS_2 in PANi) counter electrode when irradiated from both sides. This efficiency is far higher than Pt based transparent counter electrodes which showed 7.96% conversion efficiency when irradiated from both sides. Cobalt

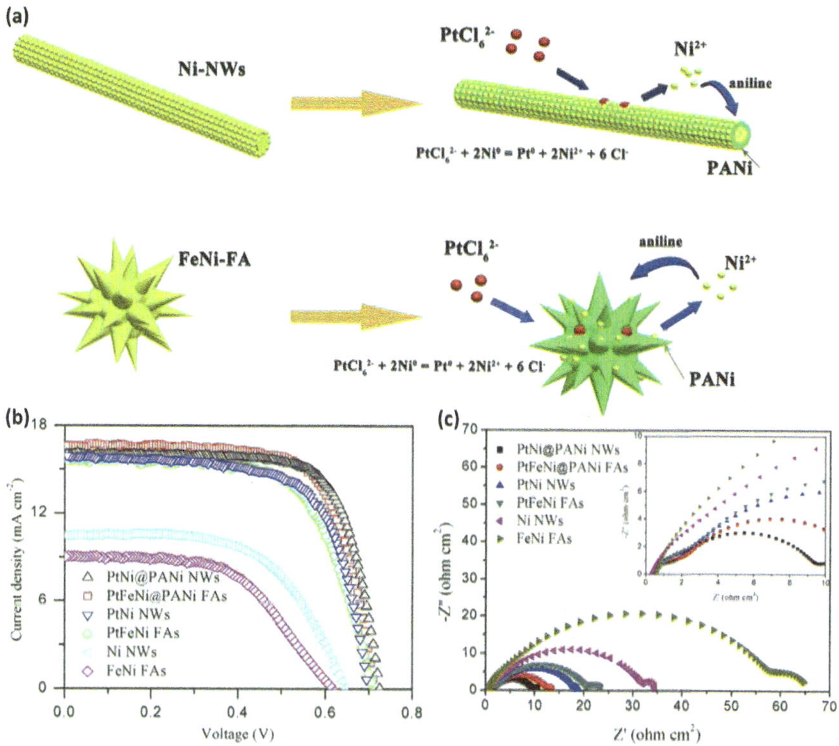

Fig. 9 **a** Schematic of synthesis route, **b** I-V curves and **c** EIS pattern [25]

sulfate (7 wt%) incorporated in PANi also displayed remarkably high conversion efficiency compared to pure Pt based counter electrode [26].

5 Poly(3,4-Ethylene-Dioxythiophene) (PEDOT) Based Nanocomposite

PEDOT due to its excellent electrical and catalytic activity is the material of choice for counter electrode in DSSC. In a study, PEDOT has been deposited onto ITO coated plastic sheet using one step electro-chemical process employing three point configuration and investigated as counter electrode. The reported efficiency was 7.18% which is close to Pt based counter electrodes (7.56%). Furthermore, Pt nanoparticles have been deposited on PEDOT layer and the observed efficiency was approximately 7.90% which is higher than all Pt based counter electrode [27]. This method reduces the amount of Pt which in turn reduces the cost of over-all cell. Furthermore, the study employed ITO coated plastic substrate which is cost effective compared to FTO glass substrate. In another study, PEDOT has been reinforced with transition

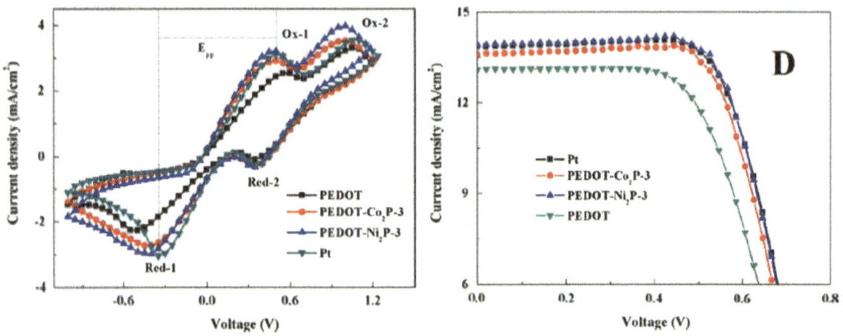

Fig. 10 **a** Cyclic voltamograms and **b** I-V curves [28]

metals (Nickel and cobalt). The transition metals due to their high conductivity and excellent catalytic properties significantly improved the over-all conversion efficiency of PEDOT counter electrode based DSSC devices. The reported efficiency of Ni and Co decorated PEDOT counter electrode was 7.14% and 6.85% respectively. The reference Pt based counter electrode displayed 7.09% efficiency which is lower compared to Ni/PEDOT nanocomposite based counter electrode [28]. Figure 10 shows catalytic properties and I-V characteristics of transition metal reinforced PEDOT based counter electrodes. Gold nanoparticles have also been incorporated in PEDOT with excellent conversion efficiency due to high catalytic and electrical properties. Also increase in surface area leads to improvement in conversion efficiency. The reported efficiency was 3.2% which is much higher compared to that of reference Pt based counter electrode showing 1.3% efficiency [29].

Carbon in the form of graphene has also been tried to improve the performance of PEDOT based counter electrodes. In a study, the electrochemically exfoliated graphene (Ex-Gr) has been prepared using two electrode configurations. The prepared graphene was incorporated in PEDOT matrix and investigated for its performance as counter electrode. The incorporation of Ex-Gr in PEDOT significantly improves the catalytic and electrical properties and in turn leads to superior conversion efficiency. The reported efficiency was 8.00% which is higher than Pt (7.70%) and bare PEDOT (4.60%) based counter electrode. Figure 11 shows TEM images of prepared graphene sheets with I-V characteristics of prepared devices. Nitrogen doped nano graphene has also been studied as possible reinforcement in PEDOT which reportedly higher conversion efficiency (8.30%) compared to Pt based counter electrode which exhibited 8.17% conversion efficiency [30].

PEDOT due to its insolubility in water is very difficult to process. Synthesis of PEDOT in the presence of polystyrenesulfonate (PSS) makes it water soluble. In the films the polycationic PEDOT chains are incorporated into a polyanionic PSS matrix to compensate the charges. Thin films deposited from an aqueous PEDOT:PSS dispersion can be utilized in a wide range of applications with much ease and comfort [31]. In a recent study, Si_3N_4/MoS_2 nanocomposite has been prepared and investigated as potential reinforcement in PEDOT:PSS matrix. Chemical route has been

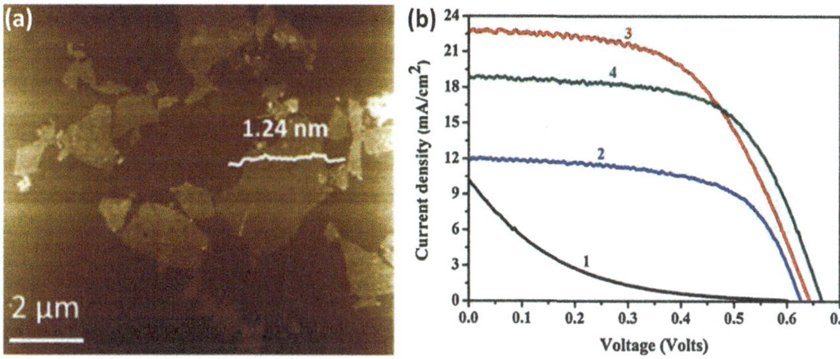

Fig. 11 **a** TEM images of Ex-Gr sheets and **b** I-V curves of (1. Ex-Gr/FTO, 2. PEDOT/FTO, 3. Ex-Gr/PEDOT/FTO and 4. Pt/FTO) [30]

adopted to prepare Si_3N_4/MoS_2 nanocomposite and its various concentrations have been incorporated in PEDOT:PSS matrix for optimization purpose. An impressive increment in conversion efficiency (7.21%) has been reported when irradiated from both sides with high transparency which is slightly less than Pt based reference electrode exhibiting 7.50% conversion efficiency when irradiated from both sides [32]. Figure 12 shows SEM images and I-V curves of reported devices.

In still another study, strontium titanate doped with cobalt has been incorporated in PEDOT:PSS for potential replacement of Pt based counter electrode. The reported conversion efficiency was 8.39% with 0.075 Co concentrations which is higher than Pt based counter electrode exhibiting 8.27% over-all photo conversion efficiency [33]. Nickel sulfide (NiS) has also been studied and optimized for PEDOT:PSS matrix as possible nanocomposite to replace Pt based counter electrode. The optimized concentration of NiS exhibited an impressive conversion efficiency of approximately 8.18% which is slightly lower than that of Pt based reference [34]. Other materials such as TiO_2, SiO_2 and Si has also been tried to improve the performance of PEDOT:PSS with remarkably impressive results [35].

6 Conclusion

The counter electrode is a key component and has a significant influence on both the device cost and photovoltaic performance of the DSSCs. It must possess high conductivity and good catalytic activity along with good stability. Platinum metal well satisfies all the requirements for counter electrode and has been the most frequently used material. However, the high cost and chemical corrosion limit its application widely. Conducting polymers in this regards, due to their cost effective nature, simple facile synthesis, excellent chemical/photo corrosion resistance and ability to combine with most of the nano architectures are gaining immense scientific interest. The improved

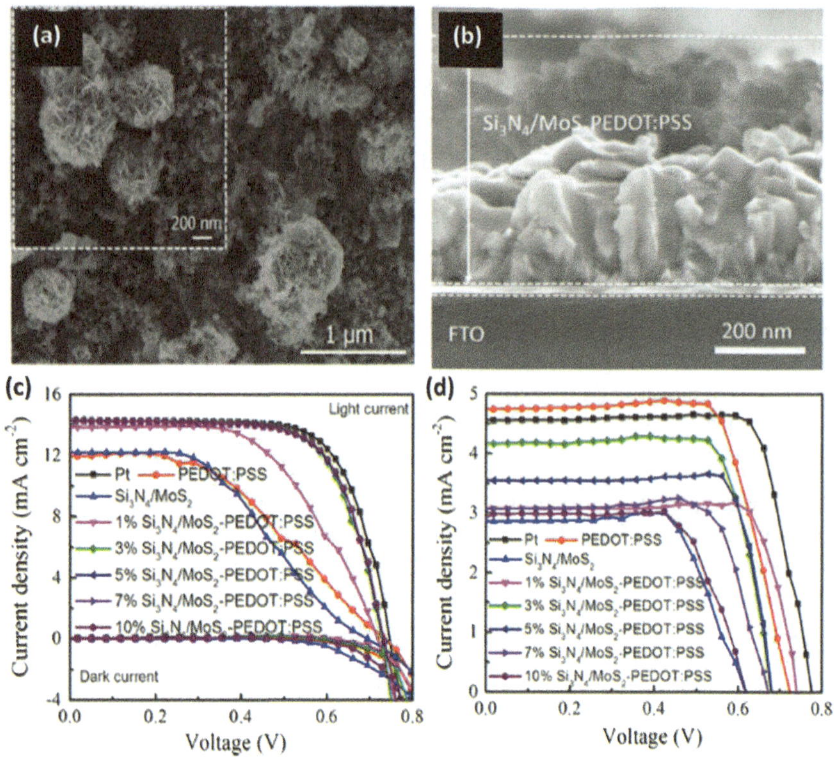

Fig. 12 **a** SEM image of Si3N4/MoS2-PEDOT:PSS, **b** cross sectional view of Si_3N_4/MoS_2-PEDOT:PSS deposited on FTO, **c** I-V curve of DSSC irradiated from front and **d** I-V curve of DSSC irradiated from rear [32]

optoelectronic and catalytic performance of their nanocomposites out performed even Pt as conventional catalyst in DSSC technology. It can be quoted with confidence now, that the age of Pt as counter electrode catalyst in DSSC has now been expired. New cost effective conducting polymers will represent the future of DSSC technology.

References

1. Ahmad, M.S., Pandey, A., Rahim, N.A.: Advancements in the development of TiO_2 photoanodes and its fabrication methods for dye sensitized solar cell (DSSC) applications. A review. Renew. Sustain. Energy Rev. **77,** 89–108 (2017)
2. Ahmed, U., et al.: A comprehensive review on counter electrodes for dye sensitized solar cells: a special focus on Pt-TCO free counter electrodes. Sol. Energy **174**, 1097–1125 (2018a)
3. Tang, Q., et al.: Recent advances in alloy counter electrodes for dye-sensitized solar cells. a critical review. Electrochim. Acta **178**, 886–899 (2015)

4. Wu, J., et al.: High-performance polypyrrole nanoparticles counter electrode for dye-sensitized solar cells. J. Power Sourc. **181**(1), 172–176 (2008)
5. Bu, C., et al.: A transparent and stable polypyrrole counter electrode for dye-sensitized solar cell. J. Power Sourc. **221**, 78–83 (2013)
6. Wang, G., et al.: Facile synthesis of hierarchical nanostructured polypyrrole and its application in the counter electrode of dye-sensitized solar cells. Mater. Lett. **214**, 158–161 (2018a)
7. Xu, Q., et al.: Polypyrrole-coated cotton fabrics prepared by electrochemical polymerization as textile counter electrode for dye-sensitized solar cells. Org. Electron. **29**, 107–113 (2016)
8. Saranya, K., Rameez, M., Subramania, A.: Developments in conducting polymer based counter electrodes for dye-sensitized solar cells–an overview. Eur. Polym. J. **66**, 207–227 (2015)
9. Gunasekera, S.S.B., I.R. Perera, and S.S. Gunathilaka, *Conducting Polymers as Cost Effective Counter Electrode Material in Dye-Sensitized Solar Cells*, in *Solar energy*. 2020, Springer. p. 345–371.
10. Hou, W., et al.: The applications of polymers in solar cells: a review. Polymers **11**(1), 143 (2019)
11. Wei, W., Wang, H., Hu, Y.H.: A review on PEDOT-based counter electrodes for dye-sensitized solar cells. Int. J. Energy Res. **38**(9), 1099–1111 (2014)
12. Gong, F., et al.: Enhanced charge transportation in a polypyrrole counter electrode via incorporation of reduced graphene oxide sheets for dye-sensitized solar cells. Phys. Chem. Chem. Phys. **15**(2), 546–552 (2013)
13. Chen, L., et al.: Graphene quantum-dot-doped polypyrrole counter electrode for high-performance dye-sensitized solar cells. ACS Appl. Mater. Interfaces. **5**(6), 2047–2052 (2013)
14. Ou, J., et al.: Highly efficient ZIF-8/graphene oxide derived N-doped carbon sheets as counter electrode for dye-sensitized solar cells. Electrochim. Acta **286**, 212–218 (2018)
15. Thuy, C.T.T., et al.: Graphene coated alumina-modified polypyrrole composite films as an efficient Pt-free counter electrode for dye-sensitized solar cells. Electrochim. Acta **205**, 170–177 (2016)
16. Hou, W., et al.: Electro-polymerization of polypyrrole/multi-wall carbon nanotube counter electrodes for use in platinum-free dye-sensitized solar cells. Electrochim. Acta **190**, 720–728 (2016)
17. Yue, G., et al.: Highly efficient and stable dye-sensitized solar cells based on nanographite/polypyrrole counter electrode. Electrochim. Acta **129**, 229–236 (2014)
18. Ghani, S., et al.: Polypyrrole thin films decorated with copper nanostructures as counter electrode for dye-sensitized solar cells. J. Power Sourc. **282**, 416–420 (2015a)
19. Mohan, K., et al.: Polyaniline nanotube/reduced graphene oxide aerogel as efficient counter electrode for quasi solid state dye sensitized solar cell. Sol. Energy **186**, 360–369 (2019)
20. Shahid, M.U., et al.: Few-layer graphene supported polyaniline (PANI) film as a transparent counter electrode for dye-sensitized solar cells. Diam. Relat. Mater. **94**, 242–251 (2019)
21. Mehmood, U., et al.: Polyaniline/graphene nanocomposites as counter electrode materials for platinum free dye-sensitized solar cells (DSSCSs). Mater. Lett. **256**, 126651 (2019)
22. Shih, Y.C., Lin, H.L., Lin, K.F.: Electropolymerized polyaniline/graphene nanoplatelet/multi-walled carbon nanotube composites as counter electrodes for high performance dye-sensitized solar cells. J. Electroanal. Chem. **794**, 112–119 (2017)
23. Wu, K., et al.: Effect of ion doping on catalytic activity of MWCNT-polyaniline counter electrodes in dye-sensitized solar cells. Mater. Des. **104**, 298–302 (2016)
24. Ghani, S., et al.: Dye-sensitized solar cells with high-performance electrodeposited gold/polyaniline composite counter electrodes. Mater. Sci. Semicond. Process. **31**, 588–592 (2015b)
25. Wang, Y., et al.: Ternary hybrid PtM@ polyaniline (M= Ni, FeNi) counter electrodes for dye-sensitized solar cells. Electrochim. Acta **291**, 114–123 (2018b)
26. He, B., et al.: Transparent molybdenum sulfide decorated polyaniline complex counter electrodes for efficient bifacial dye-sensitized solar cells. Sol. Energy **147**, 470–478 (2017)

27. Ma, J., et al.: Improvement on the catalytic activity of the flexible PEDOT counter electrode in dye-sensitized solar cells. Mater. Res. Bull. **100**, 213–219 (2018)
28. Di, Y., et al.: Electrocatalytic films of PEDOT incorporating transition metal phosphides as efficient counter electrodes for dye sensitized solar cells. Sol. Energy **189**, 8–14 (2019)
29. Koussi-Daoud, S., et al.: Gold nanoparticles and poly (3, 4-ethylenedioxythiophene)(PEDOT) hybrid films as counter-electrodes for enhanced efficiency in dye-sensitized solar cells. Electrochim. Acta **125**, 601–605 (2014)
30. Belekoukia, M., et al.: Electrochemically exfoliated graphene/PEDOT composite films as efficient Pt-free counter electrode for dye-sensitized solar cells. Electrochim. Acta **194**, 110–115 (2016)
31. Nardes, A.M., et al.: Conductivity, work function, and environmental stability of PEDOT: PSS thin films treated with sorbitol. Org. Electron. **9**(5), 727–734 (2008)
32. Ahmed, A.S., et al.: Si_3N_4/MoS_2-PEDOT: PSS composite counter electrode for bifacial dye-sensitized solar cells. Sol. Energy **173**, 1135–1143 (2018b)
33. Maiaugree, W., et al.: Influence of $SrTi_1$-$xCoxO_3$ NPs on electrocatalytic activity of $SrTi_s$-$_xCo_xO_3$ NPs/PEDOT-PSS counter electrodes for high efficiency dye sensitized solar cells. Energy **154**, 182–189 (2018)
34. Maiaugree, W., et al.: NiS (NPs)-PEDOT-PSS composite counter electrode for a high efficiency dye sensitized solar cell. Mater. Sci. Eng., B **220**, 66–72 (2017)
35. Song, D., et al.: Dye-sensitized solar cells using nanomaterial/PEDOT–PSS composite counter electrodes: effect of the electronic and structural properties of nanomaterials. J. Photochem. Photobiol. **293**, 26–31 (2014)

Basics of Dye Sensitized Solar Cell and Use of Conductive Polymer as Counter Electrode

Shahid Mehmood, S. Suresh, Sohail Ahmed, Mahdi Alizadeh, Nasrudin Abd Rahim, and Yiqiang Zhan

Abstract The photovoltaic technology is an emerging technology nowadays due to increasing energy demands and its low price, simple manufacturing process, environmental friendliness, and comparatively high conversion efficiency. Dye sensitize solar cell (DSSC) a third-generation renewable energy technology device imitates the process of photosynthesis and provides a cost-effective, easy and efficient alternative of solar to conversion of electrical energy. Like other major parts, which affect DSSC efficiency, counter-electrode catalyzes the decrease of redox by collecting electrons from external circuits. There are numerous available materials and methods that have been used to make DSSC counter electrode (CE), which significantly affects DSSC efficiency. In this chapter our focus is to provide comprehensive research on the use of conducting polymer as counter electrode for DSSC by using different synthesis techniques. It includes the use of various types of conducting polymers (CPs) employed as CE of DSSC, a replacement of conventionally used Pt CE through different means by providing an extensive overview of a wide range of techniques used on TCO to manufacture the CE.

S. Mehmood (✉) · Y. Zhan (✉)
Center of Micro-Nano System, School of Information Science and Technology (SIST), Fudan University, Shanghai 200433, China
e-mail: shahid@fudan.edu.cn

Y. Zhan
e-mail: yqzhan@fudan.edu.cn

S. Suresh
Nanotechnology and Catalysis Research Centre, University of Malaya, 50603 Kuala Lumpur, Malaysia

S. Ahmed
Faculty of Science, The University of Haripur, Haripur KPK 22620, Pakistan

M. Alizadeh
Department of Physics, Faculty of Science, University of Zanjan, 45371-38791 Zanjan, Iran

N. A. Rahim
Higher Institution Centre of Excellence (HICoE), UM Power Energy Dedicated Advanced Centre (UMPEDAC), 59990 Kuala Lumpur, Malaysia

© Springer Nature Switzerland AG 2021

S. Shahabuddin et al. (eds.), *Advances in Hybrid Conducting Polymer Technology*, Engineering Materials, https://doi.org/10.1007/978-3-030-62090-5_13

Keywords Conducting polymer · Counter electrode · Dye-sensitized solar cell (DSSC) application

1 Introduction

Energy is critical to attain nearly all sustainable development goals, from its role in poverty eradication through advances in health, education, water supply, industrialization, and climate change. The 7th goal of United Nations (UN) concerns energy resources, ensuring access for all to get benefited from affordable, reliable, viable and modern energy [1]. Energy is the single most significant challenge faced by humanity today. Worldwide excellent attention has been paid to the research and development of renewable energy resources. Renewables contributed 19.2% to global power consumption in 2014 and 23.7% to worldwide electricity manufacturing in 2015 [2], according to the 2016 study of the REN21. The following Fig. 1 shows the article published on solar cells, DSSC and CEs until 2019. In 2015, global investment in renewable technologies exceeded US$286 billion, with nations such as China and the United States investing actively in the wind, hydro, solar and biofuels [2]. As research progresses towards more efficient devices, therefore, in this chapter, an overview of polymer-based CEs for dye-sensitized solar cells application has been discussed. Particularly, conducting polymers (CPs) are the new class of materials which can be employed for the realization of a new device. Beside high conductivity and flexibility these polymers have the characteristics of having good ion transportation, adequate redox behavior, electrochemical and photo response. In DSSC the CE

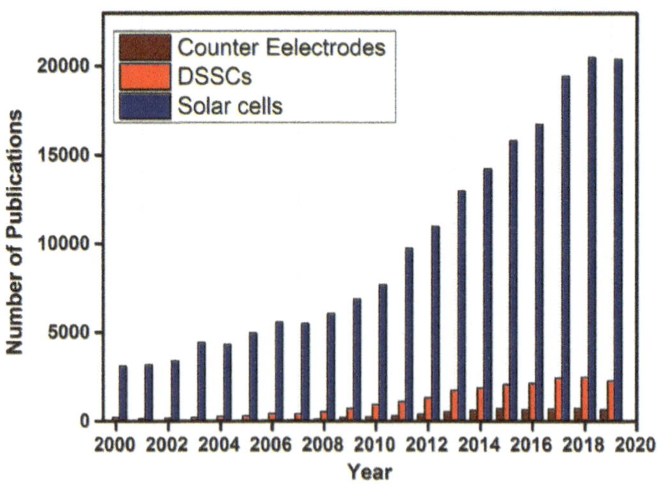

Fig. 1 Number of published articles on solar cells, DSSC and CEs of DSSC obtained from ISI Web of Science

is responsible of reducing the redox species which used as a mediator after electron injection or hole collection after the regeneration of sensitizer. Most commonly used CE of DSSC is vacuum deposited Pt, being grateful to its high conductivity and catalytic activity however, it is not only the most expensive material but has limitations like it can dissolve in electrolyte and made complex Iodide/tri-iodide of PtI_4 or H_2PtI_6 [3]. To avoid above outlined issues and to save the cost and attain the comparable efficiency of DSSCs Pt free CE were synthesized and CPs are the most promising CE so far [4–6].

2 Solar Cells

A solar cell or photovoltaic cell (PV) is a device that directly converts sunlight into electricity using the photoelectric effect. Photovoltaic technology is regarded the most promising among all renewable energy technologies [7–9]. Solar PV has become a fast-growing, multi-billion sector and the most potential of any renewable technology [10]. The abundant, clean, secure and accessible photovoltaic technology was regarded the most promising of all the new energy technologies [8]. The scientific research and industrial growth of PV technology must concentrate on high-efficiency, low-cost, and stability called the "Golden Triangle Issues." The fourth problem of environmental harmony should also be highlighted with increased government awareness of sustainable development. Up to now, three generations of solar cells have developed to satisfy these difficulties. The cells of the first generation, also known as standard, traditional or silicon wafer-based cells, are produced of polycrystalline silicon or monocrystalline silicon. Si solar cells generate efficiencies ranging from 12 to 16% and dominate the PV industry, depending on the production process and quality of the wafer. The advanced manufacturing measures and elevated environmental expenses, however, resulted in the use of polycrystalline Si rather than monocrystalline Si, which further open the ways towards environmentally friendly and low-cost alternatives. Second-generation solar cells are referred to as thin-film techniques, including cadmium telluride (CdTe), copper indium gallium selenide (CIGS), and other cells. Due to material saving, low-temperature procedures and high automation in series production, especially in flexible cells, the thin-film solar cells sandwich active components between two glass sheets and provide potential for cost reduction in the manufacturing process.

However, hard module technology and restricted stability have resulted in the thin-film cells having a small market share (about 112%). The third generation of solar cells emerged after about 30 years. Although less efficient than Si-based solar cells and thin-film cells, third-generation cells rely on low processing expenses and low environmental impacts, resulting in extensive research and development [11]. The present classification of solar cells is a bit puzzling and illogical, particularly that of the thin-film technology of the second generation and the evolving PV of the third generation. This is due to the reality that distinct coordinate systems are based on the present categories. First-generation cells, Si crystalline cells, are categorized based

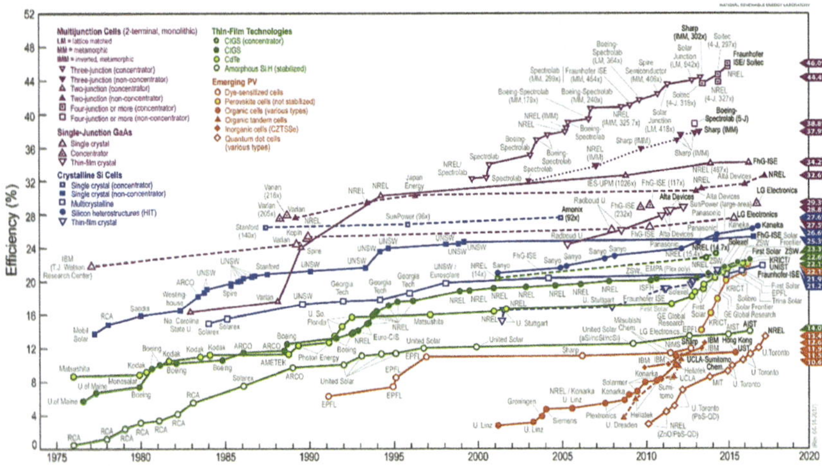

Fig. 2 Conversion efficiencies of best research solar cells worldwide for various photovoltaic technologies since 1976 (Reprinted from [12])

on active components; second-generation cells, thin-film technologies, are manufacturing technology; third-generation cells, emerging photovoltaic, are time emerging and technology mature. We believe that solar cell generations should be based on the active materials used and the order of the invention. Thus, solar photovoltaics is growing the fastest in Asia, with China and Japan accounting for half of global deployment. In the last few decades, solar photovoltaics showed an annual growth rate of more than 20%, and growth rates were even 28% in 2015 [2]. However, the current solar photovoltaic photovoltaics, contributes only 1.2% (2015) to global energy consumption, leaving a big space for industrial growth and scientific research. Figure 2 displays the growth history of the highest efficiencies in research cells from 1976 to the present [12].

3 Dye-Sensitized Solar Cells

Becquerel conducted early studies in 1839, pioneering photoelectric with non-solid-state liquid instruments. In his experiment, lighting solutions comprising silver halide generated a current in electrolytes among two platinum electrodes [13]. This impact is now acknowledged as being insensitive to the visible light because of the energy band gap of 2.7–3.2 eV of silver halide grains due to its semiconductor nature [14]. The photoelectric conversion generally depends upon energy bandgap of semiconductors. Unluckily, the semiconductors having small bandgap are sensitive to photo corrosion when absorb the visible light. The semiconductors such as TiO_2 and Nb_2O_5 which are stable under the light also displays an excessive bandgap to capture visible light efficiently. The problem was fixed by using visible-light sensitive dye molecules to modify the surface of the semiconductors [14]. The bipyridyl Ru complexes with anchoring groups to facilitate them to attach to the oxide semiconductor surface has promoted their recent development in transformation of solar energy. Initially, in dye-sensitized photoelectrochemical cells, the soft semiconductors were used as electrodes to absorb dye. It was generally ineffective to attempt to harvest more light by using multilayer dyes. The electrode of the mesoporous semiconducting material with a thickness of 10 mm and 50% of porosity has surface area more than a thousand times that of the same size of a flat electrode. It can provide sufficient space in the monomolecular layer for dye chemisorption and trap all the incident light falling on it [15]. DSSCs attracted broad attention and were thought to be one of the most promising solar cells in the PV technology of the third generation. The working principal of DSSC is demonstrated in Fig. 3, it differs significantly from the working principal of second generation and third generation solar cell. The DSSC's working principle imitate and strongly related to the photosynthesis process based on light absorption flow of charge carrier through various stages/substances.

Fundamentally the DSSC's working principle procedures include four basic steps [17–19]: (a) At first the incident light strike to the surface of the DSSC and absorbed by sensitizer due to photoexcited electron get ejected from the ground state to the excited state of dye; (b) now these electrons having life in nanoseconds are injected to the conduction band of TiO_2 photoanode lying right below the excited state of the dye, TiO_2 semiconductor will absorb a portion of these photoexcited electron and meanwhile dye will get oxidized; (c) these injected electrons from the TiO2 will diffuse to the transparent conducting oxide coated surface of substrate and travel to the counter electrode via external circuit by passing through load; (d) once the electrons reached to the CE through an external circuit, these electrons reduce the triiodide into iodide (I_3^- to I^-), the ions from the reduced electrolyte (I^-) will be injected to previously oxidized dye by regenerating it, meanwhile the redox mediator

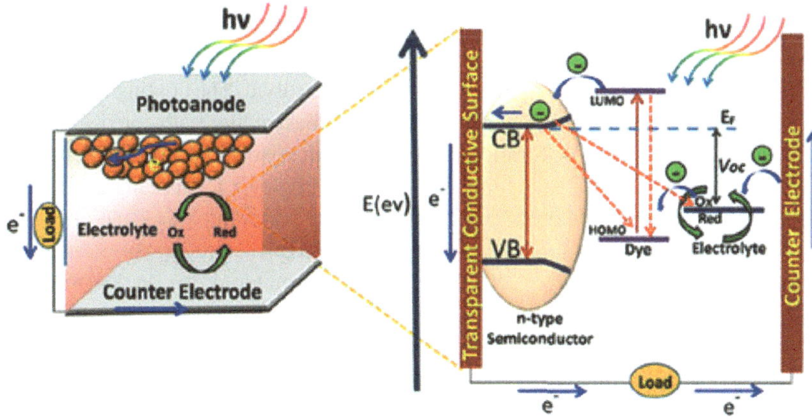

Fig. 3 Schematic representation of the working principle of a DSSC device with its components (Reprinted from [16])

I^- will be oxidize to I_3^-. Hence a cycle of DSSC's working principle complete and it will go on. Recombination by the donation of electrons to the oxidized redox pair [20, 21]. Research to optimize the photovoltaic output of DSSC focuses on modulating the physicochemical characteristics four primary parts (Fig. 3) of the devices: (i) the photosensitizer (ii) anode semiconductor; (iii) redox electrolyte; and (iv) CE. All parts involve fine-tailoring for a high-efficiency solar cell. The book chapter mainly focused on the preparation of CE as one of the vitally important components of DSSC.

4 Preparation of CEs

DSSCs contain many types of CE preparation techniques, including electrochemical deposition, chemical vapor deposition (CVD), thermal decomposition, sputter deposition chemical reduction, hydrothermal reaction and in situ polymerization, etc. The preparation techniques have significant impact on particle size, surface area, morphology and electrochemical and catalytic properties of the electrodes. The smaller particles and larger surface areas of the electrodes will generate more catalytic active sites and encourage electrode activity enhancement. The preparation techniques are diverse with the fast growth of electrode materials in latest years. The primary purpose of using various synthesis techniques is to aspire for highly conducting electrode material with higher electrocatalytic activity and efficiency. Cost effectiveness and easy fabrication of CE can also be another main objective.

4.1 Electrochemical Deposition

Electrochemical deposition, or electrodeposition in short, is commonly used to prepare the CEs of DSSCs. In this method a metallic film is grown on a base material via electrochemical reduction of metal ions in an electrolyte. The coating is prepared in this method at low temperature t achieve strong adhesive film with corrosion resistance, reduced friction and decoration of material with no residual stress of deposited layer and substrate; standardized films with different complex forms can be produced on the surface; coating thickness and chemical coated film thickness and chemical compositions can be controlled easily using this method. Electrodeposition is a very efficient method for preparation of CEs. The desired material can be deposited by selecting any instrumental method it can be Galvanostatic electrodeposition, potentiostatic electrodeposition, pulsed electrodeposition, electrophoretic deposition, etc. As-prepared CE performance depends on deposition conditions such as deposition technique, concentration of ions in solution, pH value, reaction time and temperature of reaction, deposition potential, current density, etc. [22]. The deposition technique and electrochemical parameters have a greater impact on the composition, morphology and characteristics of the resulting electrodes.

4.2 Chemical Vapor Deposition

Chemical vapor deposition (CVD) is a very fine chemical synthesis technique used to create solid products of high quality. The substrate is subjected to consist of one or more than one volatile precursors in a typical CVD; the precursors react and/or decompose to generate the required deposit on the surface of the substrate. A liquid or solid source is used for physical vapor deposition, and chemical vapor deposition technique employs vapors of chemical for deposition. The CVD technology is now a commonly employed method for the preparation of various products such as; semiconductors, synthetic diamonds, metal oxides, sulfides, nitrides, carbides, and compounds of two or more elements. Perhaps the most versatile of all application is CVD polymerization. During the deposition process [23], the physical and chemical properties of the prepared material can be precisely controlled by gas doping.

4.3 Thermal Decomposition or Pyrolysis

This method is frequently used to prepare CEs and is comparatively simple and easy. This technique included the decomposition of an organometallic precursor in a high boiling point solvent and was used by Sun et al. to synthesize Fe_3O_4 nanoparticles using a precursor $Fe(acac)_3$ and surfactant solution [24]. It can assist with thermally decomposing the precursor to obtain CE products with a porous structure. The reaction is essentially endothermic, and heat is used in the compound undergoing decomposition to break chemical bonds.

4.4 Chemical Reduction

The chemical reduction techniques is one of the most prevalent techniques has certain benefits compared to other techniques, such as its convenient procedure, easy operation, simple equipment, low temperature and less price, which make this method appropriate for large-scale productions [25]. For instance, the hydrolysis of urea at lower-temperature produces platinum hydroxide, which can be spread on the FTO substrate owing to electrostatic repulsion, and then can be lowered to Pt by EG [26]. The CE of DSSC can be obtained by using chemical reduction techniques with higher electrocatalytic activity, good conductivity, low price and higher efficiency with low charge transfer resistance between electrolyte/electrode interfaces.

4.5 Sputter Deposition

Sputter deposition is a sputtering technique of thin film deposition by physical vapor deposition (PVD). This includes ejecting of material from a "target" called a source onto a substrate like a conductive glass other conducting sheet. Sputtering is one of the primary procedures for the effective production of photovoltaic solar cells. A significant benefit of sputter deposition is that it is easy to sputter at high temperature as materials with very high melting points cannot be deposited by other deposition techniques. This method has involved no chances of impurities to include and thin films thus prepared are of high quality and uniformity. Sputter deposition technique is having high repeatability and reproducible with improved adhesion on the substrate. The loading quantity of material on substrate can be correctly controlled and monitored. Different sputtering techniques are based on the materials used and control modes which includes Radiofrequency (RF) sputtering, vacuum sputtering, magnetron sputtering, ion beam sputtering, reactive sputtering and ion-assisted sputtering, etc. This sputter deposition technique makes the conventional CE, often used for comparison, sputtered Pt electrode [27].

4.6 Hydrothermal Reaction

The hydrothermal reaction is a reaction that occurs at elevated temperature and pressure in an aqueous medium. The hydrothermal reaction technique is simpler for scientists to work and perform large-scale preparation compared with other methods. By regulating the reaction temperature, time, pressure, packing ratio and reactant ratio, this technique helps to create different materials and structures. In the preparation of CEs, the hydrothermal reaction is often used, particularly in transition metal compounds and composite CEs [28].

4.7 In Situ Polymerization

The in-situ polymerization reaction is a technique in which the nanomaterial is mixed with monomer and polymerized. This technique is used to prepare the CPs such as polyaniline (PANI), polypyrrole (PPy) and polythiophene (PT) derivatives, poly(3,4-ethylenedioxythiophene): polystyrenesulfonate (PEDOT), etc. In brief, a suitable initiator is added to the organic monomer solution, which causes the polymerization of these organic monomers and later on passed through the necessary steps of stirring/non-stirring washing and drying etc. In case of preparation of counter electrode for solar cell a conductive glass substrate is placed the reaction solution, during polymerization the polymer can be developed on the surface of the substrate, called as in situ polymerization, and a conductive polymer CE thus is in situ obtained.

Beside this nanomaterial can also be introduced during the polymerization and a nanomaterial-based polymer can be developed on the substrate. The in situ polymerization can improve DSSC photovoltaic parameters and prevent the polymer aggregation and charge transfer resistance which are the main disadvantages of standard polymerization [29]. The counter-electrode preparation technology is more advanced than the techniques listed above. Thus, there is no omnipotent technique, and the selection of methods of preparing should be based on the real scenario.

Different characterization techniques have been developed to know the relative reaction process mechanism such as charge transfer and charge recombination [30–32], to design novel materials and improve the photovoltaic efficiency of the DSSCs. Electrochemical and photoelectrochemical techniques were regarded to be the most powerful instruments to investigate the process system, analyze component interactions and evaluate DSSC performance [33–36].

4.8 Electrochemical Synthesis Method

Electrochemical method of CPs synthesis has its own importance among other different reported methods of synthesis, as it is a simple, cost effective and easy method to perform in a single glass compartment. This method is reproducible and the desired films can be synthesized with the required thickness and uniformity. The anodic oxidation method of preparation is mostly used for the preparation of ECPs by using functional monomers, cathodic reduction method is used very less often. Previously, conducting polymer synthesis and doping of counter ions on substrate as a result of oxidation happen simultaneously. There are alternative chemical and electrode reaction steps involved for polymerization of an electro active monomer such as pyrrole or thiophene [37].

4.9 Chemical Synthesis Method

It is a method of producing polymers with high chance of mass production with reasonable price. Chemical method uses process of oxidation or reduction for polymerization of monomers to obtain CPs. Extensive studies have been employed to increase the yield and quality of the polymer via oxidative polymerization route. The chemical route of polymerization has its own principle and is different form the electrochemical method [38]. A well-known CP poly (3-hexylthiophene) is generally produced by chemical route of synthesis, however, PPy and PANI can also be produced by using the same route. Whereby, electrochemical method of CP synthesis provides improved conductivity and strong mechanical properties. The important factor is to produce a stable polymer after conjugation via chemical polymerization. Successful polymerization resulting high molecular weight involves oligomers while low molecular weight polymers have good solubility and reactive to polymerize.

Unsuccessful chemical polymerization would result in an entanglement molecular weight achieved, appears in the form of mechanically unstable coating on the wall of the vessel [39].

4.10 Solid State Procedure

This is a procedure of synthesizing CPs by heat in deficiency of oxygen and water by lengthening the polymer chain. The procedure takes place in vacuum or by pushing away the by-products of reaction using an inert gas. During synthesis different parameters such as temperature, pressure and the diffusion of by-products from core to shell of the pellet are controlled. It is the key procedure used to enhance the mechanical and rheological properties of polymers after melt-polymerization prior to injection blow molding [40]. This is broadly used method in industries in production of bottle-grade PET, films and production of fiber. Solid state polymerization method focuses on advantage of simple and cheap industrial apparatus also avoiding the problem faced during conventional polymerization methods.

5 Characterization of CE

5.1 Photovoltaic Measurements

The current–voltage (I–V) characteristic is one of the backbone characterizations for DSSC from which many important parameters could be calculates as explained below. The I–V curve is generally used to determine the photovoltaic parameters of the device, it shows the relationship between the output current and voltage under standard full spectrum irradiation. As shown in Fig. 4 I–V curve is the intercept of open circuit voltage (V_{oc}) and short circuit current (I_{sc}), in the horizontal and vertical axes, respectively, it can also be denoted as J–V curve, which is the intercept of open circuit voltage (V_{oc}) and short circuit current density (J_{sc}) by dividing the I_{sc} with the area of the electrode. J_{sc} represents the current density in short circuit connection when $V = 0$ for the cell, by indicating maximum output photocurrent ability of the cell. In similar way V_{oc} are the open circuit voltage of the cell at $I = 0$ and represent the maximum output photovoltage of the cell. The intersection point corresponds to the maximum photocurrent (I_{mp}) and photovoltage (V_{mp}) is known as maximum output power (P_{max}) can be seen through rectangular areas in Fig. 4.

The fill factor of the cell can be obtained by dividing P to the product of product of V_{oc} and J_{sc} as follows in Eq. 1

$$FF = P_{max}/V_{oc} \times I_{sc} \quad (1)$$

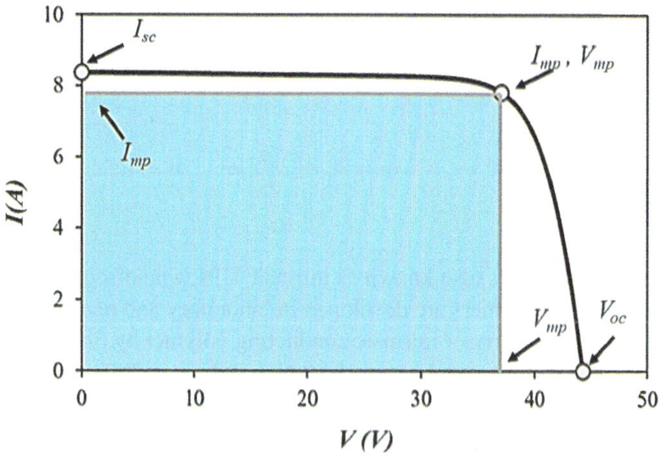

Fig. 4 I–V characteristic curves for the DSSCs

where, $P_{max} = V_{max} \times I_{max}$.

The values of *FF* will always vary from 0 to 1.

In Fig. 4 the well define rectangular shape for *I–V* curves shows the high values of *FF* and high efficiency of the device. Still *FF* can be affected by losses occurs electrically, are equivalent to the resistance in series and parallel.

The overall photo conversion efficiency (PCE) can be calculated from the formula below in Eq. 2,

$$\eta = I_{mp} V_{mp} / P_{in} A = J_{sc} V_{oc} FF / P_{in} \tag{2}$$

Here;

P_{in} = Power density of the incident light.

A = Active surface area of the cell.

The "external quantum efficiency" (EQE) also knows as incident photon to current conversion efficiency (IPCE) can be calculated by the expression given in Eq. 3

$$\text{IPCE (l)} = N_e / N_p \tag{3}$$

Here in above expression N_e referred to the number of electron calculated as photocurrent in the external circuit, while N_p is the number of monochromatic photons striking in a unit time to the solar cell. IPEC is an important parameter which helps to analyze and determine PV performance of solar cells. It can be written as in Eq. 4 from Eq. 3

$$\text{IPCE}(\lambda) = N_e / N_p = 1240 \times J_{sc}(\lambda) / \lambda \times P_{in} \tag{4}$$

where,

$J_{sc}(\lambda)$ = Short-circuit photocurrent density measure in mA/cm^2 under monochromatic light illumination with wavelength (λ, nm).

P_{in} = Power of incident light measured in mW/cm^2.

5.2 *Polymer CEs*

CPs or organic conductors also known as intrinsic CPs were discovered in 1977, as fourth generation of polymers are developed in laboratory and reached to industrial scales [41]. The rediscovery of intrinsic conducting polymer by Shirakawa, MacDiarmid, and Heeger in 2000 was honored by a Nobel Prize in 2000. CPs are the class of organic polymers which conduct electricity and their electrical properties can be tuned using different organic and dispersion synthesis techniques. CPs may have conductivity like metals or sometimes like semiconductors. The conductivity of the polymers depends upon their processability, chiefly by dispersion [42]. Conductive polymers are mostly the offshoots of many polymers such as polyacetylene, polyaniline, PPy or polythiophenes. Figure 5 shows the molecular structure of the CPs, these polymers has the double bonds for conduction. CPs has gain valuable interest as a catalyst for CE as a replacement of Pt catalysts in DSSCs. These polymers are used as potential candidate for Pt free CE of DSSCs because of their facile synthesis, spongy structure, adequate electrical conductivity, low cost, abundance and higher catalytic properties [43].

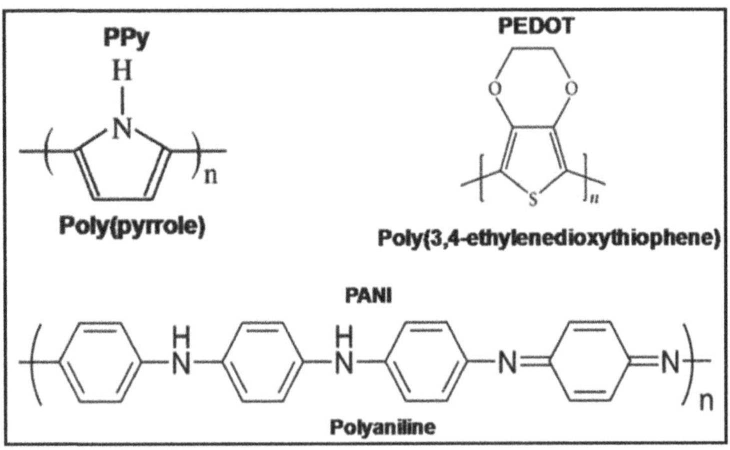

Fig. 5 Molecular structure of CPs

5.3 Poly(3,4-Ethylenedioxythiophene)

PEDOT is a polymer with excellent conductivity of 300–500 S cm^{-1}, the molecular structure of PEDOT is shown in Fig. 5. Even though PEDOT insoluble polymer, still it is widely accomplished conducting polymer amongst the polymer's library, it exhibits excellent conductivity much higher than rest of the polymers such as, polyaniline, PPy, and polythiophene. It can be doped with poly(styrene sulfonate) (PSS) to solve the in solubility issue. Ability to use in energy applications with superior conductivity, high catalytic activity and electrochemical stability are the major prospect of PEDOT. PEDOT can be prepared by various synthesis technique such as electrodeposition [4, 44], vapor phase polymerization [45, 46], chemical oxidative [47] and interfacial polymerization [48, 49]. It demonstrated and improved catalytic activity for redox couple of triiodides for DSSC applications. PEDOT was extensively explored for photovoltaic, photothermal, and thermoelectric applications [50].

As PEDOT has properties of charge transport and optical being inherited therefor, conductive film with multilayers are considered as of great interest with their nanopatterned structures. The nanopatterned conductive films can be used as CE due to several aspect including structural colors light harvesting and enhanced electrical and optical properties [51]. Several methods were adopted to prepare the highly conductive nanopatterned films of PEDOTs. Yet, rigid molecular structure is the main issue in the way of its solubility in any solvent. Still efforts are made to make a composite of nanopatterned PEDOT film with PSS was used which has soft nature as compare to PEDOT and suitable for patterning. However it results in poor conductivity, low thickness and involved complex methods [52]. Efforts are still being made to synthesis multilayered nanopattern films of PEDOT for DSSC CEs. The morphology of PEDOT affects the electrochemical properties of the CE and the photovoltaic performance of the devices. The nanoporous structure can be modified through process control. Different structure such as, nonporous films [4] nanotubes [53] nanofibers [54] has been used as CE for DDSC and efficiency of CE was improved base on the tailored morphologies are shown in Fig. 6.

5.4 Polyaniline

Polyaniline (PANI) was discovered more than a century ago but captured immense attentions by scientific community by early 1980s only. Although the compound was discovered over 150 years ago, PANI has captured intense attention from the scientific community since the early 1980s, as highly electrically conductive compound after its rediscovery. Amongst the library of many polymers PANI is mostly studied as conducting and organic semiconductors from past 50 years [55]. It is a versatile polymer due to its low cost, three distinct states of oxidation and colors, with

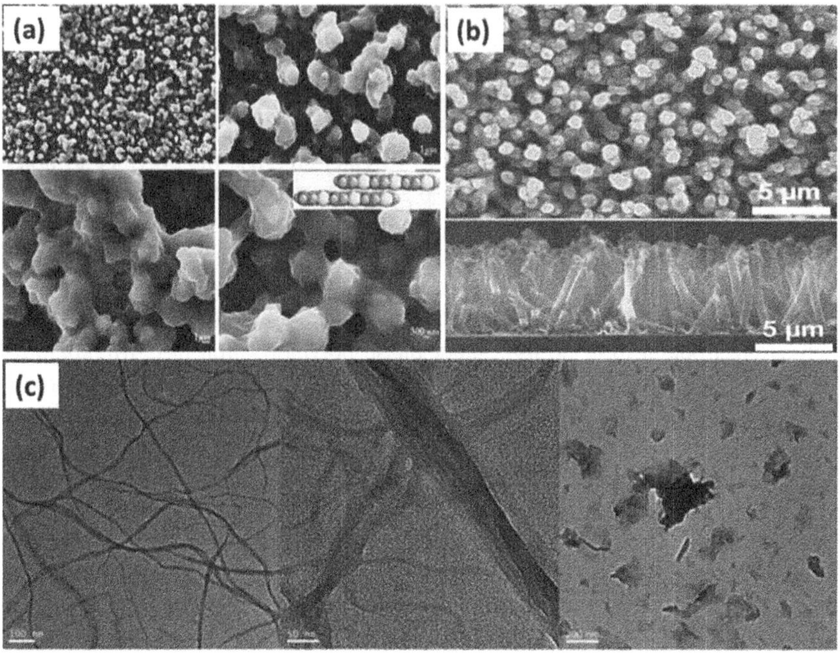

Fig. 6 FESEM images of PEDOT. **a** Nonporous films and **b** nanotubes, TEM image of PEDOT nanofibers (Reprinted from [4, 53, 54])

acid/basic doping response. The acid/basic property of PANI makes it an attractive candidate for chemical vapor sensors, biosensors and supercapacitors. It has different color, charges with multiple oxidation states which further make it an influential candidate for various technological applications such as electrochromic, supercapacitors, batteries and actuators.

PANI have been used as CE for DSSCs due to it variety of characteristics such as easy synthesis, higher conductivity, better thermal and chemical stability and its low cost [58]. There are varieties of PANI has been used as CE for DSSC. PANI has spongy structure in general with higher surface area and it can contribute well in DSSC cell performance. Furthermore, superior performance can be achieved by using electrodeposited PANI due to it adhesiveness with the substrate. It was first time reported as CE for DSSC in 2008 with a particle size of 100 nm by using aqueous oxidative polymerization method in the presence of ammonium persulfate with perchloride acid as dopant [56]. The efficiency of PANI electrode is found to be higher than the conventional Pt electrode and it exhibits higher electrocatalytic performance for the redo reaction using I_3/I. Different nanostructures of PANI are available and has been used as CE in the form of nanofibers, nanobelts, and nanotubes etc. as shown in Fig. 7 [56, 57]. Due to the excellent photoelectrical properties, easy synthesis, cost effectiveness, higher conductivity, applicability in electrocatalysis and higher surface area with various structures make PANI as credible and alternative

Fig. 7 FESEM image of **a** polyaniline nanoparticles [56] **b** micro-particles, **c** nanorods and **d** fibrils [57] (Reprinted with the permission of references—and—respectively)

candidate of Pt for DSSCs. However, PANI is not an ideal CE material because it suffers from instability, self-oxidation, and carcinogenic properties [43].

5.5 *Polypyrrole (PPy)*

PPy firstly was reported by Weiss and his colleagues in 1963 [59]. The films of PPy has yellow color but it becomes darken in air due to its oxidations, it exists in blue and black color depending upon the grade of polymerization and thickness of the firm. It is an organic polymer and synthesized by polymerization of pyrrole. PPy is insoluble in solvent same as PEDOT, it is unsolvable unlike PEDOT even in dropped form but swellable in solvents [60]. PPy is a thermally stable up to 150 °C in air [61]. Originally PPy is a non-conducting material but its derivatives are good in electrical conductions. The conductivity of PPy can be tuned by the reagent used during oxidation of pyrrole, it ranges from 2 to 100 S cm^1. It is mostly used in electronics and sensor applications [62]. It has the potential to replace the Pt CE due to its low cost, applicability in catalysis, easy synthesis good stability and higher yield (Fig. 8).

Fig. 8 The FESEM image of **a** PPy, **b** PPy nanoparticles and **c** PPy nanoparticle deposited on FTO as counter electrod for DSSC (Reprinted from [63, 64])

6 Conclusion

Dye-sensitized solar cells are becoming one of the typical representatives of the third-generation solar cells, effective in converting solar energy into low-cost electricity, simple preparation, high efficiency and environmental benignity. With a focus on the golden triangle problems of PV technology, namely efficiency, price, and stability, and based on the three main elements of DSSCs, namely photoanodes, electrolytes, and CEs, researchers have put a lot of effort into this, and significant progress has been made since 1991. The CE is the main element and has an essential impact on DSSC's photovoltaic efficiency as well as device price. As a CE, for electrolyte regeneration, it must have elevated conductivity and excellent catalytic activity, as well as excellent stability. The platinum metal meets all CE demands well and was the most commonly used CE material. However, its implementation is commonly limited by the high price and lack of stability. Conductive polymers are flexible, transparent, easy to process and easy to adjust assets among the Pt-free CEs. PEDOT shows the highest efficiency among conductive polymers, but its price is high. The PPy-based CE is cheaper, but PEDOT's output is lower. Due to its low price and comparatively better results, the PANI-based CE has more attractive opportunities.

References

1. Assembly, G.: Sustainable development goals. In: SDGs, Transforming Our World: The 2030 Agenda (2015)
2. Ren21, R.: Global status report. REN21 secretariat, Paris, 2016 (2016)
3. Tamburri, E., et al.: Structure and I2/I—redox catalytic behaviour of PEDOT–PSS films electropolymerized in aqueous medium: implications for convenient counter electrodes in DSSC. Inorg. Chim. Acta **377**(1), 170–176 (2011)
4. Ahmad, S., et al.: Dye-sensitized solar cells based on poly (3, 4-ethylenedioxythiophene) counter electrode derived from ionic liquids. J. Mater. Chem. **20**(9), 1654–1658 (2010)
5. Fan, B., et al.: Conducting polymer/carbon nanotube composite as counter electrode of dye-sensitized solar cells. Appl. Phys. Lett. **93**(14), 143103 (2008)
6. Makris, T., et al.: A quasi solid-state dye-sensitized solar cell made of polypyrrole counter electrodes. Electrochim. Acta **56**(5), 2004–2008 (2011)

7. Schiermeier, Q., et al.: Energy alternatives: Electricity without carbon. Nat. News **454**(7206), 816–823 (2008)
8. Hammond, A.L.: Solar energy: the largest resource. Science **177**(4054), 1088–1090 (1972)
9. Lewis, N.S.: Toward cost-effective solar energy use. Science **315**(5813), 798–801 (2007)
10. Lopez, A., et al.: US renewable energy technical potentials: a GIS-based analysis. NREL (2012)
11. Chu, S., Majumdar, A.: Opportunities and challenges for a sustainable energy future. Nature **488**(7411), 294 (2012)
12. Wu, J., et al.: Counter electrodes in dye-sensitized solar cells. Chem. Soc. Rev. **46**(19), 5975–6023 (2017)
13. Becquerel, M.: Mémoire sur les effets électriques produits sous l'influence des rayons solaires. Comptes Rendus Hebdomadaires Des Séances De l'Académie Des Sciences **9**, 561–567 (1839)
14. Grätzel, M.: Photoelectrochemical Cells. Nature **414**(6861), 338 (2001)
15. Desilvestro, J., et al.: Highly efficient sensitization of titanium dioxide. J. Am. Chem. Soc. **107**(10), 2988–2990 (1985)
16. Roy, A., et al.: A review on applications of Cu_2ZnSnS_4 as alternative counter electrodes in dye-sensitized solar cells. AIP Adv. **8**(7), 070701 (2018)
17. Wu, J., et al.: Electrolytes in dye-sensitized solar cells. Chem. Rev. **115**(5), 2136–2173 (2015)
18. Hagfeldt, A., Graetzel, M.: Light-induced redox reactions in nanocrystalline systems. Chem. Rev. **95**(1), 49–68 (1995)
19. Ragoussi, M.E., Torres, T.: New generation solar cells: concepts, trends and perspectives. Chem. Commun. **51**(19), 3957–3972 (2015)
20. Grätzel, M.: Solar energy conversion by dye-sensitized photovoltaic cells. Inorg. Chem. **44**(20), 6841–6851 (2005)
21. Grätzel, M.: Recent advances in sensitized mesoscopic solar cells. Acc. Chem. Res. **42**(11), 1788–1798 (2009)
22. Moharana, M., Mallik, A.: Nickel electrocrystallization in different electrolytes: An in-process and post synthesis analysis. Electrochim. Acta **98**, 1–10 (2013)
23. Nam, J.G., et al.: Enhancement of the efficiency of dye-sensitized solar cell by utilizing carbon nanotube counter electrode. Scripta Mater. **62**(3), 148–150 (2010)
24. Sun, S., Zeng, H.: Size-controlled synthesis of magnetite nanoparticles. J. Am. Chem. Soc. **124**(28), 8204–8205 (2002)
25. Dao, V.D., Choi, H.S.: Pt nanourchins as efficient and robust counter electrode materials for dye-sensitized solar cells. ACS Appl. Mater. Interfaces. **8**(1), 1004–1010 (2015)
26. Song, M.Y., et al.: High efficient Pt counter electrode prepared by homogeneous deposition method for dye-sensitized solar cell. Appl. Energy **100**, 132–137 (2012)
27. Moraes, R., et al.: Optical, electrical and electrochemical evaluation of sputtered platinum counter electrodes for dye sensitized solar cells. Appl. Surf. Sci. **364**, 229–234 (2016)
28. Xiao, Y., Han, G.: Efficient hydrothermal-processed platinum–nickel bimetallic nano-catalysts for use in dye-sensitized solar cells. J. Power Sour. **294**, 8–15 (2015)
29. Hsu, Y.C., Chen, G.L., Lee, R.H.: Graphene oxide sheet-polyaniline nanocomposite prepared through in-situ polymerization/deposition method for counter electrode of dye-sensitized solar cell. J. Polym. Res. **21**(5), 440 (2014)
30. Hagfeldt, A., et al.: Dye-sensitized solar cells. Chem. Rev. **110**(11), 6595–6663 (2010)
31. Halme, J., et al.: Device physics of dye solar cells. Adv. Mater. **22**(35), E210–E234 (2010)
32. Barnes, P.R., et al.: Interpretation of optoelectronic transient and charge extraction measurements in dye-sensitized solar cells. Adv. Mater. **25**(13), 1881–1922 (2013)
33. Mishra, A., Fischer, M.K., Bäuerle, P.: Metal-free organic dyes for dye-sensitized solar cells: From structure: Property relationships to design rules. Angew. Chem. Int. Ed. **48**(14), 2474–2499 (2009)
34. Ondersma, J.W., Hamann, T.W.: Recombination and redox couples in dye-sensitized solar cells. Coord. Chem. Rev. **257**(9–10), 1533–1543 (2013)
35. Dunn, H.K., Peter, L.M.: How efficient is electron collection in dye-sensitized solar cells? Comparison of different dynamic methods for the determination of the electron diffusion length. J. Phys. Chem. C **113**(11), 4726–4731 (2009)

36. Weiqing, L., et al.: Development and application of intensity modulated photocurrent spectroscopy and intensity modulated photovoltage spectroscopy. Prog. Chem. **21**(6), 1085–1093 (2009)
37. Guay, J., et al.: Electrooxidation of soluble. alpha.,. alpha.-coupled thiophene oligomers. Chem. Mater. **4**(2), 254–255 (1992)
38. Sharma, P.S., et al.: Electrochemically synthesized polymers in molecular imprinting for chemical sensing. Anal. Bioanal. Chem. **402**(10), 3177–3204 (2012)
39. Kumar, R., Singh, S., Yadav, B.: Conducting polymers: synthesis, properties and applications. Int. Adv. Res. J. Sci. Eng. Technol. **2**(11), 110–124 (2015)
40. Staiti, P., Lufrano, F.: Design, fabrication, and evaluation of a 1.5 F and 5 V prototype of solid-state electrochemical supercapacitor. J. Electrochem. Soc. **152**(3), A617 (2005)
41. Kirchmeyer, S., Reuter, K.: Scientific importance, properties and growing applications of poly(3, 4-ethylenedioxythiophene). J. Mater. Chem. **15**(21), 2077–2088 (2005)
42. Skotheim, T.A., Reynolds, J.: Handbook of Conducting Polymers, 2 volume set. CRC press (2007)
43. Yun, S., Hagfeldt, A., Ma, T.: Pt-free counter electrode for dye-sensitized solar cells with high efficiency. Adv. Mater. **26**(36), 6210–6237 (2014)
44. Anothumakkool, B., et al.: Electrodeposited polyethylenedioxythiophene with infiltrated gel electrolyte interface: a close contest of an all-solid-state supercapacitor with its liquid-state counterpart. Nanoscale **6**(11), 5944–5952 (2014)
45. Kim, J., et al.: The preparation and characteristics of conductive poly (3, 4-ethylenedioxythiophene) thin film by vapor-phase polymerization. Synth. Met. **139**(2), 485–489 (2003)
46. Lock, J.P., Im, S.G., Gleason, K.K.: Oxidative chemical vapor deposition of electrically conducting poly (3, 4-ethylenedioxythiophene) films. Macromolecules **39**(16), 5326–5329 (2006)
47. Anothumakkool, B., et al.: 1-Dimensional confinement of porous polyethylenedioxythiophene using carbon nanofibers as a solid template: an efficient charge storage material with improved capacitance retention and cycle stability. RSC Adv. **3**(29), 11877–11887 (2013)
48. Jang, J., Chang, M., Yoon, H.: Chemical sensors based on highly conductive poly (3, 4-ethylenedioxythiophene) nanorods. Adv. Mater. **17**(13), 1616–1620 (2005)
49. Yoon, H., Chang, M., Jang, J.: Formation of 1D poly (3, 4-ethylenedioxythiophene) nanomaterials in reverse microemulsions and their application to chemical sensors. Adv. Func. Mater. **17**(3), 431–436 (2007)
50. Park, T., et al.: Flexible PEDOT electrodes with large thermoelectric power factors to generate electricity by the touch of fingertips. Energy Environ. Sci. **6**(3), 788–792 (2013)
51. Park, C., Na, J., Kim, E.: Cross Stacking of Nanopatterned PEDOT Films for Use as Soft Electrodes. ACS Appl. Mater. Interfaces. **9**(34), 28802–28809 (2017)
52. Yang, Y., et al.: Nanoimprint of dehydrated PEDOT: PSS for organic photovoltaics. Nanotechnology **22**(48), 485301 (2011)
53. Trevisan, R., et al.: PEDOT nanotube arrays as high performing counter electrodes for dye sensitized solar cells. Study of the interactions among electrolytes and counter electrodes. Adv. Energy Mater. **1**(5), 781–784 (2011)
54. Lee, T.H., et al.: High-performance dye-sensitized solar cells based on PEDOT nanofibers as an efficient catalytic counter electrode. J. Mater. Chem. **22**(40), 21624–21629 (2012)
55. MacDiarmid, A.G.: "Synthetic metals": A novel role for organic polymers (Nobel lecture). Angew. Chem. Int. Ed. **40**(14), 2581–2590 (2001)
56. Li, Q., et al.: Application of microporous polyaniline counter electrode for dye-sensitized solar cells. Electrochem. Commun. **10**(9), 1299–1302 (2008)
57. Qin, Q., Tao, J., Yang, Y.: Preparation and characterization of polyaniline film on stainless steel by electrochemical polymerization as a counter electrode of DSSC. Synth. Met. **160**(11–12), 1167–1172 (2010)
58. MacDiarmid, A.G.: „Synthetische Metalle ": eine neue Rolle für organische Polymere (Nobel-Vortrag). Angew. Chem. **113**(14), 2649–2659 (2001)

59. McNeill, R., et al.: Electronic conduction in polymers. I. The chemical structure of polypyrrole. Aust. J. Chem. **16**(6), 1056–1075 (1963)
60. Tat'yana, V.V., Efimov, O.N.: Polypyrrole: a conducting polymer; its synthesis, properties and applications. Russ. Chem. Rev. **66**(5), 443 (1997)
61. Xia, J., Chen, L., Yanagida, S.: Application of polypyrrole as a counter electrode for a dye-sensitized solar cell. J. Mater. Chem. **21**(12), 4644–4649 (2011)
62. Huang, Y., et al.: Nanostructured polypyrrole as a flexible electrode material of supercapacitor. Nano Energy **22**, 422–438 (2016)
63. Yue, G., et al.: Application of poly (3, 4-ethylenedioxythiophene): polystyrenesulfonate/polypyrrole counter electrode for dye-sensitized solar cells. J. Phys. Chem. C **116**(34), 18057–18063 (2012)
64. Wu, J., et al.: High-performance polypyrrole nanoparticles counter electrode for dye-sensitized solar cells. J. Power Sources **181**(1), 172–176 (2008)

The manufacturer's authorised representative in the EU is Springer Nature Customer Service Centre GmbH, Europaplatz 3, 69115 Heidelberg, Germany. If you have any concerns regarding our products, please contact ProductSafety@springernature.com

Printed and bound by CPI Group (UK) Ltd, Croydon, CR0 4YY
30/09/2026
02225288-0008